APS

Advances in Pharmacological Sciences

GABA: Receptors, Transporters and Metabolism

Edited by
C. Tanaka
N.G. Bowery

Birkhäuser Verlag
Basel · Boston · Berlin

Editors:

Chikako Tanaka MD Ph.D.
Professor Emeritus, Kobe University
Director, Hyogo Institute
for Aging Brain and Cognitive Disorders
Himeji
Hyogo 670
Japan

Professor Norman G. Bowery
Head of the Department of Pharmacology
The Medical School
University of Birmingham
Edgbaston
Birmingham B15 2TT
United Kingdom

A CIP catalogue record for this book is available from the Library of Congress, Washington D.C., USA

Deutsche Bibliothek Cataloging-in-Publication Data
GABA : receptors, transporters and metabolism / ed. by C. Tanaka ;
N. G. Bowery. – Basel ; Boston ; Berlin : Birkhäuser, 1996
(Advances in pharmacological sciences)
ISBN-13:978-3-0348-9858-4 e-ISBN-13:978-3-0348-8990-2
DOI: 10.1007/978-3-0348-8990-2

NE: Tanaka, Chikako [Hrsg.]

Softcover reprint of the hardcover 1st edition 1996

Camera-ready copy prepared by the editor and authors
Printed on acid-free paper produced from chlorine-free pulp. TCF ∞

ISBN-13:978-3-0348-9858-4

9 8 7 6 5 4 3 2 1

Contents

$GABA_A$ and $GABA_B$ receptor ligands and their therapeutic application

Preface

It is just 30 years since the amino acid, GABA, was accepted as a neurotransmitter in the mammalian brain. Unequivocal evidence finally emerged in 1966 when Krnjevic and colleagues established that this substance mimicked synaptic inhibition within the cerebral cortex mediating an increase in neuronal membrane conductance to chloride ions. GABA is now believed to be responsible for at least 40% of all inhibitory synaptic events which occur in the mammalian brain. Thus, this simple amino acid appears to be a major neurotransmitter controlling neuronal function and any changes in its concentration within the brain will produce an alteration in cell function. This has been exploited successfully to provide therapeutic agents such as vigabatrin, which can provide effective treatment of epilepsy. This agent reduces the metabolism of GABA to increase its total concentration in the brain which is believed to be the basis for its anti-epileptic action.

However, the aspect of GABA physiology and pharmacology which has probably received most attention over the years is the site or sites at which it acts. The receptor for GABA was initially considered to be a single site coupled directly to the chloride ion channel. Discovery of the antagonistic action of bicuculline by Curtis, Johnston and colleagues in 1971 supported this concept. But we now know that GABA receptors are not homogeneous and can be readily sub-divided into ionotropic ($GABA_A$) and metabotropic ($GABA_B$) classes although further divisions within the ionotropic group e.g. isoforms including $GABA_C$ sites have been suggested.

In recent years, molecular biology has enabled the structure of the $GABA_A$ receptor to be discerned such that we now know that the receptor is part of a super family of fast ion channel receptors. The receptor comprises a pentameric structure with each protein subunit linking to form a central pore through which Cl^- selectively permeates. Evidence indicates that the subunit composition can vary and whilst many thousands of combinations could exist, in practice the forms normally expressed are probably fewer than 20. Even so this would suggest that there is considerable heterogeneity within the $GABA_A$ receptor family but this may provide distinct and specific drug target sites.

The $GABA_A$ receptor complex has already proved to be an important site for drug action since many of our existing centrally-active drugs exert their effects by modification of the response to GABA. Probably the most widely prescribed drugs in this category are the benzodiazepines and whilst they are effective their specificity could still be improved. Subunit site-directed compounds might provide an improvement over presently available compounds. Similarly, the specificity of compounds acting at other sites on the $GABA_A$ receptor complex e.g. neurosteroids may benefit from our increasing knowledge of the $GABA_A$ subunit structures and their biochemical manipulation.

The metabotropic receptor for GABA emerged in 1981 and now this G-protein linked receptor may also prove to be a heterogeneous group. Like the $GABA_A$ receptor, the $GABA_B$ site also contributes to synaptic transmission but seems to be responsible for slower postsynaptic events mediated by an increase in K^+ channel conductance as well as a suppression of Ca^{++} conductance in nerve terminals. Whilst it is already clear that a $GABA_B$ agonist, baclofen, is important in therapeutics as a muscle relaxant, there exists considerable potential for effective drugs emerging from antagonists of the $GABA_B$ receptor. These include cognitive enhancers and anti-absence epilepsy drugs.

Much of the present volume is devoted to recent advances in our knowledge of receptors for GABA obtained by many experts in this research area.

Fundamental knowledge about the high affinity transporter mechanisms responsible for inactivation of GABA has also progressed significantly in recent years. Multiple forms of the transporter (GAT) proteins have been described and recent research on their characteristics and significance makes an important contribution to the book.

Of course, when considering GABA mechanisms in mammals it is not unusual to assume that the central nervous system is the only locus of its action. However, the concentration of GABA and activity of its forming enzyme, glutamic acid decarboxylase (GAD) in certain peripheral organs is at least as high as in the CNS. Particularly notable is the pancreas which contains very high concentrations and recent data suggest that GAD may even be implicated in the production of diabetes. This is considered in some detail by expert contributors to the book.

Overall this volume provides a wide but detailed coverage of information obtained from GABA research currently in progress around the world. The contributors are recognised experts who met to discuss their findings in Kyoto, Japan in July 1995. The original venue was to be Kobe but the earthquake which occurred earlier in the year demolished the venue. Despite this, the meeting was still held. This says a great deal about our Japanese hosts, Professor Tanaka and her colleagues, who managed to rearrange the venue against all the odds and at very short notice. May I take this oppurtunity, on behalf of all the participants, to express my gratitude to Professor Tanaka and her group.

We believe that chapters herein, which are based on the speakers' contributions, provide not only new information but have captured some of the excitement and enthusiasm which was engendered in their oral presentations.

N. G. Bowery
November 1995

GABA: Receptors, Transporters and Metabolism
ed. by C. Tanaka & N.G. Bowery

ESTABLISHMENT OF GABA AS AN INHIBITORY NEUROTRANSMITTER AT CRUSTACEAN NEUROMUSCULAR JUNCTION AND IN THE MAMMALIAN CENTRAL NERVOUS SYSTEM

M. Otsuka

Institute of Bio-Active Science, Nippon Zoki Pharmaceutical Co., Ltd., Yashiro-cho, Hyogo 673-14, Japan

Summary: In the late 1950s, Kuffler, Kravitz and Potter started a systematic search for an inhibitory transmitter in crustacean nervous system, and found in 1963 that inhibitory axons of lobsters contain about 200 times higher concentration of GABA than excitatory axons, which strongly suggested that GABA is an inhibitory transmitter. When I joined the Kuffler's group in 1964, one crucial piece of evidence was missing, namely the demonstration of the release of GABA from an inhibitory axon by stimulation. This was successfully achieved in 1966 by Otsuka, Iversen, Hall and Kravitz.

I would like to thank Prof. Chikako Tanaka and the Organizing Committee of the GABA Symposium '95 for inviting me to give this opening lecture. It is already a long time ago that I worked in the field of GABA, and I hardly expected that people still remember the work we did 30 years ago. Indeed, when I was a medical student if I heard that a study had been done 30 years before, I tended to think that the investigators must be dead!

This morning I would like to tell you about the time when GABA was established as an inhibitory transmitter, and I hope you permit me to tell you about my own participation. It is now generally believed that the nervous system uses many different neurotransmitters, the number of which may exceed 50. In the early 1960s this was not the prevailing thought. Only two kinds of neurotransmitters, acetylcholine and catecholamines, namely adrenaline and noradrenaline, were known, and it was thought that the nervous system uses a small number, say 5 and certainly less than 10, of neurotransmitters. As I will later tell you, GABA was established as an inhibitory neurotransmitter at the crustacean neuromuscular junction in 1966, and in retrospect this was the starting point of the rapid expansion of the list of neurotransmitters.

It was 1950 when GABA appeared in the field of Neuroscience. Eugene Roberts in St Louis was analyzing the amino acid contents of neoplastic tissues by means of paper chromatography in the hope of finding some difference between normal and neoplastic tissues[1]. Dr. Roberts did not find any difference between normal and neoplastic tissues, but to his surprise he found a big difference between brain and non-brain tissues. When he analyzed the extract of mouse brain by two dimensional paper chromatography, he found a large ninhydrin positive spot which later turned out to be GABA. The absolute identification of GABA in the extract was done by Sidney Udenfriend by means of the isotope derivative method.

Roberts and Frankel submitted an abstract to the Federation meeting, and when Dr. Roberts was registering at the desk, he was asked to share a room with someone who was unable to reserve a room. On his arrival to the hotel, Roberts found his roommate to be Jorge Awapara, who was, to the great surprise of Roberts, also going to report on the presence of GABA in the brain.

The significance of the presence of GABA specifically in the brain was not immediately understood. It is surprising that no one mentioned the neurotransmitter role of GABA, and no one examined the action of GABA on neurons for the next few years. In 1956 however, at the International Congress of Physiology in Brussels, Hayashi and Nagai reported that GABA exerts an inhibitory action on dog cortical neurons[2)]. Hayashi was a professor of physiology at Keio University, and by applying amino acids to dog's motor cortex he discovered that GABA and glutamate exert inhibitory and excitatory actions, respectively, on mammalian central neurons. This brilliant deduction of Hayashi based on relatively simple experiments may have been related to the fact that Hayashi was a famous and successful writer of detective stories.

In the same year Bazemore, Elliott and Florey working in Montreal published a paper in Nature, that reported an inhibitory action of GABA on crustacean stretch receptors, and they later suggested the hypothesis that GABA may be an inhibitory transmitter[3)].

In the 1950s and 1960s several investigators compared the inhibitory action of GABA to the endogenous, unknown inhibitory neurotransmitter at crustacean neuromuscular junction and stretch receptors, and they showed that GABA faithfully reproduced the action of the unknown inhibitory neurotransmitter[4,5,6)].

In the late 1950s Prof. Stephen Kuffler together with Ed Kravitz and Dave Potter started a project of systematically searching for the inhibitory transmitter in the crustacean nervous system. They started the project in Baltimore and soon moved to Boston, and as you know both cities are famous for seafood. Large crates of lobsters arrived in the laboratory. Some 500 lobsters were sacrificed, and the extract of their nervous system was subjected to a long series of fractionations and bioassays at crustacean neuromuscular junction. As a result, ten different substances producing inhibitory effects were found. Among them, GABA was by far the most potent[7)].

They then dissected single excitatory and inhibitory axons of lobsters and analyzed their GABA contents. They found that inhibitory axons contain a high concentration, i.e. 100mM of GABA whereas excitatory axons contain 200 times lower concentration of GABA. Of course this strongly suggested that GABA is the inhibitory transmitter[8].

In memory of Prof. Kuffler, Sir Bernard Katz stated in 1982[9]:

"Although Kuffler and his collaborators expressed themselves very cautiously concerning the physiological significance of their observations, to most of us their result together with previous findings on the actions of GABA was convincing evidence for its identification as the inhibitory transmitter."

Sir Bernard is entirely right in retrospect, but at the time, this was only clear to very wise people like Sir Bernard. After the work of Kuffler, Kravitz, Potter and colleagues were published in 1963, many people were still skeptical or even opposed to the transmitter role of GABA. One of the main reasons was that a crucial piece of evidence was still missing, namely the demonstration of the release of GABA from an inhibitory axon upon stimulation.

In 1964 I was fortunate enough to be allowed to work in Prof. Kuffler's laboratory. On my arrival I was told by Prof. Kuffler that I should work with Dave Potter, and Dave suggested that we should analyze the contents of single excitatory and inhibitory cell bodies. We showed that the inhibitory cell bodies contain high concentration of GABA whereas the excitatory cell bodies contain low concentration of GABA. We made a map of abdominal ganglion of the lobster and showed that the locations of individual excitatory and inhibitory cell bodies are constant from one lobster to the next[10].

After this extremely enjoyable year, I had another year to work, and I said to myself that since I had already completed a study before my going back home, I could try a more risky project. I decided to tackle the last missing piece of evidence that GABA may be an inhibitory transmitter, namely, the demonstration that GABA is released from an inhibitory axon upon stimulation. I thought that this must be a very difficult project which had been tried by previous investigators without success, and that I should try some complicated experiments because simple experiments must have been done in the past without success. I prepared hot GABA, injected it into an inhibitory cell body, and waited for some time so that the labelled GABA could be transported to the neuromuscular junction. When I stimulated the inhibitory axon and measured the radioactivity of the perfusate, I could obtain just low, negligible count. Then I soaked an isolated nerve-muscle preparation in a radioactive GABA solution, waited for some time so that the hot GABA would be taken up by nerve terminals and then stimulated the inhibitory axon. I obtained some increase in radioactivity in the perfusate upon stimulation. I was excited and showed the result to my colleagues, but to my disappointment they did not show much interest.

Since only a few months were left for me before I had to go back to Tokyo, I consulted my friend Dick Orkand working next door about whether I should give up and change my project. Dick told me I had invested too much time, and it was too late to change the project. One day in the corridor Prof. Kuffler asked me about the progress of my experiments. When I told him that I could not get an increase in GABA release upon stimulation, he said "May be GABA is not the transmitter". It was clear that he was joking, but I was slightly upset because when I was desperate, I often said to myself, "This is Steve's project, and I know that his ideas are always right".

Around this time, Les Iversen who had been working on this project with me, told me that he and Ed Kravitz succeeded in developing a new method of determination of GABA using ion exchange resins to separate GABA from salts. He suggested that I should simply perfuse a nerve-muscle preparation, stimulate the inhibitory nerve and give the perfusates to them. Within a few days we obtained a clear indication of the release of GABA. We changed our preparation from abdominal superficial flexor muscle to the opener muscle of the big claw, and the results became much much better[11). Les, Ed Kravitz, Zach Hall and I worked hard because we had very little time left. Les had to go back to Cambridge. Zach had to go to San Francisco, and I had to return to Tokyo. Since we never had a chance to get together to take a picture at that time, a few years ago at the occasion of Ed's 60th birthday, we decided finally to take a picture of us.

When Dave and Ed reported the results at an International Symposium in Stockholm in 1966[12,13)], at the end of the session Dr. Dudel gave a summary as follows[14)]:

"As you all know there are only a few instances of synaptic transmission in which the transmitter is fully identified according to the criteria cited above".

and, after describing the role of acetylcholine as a transmitter, Dudel continued:

"After this session I feel we can add another substance to the list of approved transmitter, namely GABA as an inhibitory transmitter in the crustacean neuromuscular junction".

I am afraid that I have no time to describe the work done by K. Obata, M. Ito, Y. Okada and myself on the role of GABA as an inhibitory transmitter of Purkinje neurons in mammalian CNS. While we were working on lobster nervous system in the United States, Ito, Obata and colleagues in Tokyo were obtaining many important lines of evidence showing that GABA is the inhibitory transmitter of Purkinje neurons innervating vestibular Deiters cells[15)], and after I returned to Tokyo I helped Obata and contributed to their project. We measured the contents of GABA in isolated dorsal Deiters neurons by using the enzymatic cycling method of Lowry. We found that isolated dorsal Deiters cells contained high concentrations of GABA, which, however, were dramatically reduced by removal of cerebellar vermis. This suggested that GABA is concentrated in axon terminals of Purkinje neurons making synapses with Deiters neurons[16)]. That this is the case was shown by Okada by direct measurement of GABA contents of presynaptic terminals attached to Deiters cells[17)].

I left the field of GABA in 1975 and changed my field to substance P. When I saw the prosperity of the GABA field from a distance, I consoled myself by saying to myself: "If we manipulate a system of a major neurotransmitter like GABA, we get convulsions or death of the individual, whereas, if we manipulate a system of a minor neurotransmitter like substance P or enkephalins, we may get a useful drug effect like the analgesia by morphine." This is evidently mistaken comfort because the GABA-mediated transmission is related to the effects of many useful drugs, such as benzodiazepines and barbiturates, and possibly also alcohol and general anesthetics. Therefore, I feel some regret that I left the GABA field so early.

References

1. Roberts E. Introduction. In: GABA in Nervous System Function. Roberts E, Chase TN, Tower DB, editors. New York: Raven Press, 1976: 1-6.
2. Hayashi T, Nagai K. Action of ω-amino-acids on the motor cortex of higher animals, especially γ-amino-β-oxy-butyric acid as the real inhibitory principle in brain. XXth International Physiological Congress, Brussels, 1956: 410.
3. Bazemore AW, Elliott KAC, Florey E. Isolation of Factor I. J Neurochem 1957; 1: 334-339.
4. Boistel J, Fatt P. Membrane permeability change during inhibitory transmitter action in crustacean muscle. J Physiol 1958; 144: 176-191.
5. Kuffler SW. Excitation and inhibition in single nerve cells. The Harvey Lectures, 1958-1959. New York: Academic Press, 1960: 176-218.
6. Takeuchi A, Takeuchi N. Localized action of gamma-aminobutyric acid on the crayfish muscle. J Physiol 1965; 177: 225-238.
7. Kravitz EA, Kuffler SW, Potter DD, van Gelder NM. Gamma-aminobutyric acid and other blocking compounds in Crustacea. II Peripheral nervous system. J Neurophysiol 1963; 26: 729-738.
8. Kravitz EA, Kuffler SW, Potter DD. Gamma-aminobutyric acid and other blocking compounds in Crustacea. III. Their relative concentrations in separated motor and inhibitory axons. J Neurophysiol 1963; 26: 739-751.
9. Katz B. Stephen William Kuffler 1913-1980. Biographical Memoirs of Fellows of the Royal Society. 1982; 28: 225-259.
10. Otsuka M, Kravitz EA, Potter DD. Physiological and chemical architecture of a lobster ganglion with particular reference to gamma-aminobutyrate and glutamate. J Neurophysiol 1967; 30: 725-752.
11. Otsuka M, Iversen LL, Hall ZW, Kravitz EA. Release of gamma-aminobutyric acid from inhibitory nerves of lobster. Proc Natl Acad Sci U.S.A. 1966; 56: 1110-1115.
12. Potter DD. The chemistry of inhibition in crustaceans with special reference to gamma-aminobutyric acid. In: Structure and Function of Inhibitory Neuronal Mechanisms. von Euler C, Skoglund S, Söderberg U, editors. Oxford: Pergamon Press, 1968: 359-370.

13. Kravitz EA, Iversen LL, Otsuka M, Hall ZW. Gamma-aminobutyric acid in the lobster nervous system: Release from inhibitory nerves and uptake in nerve-muscle preparations. In: Structure and Function of Inhibitory Neuronal Mechanisms. von Euler C, Skoglund S, Söderberg U, editors. Oxford: Pergamon Press, 1968: 371-376.

14. Dudel J. Criteria for identification of transmitter substances. In: Structure and Function of Inhibitory Neuronal Mechanisms. von Euler C, Skoglund S, Söderberg U, editors. Oxford: Pergamon Press, 1968: 523-525.

15. Obata K, Ito M, Ochi R, Sato N. Pharmacological properties of the postsynaptic inhibition by Purkinje cell axons and the action of γ-aminobutyric acid on Deiters neurones. Exp Brain Res 1967; 4: 43-57.

16. Otsuka M, Obata K, Miyata Y, Tanaka Y. Measurement of γ-aminobutyric acid in isolated nerve cells of cat central nervous system. J Neurochem 1971; 18: 287-295.

17. Okada Y. Fine localization of GABA (γ-aminobutyric acid) and GAD (glutamate decarboxylase) in a single Deiters' neuron—significance of the uneven distribution of GABA and GAD in the CNS. In: Problems in GABA Research from Brain to Bacteria. Okada Y, Roberts E, editors. Amsterdam: Excerpta Medica, 1982: 23-29.

GABA: Receptors, Transporters and Metabolism
ed. by C. Tanaka & N.G. Bowery

PHYSIOLOGICAL ASPECTS OF GAD-GABA SYSTEM IN THE PANCREATIC ISLET

Hiroshi Taniguchi, Junichi Katoh*, Motoyoshi Sakaue† and Yasuhiro Okada§

Department of Metabolism and Community Health Science, Faculty of Health Science,
†Second Department of Internal Medicine and §First Department of Physiology,
Faculty of Medicine, Kobe University School of Medicine, and
*Department of Internal Medicine, Hyogo Rehabilitation Center, Kobe, Japan

Summary: In the pancreatic islet, which is non-neuronal, γ-aminobutyric acid (GABA) and its synthesizing enzyme, L-glutamate decarboxylase (GAD), are so high in concentration and activity respectively as in the brain, where GABA functions as an inhibitory neurotransmitter. This GABA may exert an inhibitory action on the secretion of somatostatin and glucagon possibly via GABAergic peri-insular neurons. Besides, the GAD-GABA system appears to be involved in insulin synthesis in the islet.

γ-Aminobutyric acid (GABA) is an inhibitory neurotransmitter and present in the brain abundantly. The major pathway of GABA synthesis is decarboxylation of glutamate catalyzed by L-glutamate decarboxylase (GAD), while GABA degradation is catalyzed by GABA-α-ketoglutarate transaminase (GABA-T). The activity of GAD is also high in the neuronal tissues. Some non-neuronal tissues also have high concentration of GABA and high GAD activity, which are comparable to those of the brain. One of such tissues is the pancreatic islet. Especially B cells of the islet possess extremely high concentration of GABA and GAD activity (1,2). This high activity of GAD indicates that GABA is synthesized in B cells rather than taken up from the circulation. GABA is present not only in the islet of the mammals but in the Brockmann body of the fish such as Seriora Quinqueradiata, which corresponds to the pancreatic islet of the mammals. Thus, GABA exists in a variety of species (Table 1). Two isoforms of GAD, i.e., GAD65 and GAD67, are present in the brain as well as the islet (3).

Table 1. GABA content in the pancreatic islet of different species.

Source	GABA content (mmol/kg dry weight) Mean ± SEM
Rat (islet)	16.3 ± 1.4 (n=27)
Mouse (islet)	3.0 ± 0.1 (n=19)
Human (islet)	8.3 ± 1.2 (n=5)
Fish (Brockmann body)	0.67 ± 0.1 (n=5)*
Human insulinoma case 1	25.6 ± 0.8 (n=6)
case 2	10.6 ± 1.7 (n=10)
Rabbit (Deiters' nucleus) dorsal part	17.5 ± 1.6 (n=43)
ventral part	8.4 ± 0.4 (n=6)
Rat (acinus)	2.0 ± 0.2 (n=52)

* The value is expressed as mmol/ kg wet weight. Approximate value per kg dry weight is obtained through multiplication by 5.

Though GABA is highly concentrated in the brain and the pancreatic islet, GABA is not taken up in B cells of the islet, whereas it is incorporated in the neuronal tissues abundantly. However, some of D cells and the acinar cells incorporate GABA (4). Immunohistochemical studies show that only a small component of B cell GABA is present in insulin secretory granules. GABA is associated with a vesicular component distinctly different from insulin secretory granules. GAD is found in a unique extensive tubular cisternal network (GAD complex) (5-7). GABA- and GAD-containing neuronal cell bodies send extremely fine processes into islets, which seem to surround most of the cells of the islet mantle. Few processes are found in the B cell region of the islet (7). This suggest that the neuronal GABA may have a regulatory role in secretion of glucagon and somatostatin from A and D cells situated in the periphery of the islet. The cellular and subcellular localization of GABA, GABA-T and GAD as well as GABA uptake in the rat pancreatic islet and acinus is summarized in Table 2 (8).

GABA content of the islet changes in parallel to that of the superior colliculus, the hippocampus and the striatum of the brain in response to 3-mercaptopropionic acid which competes with glutamate and blocks GAD, as well as sodium valproate which competes with GABA-T in vivo. In these non-neuronal as well as neuronal tissues, the former agent lowers the GABA content in a dose-dependent fashion, whereas the latter elevates it. The similar alterations are also induced by these agents in vitro (9,10). Thus, the properties of GAD and GABA-T in terms of response to their inhibitors do not appear to be different between the neuronal tissues and the non-neuronal islet.

The function of the GAD-GABA system in the islet has not been clarified. However, several suggestive observations have been reported in recent years. Studies with the perifusion and

Table 2. Cellular and subcellular localization of GABA, GABA-T and GAD as well as GABA uptake in the rat pancreatic islet and acinus.

Cell type	GABA	GABA-T	GAD	GABA uptake
A cells	-	-		
	+*			
B cells				+†
membrane fraction			+++	
cytoplasm	+++		+++	
synaptic-like microvesicles	+++			
mitochondria	++	+++		
secretory granules	-			
RINm5F cells			-	
D cells	-	-		+
	+*			
Acinar cells	+		+	+‡
zymogen granules	+++	-	+	
cytoplasm	+	-	+++	

* Measurement was done in the outer part of the pancreatic islet consistent with A and D cells.
† Immature B cells only.
‡ In fetal and perinatal period; controversial results were obtained in adult animals.
(Cited from reference 8)

static incubation system of rat islets indicate that glucose-stimulated somatostatin secretion is suppressed by GABA as well as its agonist, muscimol (11), whereas insulin secretion is unchanged (12). This suppression of the somatostatin secretion is reversed by an addition of bicuculline, antagonist of GABAA receptor (11). In studies with the static incubation of guinea pig islets, GABA significantly suppresses the arginine-stimulated glucagon secretion, and this

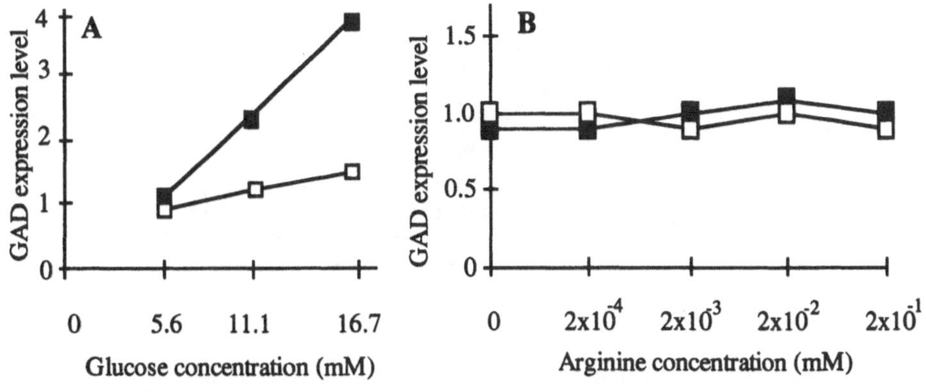

Figure 1. Effect of glucose (A) and arginine (B) on the level of GAD expression in isolated rat islets. (■: GAD65, □: GAD67) (Cited from reference 14)

is blocked by bicuculline (13), though muscimol, agonist of GABAA receptor, does not affect this arginine-stimulated glucagon secretion in studies with isolated perfused rat pancreas (7).

Our studies indicate that ^{3}H-leucine incorporation into (pro-)insulin immunoreactivity is enhanced by γ-vinyl GABA which is a selective GABA-T inhibitor, and aminooxyacetic acid which inhibits GABA-T more strongly than GAD (12). This potentiation of insulin synthesis agrees with the increased expression of GAD by elevation of glucose concentration which stimulates not only secretion but synthesis of insulin. However, this GAD expression is unchanged by elevation of arginine concentration which stimulates insulin secretion, but does not change its synthesis (Fig.1) (14). Insulin secretion stimulated by 3-phenylpyruvate, non-nutrient secretagogue, is reported to be suppressed by γ-acetylenic GABA, an irreversible GABA shunt inhibitor selective for GABA-T and GAD. This may imply that the GABA shunt pathway has a role in insulin secretion (7).

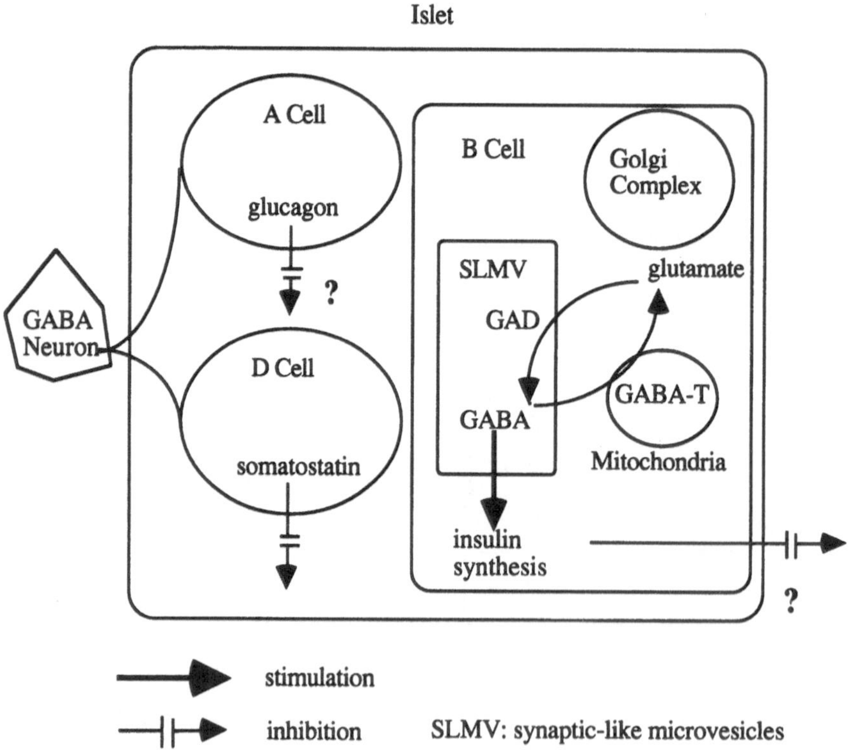

Figure 2. Hypothetical physiological function of GABA-GAD system in pancreatic islets.

In conclusion, the presence of high concentration of GABA and high activity of GAD is a consistent observation in the islet of all species from the fish to the human. The inhibition of somatostatin secretion from D cells by GABA is consistent in most studies reported, while that of glucagon secretion is still conflicting. GABA in B cells of the islet potentiates insulin synthesis and the GABA shunt pathway may be related to insulin secretion. Thus, the GAD-GABA system in the islet is strongly suggested to be involved in the regulation of hormone secretion as well as insulin synthesis (Fig. 2).

References

1. Okada Y, Taniguchi H, Shimada C. High concentration of GABA and high glutamate decarboxylase activity in rat pancreatic islets and human insulinoma. Science 1976; 194: 620-622.

2. Taniguchi H, Okada Y, Seguchi H, Shimada C, Seki M, Tsutou A, Baba S. High concentration of gamma-aminobutyric acid in pancreatic beta cells. Diabetes 1979; 28: 629-633.

3. Christgau S, Schierbeck H, Aanstoot H-J, Aagaard L, Begley K, Kofod H, Hejnaes K, Baekkeskov S. Pancreatic β cells express two autoantigenic forms of glutamic acid decarboxylase, a 65-kDa hydrophilic form and a 64-kDa amphiphilic form which can be both membrane-bound and soluble. J. Biol. Chem. 1991; 266: 21257-21264.

4. Okada Y, Hosoya Y, Taniguchi H. Uptake of ^{3}H-GABA(gamma-aminobutyric acid) and ^{3}H-leucine in the pancreatic islets and substantia nigra of the rat. Experientia 1980; 36: 131-133.

5. Reetz A, Solimena M, Matteoli M, Folli F, Takei K, De Camilli P. GABA and pancreatic β-cells: colocalization of glutamic acid decarboxylase (GAD) and GABA with synaptic-like microvesicles suggests their role in GABA storage and secretion. EMBO J. 1991; 10: 1275-1284.

6. Garry DJ, Appel NM, Garry MG, Sorenson RL. Cellular and subcellular immunolocalization of L-glutamate decarboxylase in rat pancreatic islets. J. Histochem. Cytochem. 1988; 36: 573-580.

7. Sorenson RL, Garry DJ, Brelje TC. Structural and functional considerations of GABA in islets of Langerhans — β-cells and nerves. Diabetes 1991; 40: 1365-1374.

8. Michalik M, Erecinska M. GABA in pancreatic islets: metabolism and function. Biochem. Pharmacol. 1992; 44: 1-9.

9. Katoh J, Taniguchi H, Ogura M, Kasuga M, Okada Y. The effect of a convulsant, 3-mercaptopropionic acid, on the GABA level in rat pancreatic islets and brain, in-vivo and in-vitro. Pharmaceut. Sci. 1995;1: 23-25.

10. Katoh J, Taniguchi H, Kasuga M, Okada Y. The effect of an anticonvulsant, sodium valproate, on the GABA level in rat pancreatic islets and brain, in-vivo and in-vitro. Pharmaceut. Sci. 1995; 1: 239-241.

11. Taniguchi H, Yoshioka M, Ejiri K, Ishihara K, Tamagawa M, Hirose Y, Ishihara K, Utsumi M, Baba S, Okada Y. Suppression of somatostatin release and increase of somatostatin content in pancreatic islets by GABA. In: Problems in GABA Research — from Brain to Bacteria. Okada Y, Roberts E, editors. Amsterdam: Excerpta Medica, 1982: 406-411.

12. Taniguchi H, Murakami K, Yoshioka M, Ejiri K, Ishihara K, Baba S, Okada Y. GABA and insulin in pancreatic islets. In: Problems in GABA Research — from Brain to Bacteria. Okada Y, Roberts E, editors. Amsterdam: Excerpta Medica, 1982: 387-405.

13. Rorsman P, Berggren P-O, Bokvist K, Ericson H, Möhler H, Östenson C-G, Smith PA. Glucose-inhibition of glucagon secretion involves activation of GABAA-receptor chloride channels. Nature 1989; 341: 233-236.

14. Katoh J, Taniguchi H, Kasuga M. Response of glutamic acid decarboxylase to glucose but not arginine in islets. Life Sci. 1995; 56: 1799-1805.

GABA: Receptors, Transporters and Metabolism
ed. by C. Tanaka & N.G. Bowery

MEMBRANE ASSOCIATED L-GLUTAMATE DECARBOXYLASE AND INSULIN-DEPENDENT DIABETES MELLITUS (IDDM)

[1]J.-Y. Wu, [1]B.Nathan, [1]C.-C. Hsu, [2]C.-Y. Kuo, [2]G.A. Burghen, [1]R. Wu and [1]X.-W. Tang

[1]Department of Physiology & Cell Biology, University of Kansas, Lawrence, KS 66045-2106, USA and [2]Department of Pediatrics, University of Tennessee, Memphis, TN 38163, USA

Summary: Recently, three novel forms of membrane associated L-glutamate decarboxylase (M-GAD) referred to as M-GADI, M-GADII and M-GADIII were isolated and purified from porcine brain. M-GADI and M-GADII appear to be integral membrane proteins whereas M-GADIII is a peripheral protein attached to membranes in a Ca^{2+} dependent manner. In addition, M-GADI and M-GADII were found to be the major autoantigens in insulin-dependent diabetes mellitus (IDDM), also known as type 1 diabetes, even more potent than GAD_{65}/GAD_{67}, as demonstrated in immunoprecipitation and immunoblotting tests. M-GADI has a native molecular weight of 96 kDa and consists of two identical subunits of 48 kDa whereas both M-GADII and M-GADIII are homodimers of 60 kDa. M-GADI and M-GADII have fulfilled all three criteria routinely used to determine whether a protein is an integral membrane protein including partitioning in the detergent phase in Triton X-114 partitioning assay, shift in electrophoretic mobility in the presence of ionic detergent and elution by low ionic strength buffer in hydrophobic interaction chromatography. Based on these observations, it is hypothesized that M-GADI and M-GADII may be involved in the pathogenesis of IDDM.

Introduction:

Although there are many lines of evidence to support the hypothesis that insulin-dependent diabetes mellitus [(IDDM), also known as type 1 diabetes or juvenile diabetes] is caused by autoimmune destruction of insulin secreting pancreatic β-cells (1), the underlying mechanism that triggers the initial "insult" to the β-cells remains unclear. The earliest detectable β-cell autoantibodies in IDDM have been shown to be directed against 64K (MW = 64,000 kDa) and 38K islet protein (2). The 64K autoantigen in IDDM was subsequently identified as the GABA-synthesizing enzyme, L-glutamate decarboxylase (GAD) (3). Later, it has been shown that both forms of GAD, GAD_{64} and GAD_{65} [later changed to as GAD_{65} and GAD_{67} from sequence information (4), for reveiw see (5)] were identified as autoantigens in IDDM with GAD_{64} (now referred to as GAD_{65}) as the major form (6). Recently, two groups have independently shown that loss of tolerance to GAD_{65} is an early and necessary step towards the development of diabetes in NOD mice (7,8). Furthermore, these two studies also show that intravenous (7) or intrathymic (8) injection of GAD_{65} in three-week-old NOD mice markedly reduces T-cell proliferative responses both to GAD and to other IDDM autoantigens and prevents both insulitis and diabetes suggesting that GAD_{65} might be involved in the initiation of autoimmune process.

If GAD_{65} is indeed to be the earliest antigen for the initiation of autoimmune reaction as proposed (7,8), it would be expected that the target antigen, in this case GAD_{65}, should be localized on the β-cell surface to trigger the cell-mediated destruction. However, several laboratories have failed to detect GAD_{65} on the β-cell surface (9-11). The reason for the failure of detecting GAD_{65} on the β-cell surface could be due to technical reasons as well as due to the intrinsic properties of GAD_{65}. From sequence information, it is clear that both GAD_{65} and GAD_{67} are not integral membrane proteins since none of them contains a stretch of hydrophobic amino acids long enough to span the membrane (about 20 residues), a typical feature for integral membrane proteins (4,5).

Our recent findings of a new class of GAD, namely, membrane associated GAD (M-GAD), which are major autoantigens in IDDM may provide an alternative mechanism regarding the role of GAD in the etiology of IDDM (12-15). Evidence is presented to support the hypothesis that M-GADI and M-GADII are integral membrane proteins and may be involved in the pathogenesis of IDDM.

Materials and Methods

Preparation and purification of membrane associated GAD - In a typical preparation, a 10% porcine brain homogenate was made in doubly distilled water containing 2 mM S-(2-aminoethyl)isothiouronium bromide hydrobromide (AET) and 0.4 mM PLP. The homogenate was centrifuged at 100,000 x g for 1 hr and the pellet was rehomogenized and centrifuged as before. The pellet after the second centrifugation was solubilized in 0.05 M standard buffer (0.05 M KP_i, pH 7.2/0.4 mM PLP/2 mM AET/1 mM EDTA) containing 0.5% Triton X-100. The solubilized M-GAD was purified to homogeneity by a combination of column chromatographies on DE-52, gel filtration, hydroxylapatite, Sephadex-200 and non-denaturing gradient polyacrylamide gel electrophoresis (PAGE) as previously described (12-15).

Immunoblotting with non-denaturing- and SDS-PAGE - Briefly, blotting of proteins from non-denaturing gradient PAGE or SDS-PAGE was carried out at 4°C in an LKB 2005 transfer unit, and the tank buffer contained 25 mM Tris Cl (pH 8.3), 0.192 M glycine, 0.5%

SDS (or 0.05% SDS in the nondenaturing system), and 20% methanol. Blotting time varied from 3-12 hr based on the number of gels. Immunodetection was carried out with the ECL Western blot system (Amersham) as described (12).

Immunoprecipitation. Partially purified M-GAD or soluble GAD (S-GAD) in standard buffer was incubated with 5 µl of normal human serum, preimmune rabbit serum, IDDM serum, or standard buffer at 4°C for 24 hr. After incubation, samples were centrifuged at 10,000 x g for 3 min, and the resulting pellets were washed once in standard buffer. The pellets from the second wash and the two supernatant solutions were assayed for GAD activity. The activity in the two supernatants was combined to obtain the total GAD activity in the supernatant.

Hydrophobic interaction chromatography - Hydrophobic interaction chromatography was conducted as described (12) using a 5PW Phenyl-Sepharose HPLC column.

Triton X-114 Phase-Partitioning Assay was carried out as described (16), except that both the sample and the sucrose cushion contained 150 mM KP_i (pH 7.2) 2 mM AET, and 0.4 mM PLP instead of 10 mM Tris·HCl (pH 7.4) and 150 mM NaCl as reported previously (12).

Charge-Shift Electrophoresis. Charge-shift electrophoresis was carried out (17) with modifications (12). The buffer in the gel was 0.05 M glycine·NaOH, pH 8.4/0.1 M NaCl/1 mM AET/0.2 mM PLP/0.5% Triton X-100 with or without 0.25% sodium deoxycholate. The running buffer was the same except that Triton X-100 was omitted. Electrophoresis was carried out at 5 V/cm until the tracking dye reached within 1 cm of the bottom of the gel. After electrophoresis, lanes loaded with enzymes were sliced into 1 cm squares and assayed for GAD activity.

Results and Discussion

Evidence of multiple forms of M-GAD in porcine brain - Similar to S-GAD, M-GAD was also found to be present in the brain in multiple forms. When solubilized M-GAD was separated on a DE-52 column, it was fractionated into three activity peaks, namely M-GADI, II and III as shown in Fig. 1.

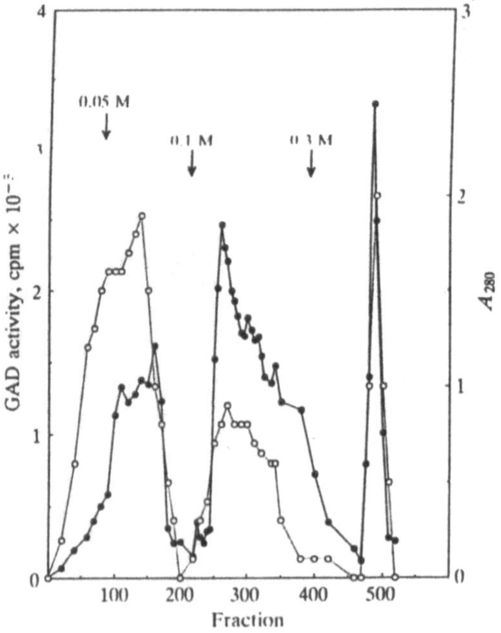

Fig. 1. Purification of solubilized M-GAD on DE-52 column.
Three GAD activity peaks, referred to as M-GADI, II and III were obtained after elution with 0.05 M, 0.1 M and 0.3 M standard GAD-buffer containing 0.5% Triton X-100, respectively.

Purification of M-GAD from porcine brain. – M-GADI, II and III were further purified by a combination of column chromatographies including AcA34 gel filtration, hydroxylapatite (HTP) and Sephadex G-200 gel filtration and native gradient PAGE as summarized in Table 1. M-GADI thus obtained migrated as a single Protein band corresponding to a molecualr weight of 96 ± 5 kDa on a native 5-25% gradient PAGE (Fig. 2C), which represented an overall purification of 2083-fold over DEAE-52 preparation. Furthermore, the protein band correlated with GAD activity. When M-GADI was analyzed on SDS-PAGE, a single protein band corresponding to a molecular weight of 48 ± 3 kDa was obtained suggesting that M-GADI is a homodimer of 48 kDa (Fig. 2E). These results suggest that M-GADI is a new form of GAD which differs in its molecular structure from any other forms of GAD that have been reported in the literature (5). Both M-GADII and M-GADIII have also been purified to apparent homogeneity and found to have a native molecular weight of 120 kDa consisting of two identical subunits of 60 kDa (Figs. 3 and 4). However, M-GADII differs fromM-GADIII in that M-GADII is an integral membrane protein while M-GADIII is a peripheral membrane protein (see followng section).

Table 1. Purification of M-GADI, M-GADII and M-GADIII from porcine brain

Sample	Specific activity, (units/mg) x 10^3			Purification, fold		
	M-GADI	M-GADII	M-GADIII	M-GADI	MGADII	MGADIII
Brain homogenate	0.16	0.160	0.16	--	--	--
Membrane (P_2M)	0.13	0.130	0.13	--	1	1
Solubilized M-GAD	0.19	0.191	0.19	--	1	1
DE-52	0.06	1.08	0.07	1	7	4
Ultrogel AcA 34	0.19	2.06	1.5	3	13	9
Hydroxylapatite	1.69	4.71	4	28	30	25
Sephadex G-200	2.97	12.8	8	50	80	50
Nondenaturing gradient PAGE	125	1095	462	2083	6844	2,893

One unit of activity forms 1 µmol of product per minute at 37°C. The specific activity of MGADI after DE-52 was used as reference for calculation of fold purification.

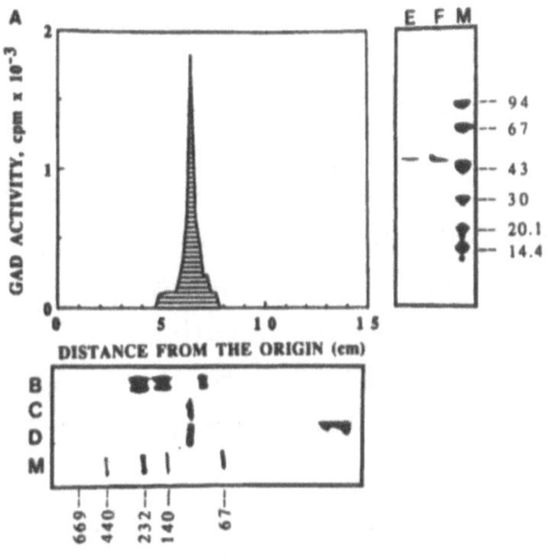

Fig. 2. Analysis of M-GADI on various gel electrophoresis and immunoblotting systems using IDDM serum.

(A) GAD activity profile of M-GADI on a native 5-25% gradient PAGE. (B) Protein profile of M-GADI from Sephadex G-200 column on a native 5-25% gradient PAGE. (C) Protein profile of purified M-GADI on a native 5-25% gradient PAGE. *(D) Immunoblotting of crude M-GADI preparation with IDDM serum on a native 5-25% gradient PAGE. (E) Protein profile of purified M-GADI on a SDS-PAGE. *(F) Immunoblotting of crude M-GADI preparation with IDDM serum on a SDS-PAGE. (M) Molecular weight markers in kDa. *IDDM serum used in immunoblotting tests was diluted at 1:6000. Note: GAD_{65}/GAD_{67} could not be detected by IDDM serum at such diluted condition.

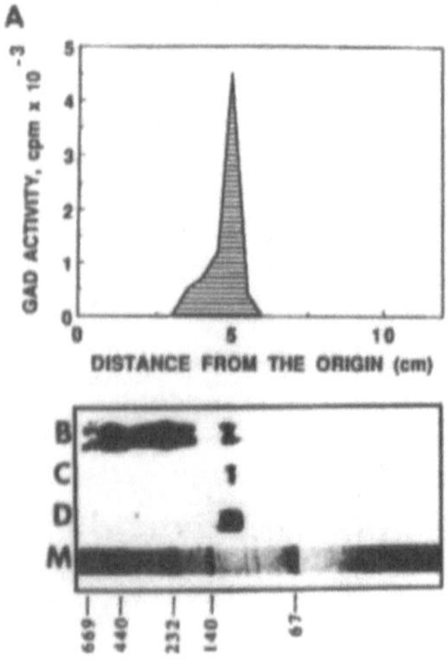

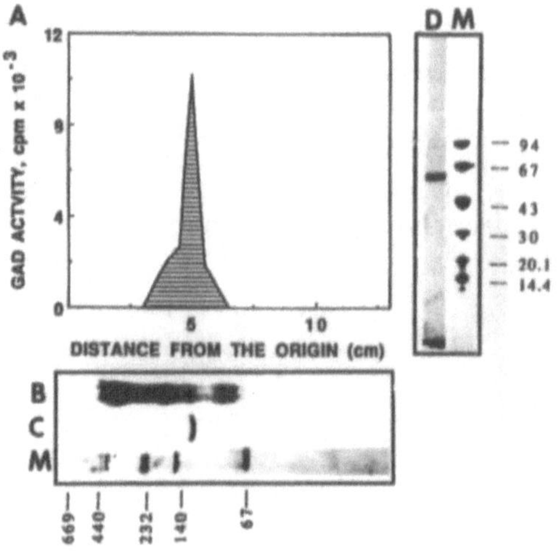

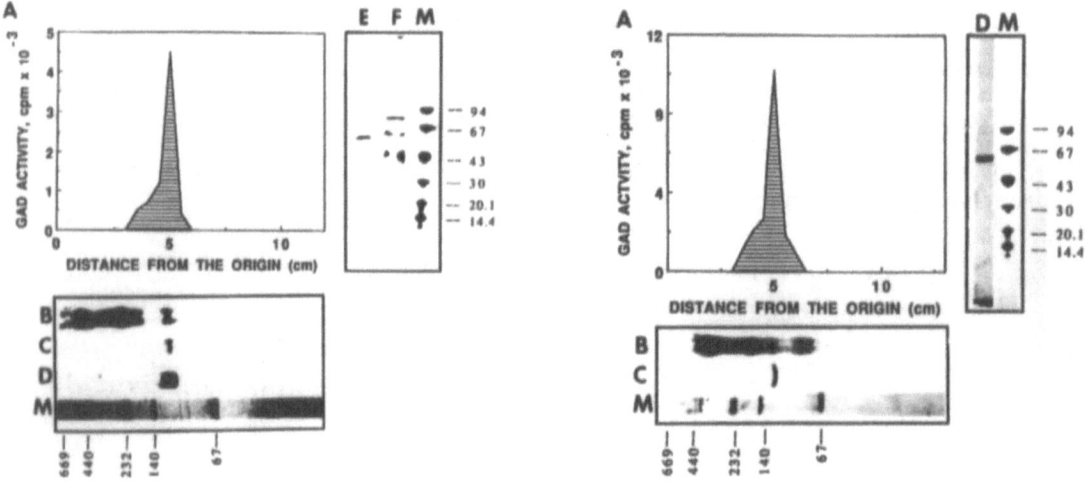

Fig. 3. Analysis of M-GADII on various gel electrophoresis and immunoblotting systems using IDDM serum.
(A) GAD activity profile of M-GADII on a native 5-25% gradient PAGE. (B) protein profile of M-GADII from Sephadex G-200 column on a native 5-25% gradient PAGE. (C) protein profile of purified M-GADII on a native 5-25% gradient PAGE. (D) immunoblotting of M-GADII with IDDM serum (1:6000 dilution) on a native 5-25% gradient PAGE. (E) protein profile of purified M-GADII on a 10% SDS-PAGE. (F) immunoblotting of M-GADII with IDDM serum (1:6000 dilution) on a 10% SDS-PAGE. (M) molecular weight markers in kDa.

Fig. 4. Analysis of M-GADIII on various gel electrophoresis systems.
(A) GAD activity profile of M-GADIII on a native 5-25% gradient PAGE. (B) protein profile of M-GADIII from Sephadex G-200 column on a native gradient 5-25% PAGE. (C) protein profile of purified M-GADIII on a native 5-25% gradient PAGE. (D) protein profile of purified M-GADIII on a SDS-PAGE. (M) molecualr weight markers in kDa.

Identification of M-GADI and M-GADII as major autoantigens in IDDM – (i) Immunoprecipitation tests using IDDM serum -- In immunoprecipitation tests, IDDM serum precipitated almost 100% of M-GADI and M-GADII activity as compared to normal human serum (Table 2).

Table 2. Immunoprecipitation test of M-GADI and M-GADII using IDDM serum.

Serum	GAD Activity (CPM)‡ M-GADI Supernatant	M-GADI Pellet	M-GADII Supernatant	M-GADII Pellet
Normal Human Serum	7224±111	153±36	7558±135	180±46
IDDM Serum	197±21	7848±220	211±38	8055±288
Control*	8621±162	164±17	9018±319	193±22

‡GAD actvity is the average from five different samples.

*Controls are the samples containing standard GAD buffer without serum.

(ii) Immunoblotting test using IDDM serum -- When M-GADI preparation was used as antigen in immunoblotting tests with IDDM serum (1:6000 dilution), a single protein band corresponding to a molecular weight of native M-GADI, 96±3kD and subunit of M-GADI, 48±2kD (Fig. 2D and 2F), were obtained in the native gradient PAGE and SDS-PAGE, respectively. M-GADII also crossreacted strongly with IDDM serum in immunoblotting test as shown in Fig. 3D and 3F. Similar results were obtained with other IDDM sera except that more concentrated sera (1:1000 dilution) were used.

(iii) Immunoblotting test with crude pancreas extract -- When crude rat pancreas extract was tested with different dilutions of IDDM serum, M-GADI was found to react with IDDM serum much stronger than GAD_{65} as shown in Fig. 5. No GAD_{65} or GAD_{67} was detected under such diluted conditions suggesting that either M-GAD is much more antigenic or is more abundant than GAD_{65} in the pancreas.

Characterization of M-GADI, M-GADII and M-GADIII – Three different methods, namely phase partitioning assay in Triton X-114, charge shift electrophoresis and chromatography on a hydrophobic column have been widely used to determine whether a protein is an integral membrane protein. Purified M-GAD and S-GAD from porcine brain were used for the experiments described above. In phase partitioning assay (16), it was found that approximately 58±10% of M-GADI and 3±1% S-GAD partitioned in detergent rich phase. In charge shift electrophoresis experiment (17), the electrophoretic mobility of M-GADI treated with a mixture of nonionic detergent, Triton X-100 (TX 100), and anionic detergent, sodium deoxycholate, increased towards anode as compared to the mobility of M-GADI treated wth non-ionic detergent (TX 100) alone. While in the case of S-GAD no significant shift in the electrophoretic mobility was observed when S-GAD was treated under the same conditions as M-GADI. In hydrophobic interaction column chromatography (HPLC phenyl 5PW), M-GADI was eluted in low ionic strength buffer similar to many other integral membrane proteins (18), whereas S-GAD was eluted in high ionic strength buffer. M-GADII and M-GADIII is similar to M-GADI and S-GAD, respectively, in all three tests described above. Taking all these results into consideration we conclude that M-GADI and M-GADII may be an integral membrane protein whereas M-GADIII is a peripheral protein.

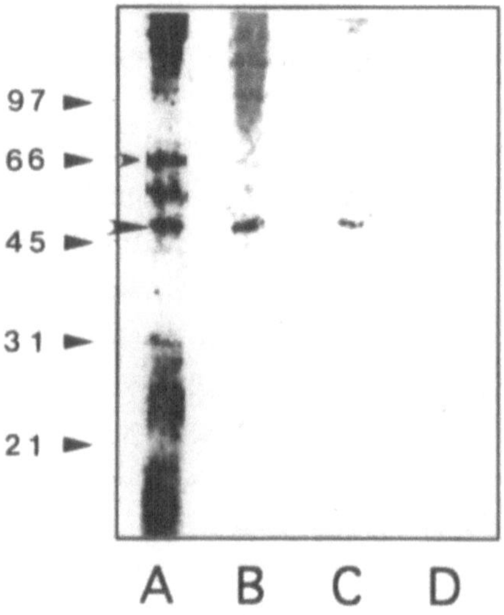

Fig. 5. Immunoblotting test of crude rat pancreas extract with IDDM serum.
Lane A, treated with IDDM serum at 1:1000 dilution. Large arrowhead and small arrowhead indicate the position of M-GADI (48 kDa) and GAD_{65} (65 kDa), respectively. Lane B and Lance C, same as Lane A except IDDM serum used was 1:2000 and 1:6000 dilution, respectively. Lane D, same as Lane A, except IDDM serum was replaced by a normal human serum. Molecular weight standards are indicated.

Based on the above observations, we hypothesize that M-GADI and M-GADII are integral membrane proteins which span the membrane with the extracellular domains exposed on the surface of β-cells and therefore they are capable of triggering the initial event in T-cell mediated destruction of β-cells as seen in IDDM. This hypothesis is supported from several lines of evidence. First of all, the molecular structure of M-GADs and GAD_{65}/GAD_{67} are distinctly different. M-GADI has a native molecular weight of 96 kDa and consists of two identical subunits of 48 kDa whereas both M-GADII and M-GADIII are homodimers of 60 kDa. GAD_{65} and GAD_{67} are homodimers of 65 kDa and 67 kDa, respectively (4, 5). Secondly, M-GADI and M-GADII have fulfilled all three criteria routinely used to determine whether a protein is an integral membrane protein while GAD_{65} and GAD_{67} are not integral membrane proteins (5). These criteria include partitioning in the detergent phase in Triton X-114 partitioning assay (16); shift in electrophoretic mobility in the presence of ionic detergent (17); and elution by low ionic strength buffer in hydrophobic interaction chromatography (18), e.g., phenyl Sepharose hydrophobic column. However, it must be stressed that the final verification of M-GADI and M-GADII being integral membrane proteins has to come from the sequence information. Thirdly, our recent observations that in immunoblotting test, M-GADI but not GAD_{65}/GAD_{67} could be easily detected with diluted IDDM serum suggest that M-GADI is either more antigenic or more abundant than GAD_{65}/GAD_{67} in the pancreas. Although these findings may be of considerable interest and importance, more studies are needed to further determine the relevance of M-GAD in the pathogenesis of IDDM and its potential application in the diagnosis and therapeutic intervention of this disorder.

References

1. Bottazzo GF, Pujol-Borell, R, Gale, E. Type 1 diabetes: role of autoimmune mechanisms. In Diabetes Annual/1. Alberti KGMM, Krall LP, Eds. Amsterdam, Elsevier 1985; pp. 16-52.

2. Baekkeskov S, Nielsen, JH, Marner, B, Bilde, T, Ludvigsson, J, Lernmark, A. Autoantibodies in newly diagnosed diabetic children immunoprecipitate human pancreatic islet cell proteins. Nature 1982; 298:167-169.

3. Baekkeskov S, Aanstoot, HJ, Christgau, S, Reetz, A, Somilena, M, Cascaiho, M, Foili, F, Richter-Oelsen, H, De Camilli, P. Identificationof the 64K autoantigen in insulin-dependent diabetes as the GABA-synthesizing enzyme glutamic acid decarboxylase. Nature 1990; 347:151-156.

4. Bu DF, Erlander, MG, Hitz, BC, Tillakaratne, NJK, Kaufmann, DL, Wagner-McPherson, CB, Evans, GA, Tobin, AJ. Two human glutamate decarboxylases (GAD_{65} and GAD_{67} are each encoded by a single gene. Proc. Natl. Acad. Sci. (USA) 1992; 89:2115-2119.

5. Erlander MG, Tobin, AJ. The structure and functional heterogeneity of glutamic acid decarboxylase: A review. Neurochem. Res. 1991; 16:215-226.

6. Christgau S, Schiebeck, H, Aanstoot, H-J, Aagaard, L, Begley, K, Petersen, JS, Kofoed, H, Hejnaes, K, Baekkeskov, S. Pancreatic β cells express two autoantigenic forms of glutamic acid decarboxylase, A 65 kDa hydrophilic form and a 64 kDa

amphiphilic form which can be both membrane-bound and soluble. J. Biol. Chem. 1991; 266:21257-21264.

7. Tisch R, Yang, X-D, Singer, SM, Liblau, RS, Fugger, L, McDevitt, HO. Immune response to glutamic acid decarboxylase correlates with insulitis in nonobese diabetic mice. Nature 1993; 366:72-75.

8. Kaufman DL, Clare-Salzler, M, Tian, J, Forsthuber, T, Ting, GS, Robinson, P, Atkinson, MA, Sercarz, EE, Tobin, AJ, Lehmann, PV. Spontaneous loss of T-cell tolerance to glutamic acid decarboxylase in murine insulin-dependent diabetes. Nature 1993; 366:69-72.

9. Colman PG, Campbell, IL, Kay, TWH, Harrison, LC. 65,000-Mr. Autoantigen in type 1 diabetes. Evidence against its surface location on human islets. Diabetes 1987; 36:1432-1440.

10. Vives M, Somoza, N, Soldevila, G, Gomis, R, Lucas, A, Sanmarti, A, Pujol-Borrell, R. Reevaluation of autoantibodies to islet cell membrane in IDDM. Failure to detect islet cell surface antibodies using human islet cells as substrate. Diabetes 1992; 41:1624-1631.

11. Formby B, Miller, N. Autoantibodies in the sera from newly diagnosed diabetic NOD mice: evidence against cross-reactivity with a putative β-cell surface autoantigen. Diabetes Res. 1991; 13:13-17.

12. Nathan B, Bao, J, Hsu, C-C, Aguilar, P, Wu, R, Yarom, M, Wu, J-Y. A novel brain glutamate decarboxylase is an autoantigen in insulin dependent diabetes. Proc. Natl. Acad. Sci. (USA) 1994; 91:242-246.

13. Nathan B, Bao, J, Hsu, C-C, Yarom, M, Deupree, DL, Lee, Y-H, Tang, X-W, Kuo, CY, Burgeon, GA, Wu, J-Y. An integral membrane protein form of brain L-glutamate decarboxylase-purification and characterization. Brain Res. 1994; 642:297-302.

14. Nathan B, Floor, E, Kuo, C-Y, Wu, J-Y. Synaptic vesicle associated glutamate decarboxylase - Identification and relationship to insulin dependent diabetes mellitus. J. Neurosci. Res 1995; 40:134-137.

15. Nathan B, Hsu, C-C, Bao, J, Wu, R, Wu, J-Y. Purification and characterization of a novel form of brain L-glutamate decarboxylase - A Ca^{2+}-dependent peripheral membrane protein. J. Biol. Chem. 1994; 269:7249-7254.

16. Bordier C. Phase separation of integral membrane proteins in Triton X-114 Solution. J. Biol. Chem. 1981; 256:1604-1607.

17. Helenius A, Simons, K. Charge shift electrophoresis: Simple method for distinguishing between amphiphilic and hydrophobic proteins in detergent solution. Proc. Natl. Acad. Sci. (USA) 1977; 74:529-532.

18. Goheen SC, Englehorn, SC. Hydrophobic interaction high performance liquid chromatography of proteins. J. Chromatogr. 1984; 317:55-65.

GABA: Receptors, Transporters and Metabolism
ed. by C. Tanaka & N.G. Bowery

GLUTAMATE DECARBOXYLASE, GABA AND AUTOIMMUNITY

Allan J. Tobin* and Daniel L. Kaufman†§

*Departments of Neurology and Physiological Science and †Molecular and Medical Pharmacology, University of California, Los Angeles, CA 90095

GAD and GABA in the central nervous system

GABA is a widely utilized neurotransmitter throughout the central and peripheral nervous systems (reviewed in 1-4). Its synthesis is largely controlled by the activity of glutamate decarboxylase which converts the amino acid glutamate directly into GABA. Alterations in GAD and GABA have been associated with a number of neurogenetic diseases and pharmacological interference with GAD or GABA rapidly induces seizures in both man and animals. Growing evidence also suggests that GABA acts as a neurotrophic factor during development.

We have isolated two genes which encode GADs (5,6). The genes are about 65% homologous and lie on different chromosomes. They encode polypeptides of 67 and 65 kD and we refer to them as GAD67 and GAD65 respectively. Since our original description, GAD cDNAs have been isolated from a number of different species and tissues (reviewed in 7-9). There are no reported sequence differences found between the GADs expressed in the brain and those expressed in peripheral tissues.

Why do neurons express two homologous forms of a GABA synthesizing enzyme and how might their different properties contribute to the regulation of GABA production? Using antibodies that were specific to each form of GAD, we found that while both forms of GAD are generally co-expressed in neurons, the two forms of GAD have quite different intraneuronal distributions and cofactor interactions (6,10, see also 11-13). While GAD67 is widely distributed throughout the neuron, GAD65 is transported to axon terminals and associates with membranes. And, while most of GAD67 is in an active holoenzyme form (saturated with its cofactor, pyridoxal phosphate), less than half of GAD65 exists as active holoenzyme. Extending the observations of Bayon and Tapia (14), we suggested that the modulation of GAD65 apo/holoenzyme levels in synaptic terminals may couple GABA production to neuronal activity (10).

Since GABA is produced by GAD in a single step, rather than a multi-step pathway, GABA synthesis can be readily genetically manipulated *in vitro* and *in vivo* to examine the regulation of GABA production and its action as a neurotrophic factor. We have engineered cell lines to produce different forms of GAD, which can be used to create novel neuronal interactions in cell culture and in the developing CNS. In one set of experiments, we have engineered the AtT-20 pituitary cell line to make GAD and secrete GABA in response to a cyclic AMP analog. In another set of experiments, Jose Segovia, in collaboration with Fred Gage (UCSD), has injected fibroblasts, engineered to express GAD and secrete GABA, into the striatum of rats that had received ibotenic acid lesions. One can readily imagine that in the future, such engineered somatic cells will be used in grafts to deliver proteins of therapeutic value to diseased tissues.

Autoimmunity to GAD

Insulin dependent diabetes mellitus (IDDM, also known as juvenile diabetes or type 1 diabetes) is an autoimmune disease in which T cells mistakenly destroy the ß-cells of the pancreas which are the sole producers of insulin (reviewed in 15,16). The ß-cells share many properties with neurons (17-20). They express GAD and release GABA which is thought to bind to $GABA_A$ receptors on the neighboring alpha cells and inhibit their secretion of glucagon. ß-cells produce neuronal antigens such as GABA-transaminase, tyrosine hydroxylase, PNMT and carboxypeptidase H. They also extend neurites in culture and make synaptic-like vesicles.

A key issue in understanding the pathogenesis of IDDM is identifying the initial target of the autoimmune response. Diabetologists have known that the presence of autoantibodies to islets, as revealed by immunohistochemistry, was predictive of impending IDDM. Such autoantibodies could be detected several years before IDDM onset, providing a window of opportunity in which the disease process, if detected could be inhibited. Years of effort by diabetologists has led to the identification of a number of ß-cell proteins, including GAD, heat shock protein, carboxypeptidase H and insulin, as targets of autoantibodies (21-25).

The availability of cloned autoantigens has recently allowed the design of inexpensive, rapid pre-diagnostic clinical assays for autoantibodies, replacing the labor intensive immunohistochemical assays. Using recombinant GADs, we have shown that GAD autoantibodies can be detected years prior to actual disease onset (26). Our studies, as well as those of numerous other laboratories, suggest that the presence of GAD autoantibodies may be the most sensitive and predictive marker of impending IDDM. Clinical trials are currently underway that will test the efficacy of these GAD autoantibody assays in prospective studies.

While autoantibodies serve as hallmarks of the autoimmune process and are valuable prediagnostic markers, autoreactive T cells, and not autoantibodies, are the underlying cause of IDDM. In order to test the relevance of T cell autoimmunity to different autoantigens in the pathogenesis of IDDM, we characterized the development of T cell responses to GAD65 and other ß-cell autoantigens in a genetically prone diabetic strain of mice (27). We found that T cell responses to ß-cell antigens develop spontaneously in a defined chronological order. T cell autoimmunity arose first to GAD65, concurrent with the onset of insulitis. Subsequently, T cell responses spread to other ß-cell autoantigens.

GAD-reactive T cells were the first to arise among the autoantigens tested, but did the anti-GAD response itself develop as a secondary consequence of the spreading of autoimmunity, after a T cell response to a yet unidentified ß-cell antigen? If the development of T cell reactivity to GAD is a primary event in the pathogenesis of ß-cell , the early inactivation of GAD-reactive T cells before their spontaneous activation should prevent the cascade of T-cell responses to other ß-cell antigens, insulitis and diabetes. We injected mice at 3 weeks of age (before the onset of insulitis) with various autoantigens intravenously, a treatment that has been shown to inactivate antigen-specific T cells. We showed that induced tolerance to GAD65 (but not other autoantigens) could completely prevent insulitis and IDDM, demonstrating the crucial role GAD autoimmunity plays in the initiation and propagation of ß-cell autoimmunity in NOD mice (27). These data demonstrated that early tolerance induction to initial target antigens can circumvent spontaneous autoimmune disease. Unlike immunosuppressants or monoclonal antibodies which target proteins on T cells, such antigen specific immunotherapies should not compromise immune system function.

Molecular mimicry hypothesis

In several autoimmune diseases, the conversion from a state of self tolerance to that of autoimmunity is thought to involve "molecular mimicry" in which a bacterial or viral antigen triggers an immune response which cross-reacts with a similar self-antigen (28). We found a region of amino acid sequence similarity between GAD and Coxsackievirus (26). Although controversial, Coxsackievirus has been epidemiologically associated with IDDM and can induce diabetes in experimental animals (see references within 26). We demonstrated in mice that a T cell response against Coxsackievirus could indeed cross-react against GAD65, but only in the context of a diabetes MHC susceptibility allele (29)--a finding which may help explain the association of autoimmune disease with certain MHCs. Working with Dr. Mark Atkinson, we found T cell reactivity to GAD65 in individuals at risk, or with IDDM (30). Next, we showed that the major GAD65 determinant recognized by T cells from individuals at high risk for IDDM is the small region of GAD65 which shares a sequence similarity with Coxsackievirus (31), consistent with the molecular mimicry hypothesis.

Conclusion

The characterization of GAD which was begun by neuroscientists decades ago has led to broad new areas of research in the neurosciences, molecular biology and immunology. In particular, GAD based therapeutics may have clinical potential for the treatment of human disease. We envision that GAD engineered cell lines will one day be used to provide GABA for the treatment of neurological disease. In the field of diabetes, GAD research has lead to a new generation of pre-diagnostic tests for IDDM, a better understanding of the basic mechanisms that initiate and propagate autoimmunity, and the first antigen-specific immunotherapies that may be able to prevent human IDDM. Thus, GAD research has grown over the decades by building on results of investigators in many different fields and demonstrates the synergism that can take place when different scientific disciplines converge.

REFERENCES

1. Cooper, J.R., Bloom, F.E. and Roth, R.H. The Biochemical Basis of Neuropharmacology 5th ed. (Oxford: Oxford University Press), 1986.

2. Hertz, L., Kvamme, E., McGeer, G. and Schousboe, A. Eds., Glutamine, Glutamate and GABA in the Central Nervous System (Liss, New York) 1983.

3. Mugnaini, E., and Oertel, W.H. In Handbook of Chemical Neruoanatomy, Volume 4 Björklund, A., and Hökfelt, T. eds. (Amsterdam:Elsevier), 436-622, 1985.

4. Erdo, S.L. and Wolff, J.R. Journal of Neurochemistry 54, 363-372, 1990.

5. Kaufman, D.L., McGinnis, J.F., Krieger, N.R. and Tobin, A.J. Science 232, 1138-1140, 1986.

6. Erlander, M.G., Tillakaratne, N.J.K., Feldblum, S., Patel, N. and Tobin, A.J. Neuron 7, 91-100, 1990.

7. Erlander, M.G. and Tobin, A.J. Neurochemical Research 16, 215-226, 1991.

8. Dingfang, B., Erlander, M.G., Hitz, B.,Tillakaratne, N.J.K., Kaufman, D.L., Wagner-McPherson, C.B., Evans, G.A. and Tobin, A.J. Proc. Nat. Acad. Sci. (USA) 89, 2115-2119, 1992.

9. Kaufman, D.L. and Tobin, A.J. Trends in Pharmacological Sciences 14, 107-109, 1993.

10. Kaufman, D.L., Houser, C.R. and Tobin, A.J. Journal of Neurochemistry 56, 720-723, 1991.

11. Martin, D.L. Cellular and Molecular Neurobiology 7, 237-253, 1987.

12. Chang, Y.-C. and Gottlieb, D.I. Journal of Neuroscience 8, 2123-2130, 1988.

13. Martin, D.L., Wu, S.J. and Martin, S.B. Journal of Neurochemistry 55, 524-532, 1990.

14. Bayon, A., Possani, L.D. and Tapia, R. Journal of Neurochemestry 29, 513-517, 1977.

15. Castano, L. and Eisenbarth, G.S. Annual Review of Immunology 8, 647-679, 1990.

16. Atkinson, M.A. and Maclaren, N.K. Scientific American July, 62-67, 1990.

17. Rorsman, P., P.-O. Berggren, K. Bokvist, H. Ericson, H. Mîhler, C.-G. ôstenson, and
P.A. Smith. Nature 341: 233-236, 1989.

18. Reetz, A., Solimena, M., Matteoli, M., Folli, F., Takei, K. and De Camilli, P. EMBO J. 10, 1275-1284, 1991.

19. Sorenson, R.L., Garry, D.G. and Brelje, T.C. Diabetes 40, 1365-1374, 1991.

20. Teitelman, G. and Evinger, M. In Persectives of the Molecular Biology and Immunology of the Pancreatic Beta Cell. D. Hanahan, H. O. McDevitt and Cahill G.F. Eds. (Cold Spring Harbor Laboratory) 1989.

21. Rossini, A.A., Greiner, D.L., Friedman, H.P., Mordes, J.P. Diabetes Reviews 1, 43-75, 1993.

22. Palmer J.P. Diabetes Reviews 1, 104-115, 1993.

23. Baekkeskov, S., Nielsen, J.H., Marner, B., Bilde, T., Ludvigsson, J. and Lernmark, A. Nature 298, 167-169, 1982.

24. Baekkeskov, S., Aanstoot, H.-J., Christgau, S., Reetz, A., Solimena, M., Cascalho,M.,
Folli, F., Richter-Olesen, H. and De Camilli, P. Nature 347, 151-156, 1990.

25. Solimena, M. and De Camilli, P. Trends in Neuroscience 14, 452-457, 1991.

26. Kaufman, D.L., Erlander, M.G., Clare-Salzler, M., Atkinson, M.A., Maclaren, N.K. and Tobin, A.J. Journal of Clinical Investigation 89, 283-292, 1992.

27. Kaufman, D.L., Clare-Salzler, M., Tian, J., Forsthuber, T., Ting, G.S.P., Robinson, P., Atkinson, M.A., Sercarz, E.E., Tobin, A.J. and Lehmann, P.V. Nature 366, 69-72, 1993.

28. Oldstone, M.B.A. Cell 50, 819-820, 1987.

29. Tian, J., Lehmann, P. and Kaufman, D.L. Journal of Experimental Medicine 180, 1979-1984, 1994.

30. Atkinson, M.A., Kaufman, D.L., Gibbs, K.A., Campbell, L., Shah, S.C., Tobin, A.J. and Maclaren, N.K. Lancet 339, 458-459,1992.

31. Atkinson, M.A., Bowman, M.A., Campbell, L., Kaufman, D.L. and Maclaren, N.K. Journal of Clinical Investigation 94, 2125-2129, 1994.

GABA: Receptors, Transporters and Metabolism
ed. by C. Tanaka & N.G. Bowery

STIFF-MAN SYNDROME AND GLUTAMIC ACID DECARBOXYLASE: AN UPDATED VIEW.

M. Solimena*, R. Dirkx*, M. Butler‡, J.-M. Hermel*, J. Guernaccia$^{\Delta}$, K. Marek$^{\Delta}$, C. David‡ and P. De Camilli‡.

*Department of Internal Medicine, Section of Endocrinology, ‡ Department of Cell Biology and the Howard Hughes Medical Institute, Department of Neurology$^{\Delta}$, Yale University School of Medicine, New Haven, CT 06510.

Summary: Evidence collected during the last decade suggests that Stiff-Man syndrome (SMS), a neuromuscular disorder characterized by continuous rigidity of the body musculature, results from an autoimmune impairment of inhibitory neurons controlling motor neuron activity. The dominant autoantigen is glutamic acid decarboxylase (GAD), the enzyme responsible for the synthesis of GABA. In this chapter, we summarize some of the relevant features of SMS and GAD, with an emphasis on the contribution of our group to this field.

Clinical Features of Stiff-Man Syndrome

Stiff-man syndrome (SMS) is a rare disease of the central nervous system characterized by muscle rigidity and painful spasms of limbs, trunk and abdominal muscles (1-5). The disease develops in adulthood and affects mostly women. SMS resembles a chronic form of tetanus, but trismus does not occur. Electromyography shows persistent motor unit activity which is, however, indistinguishable electrophysiologically from normal voluntary muscle activity except that patients are not able to relax. Rigidity and spasms result from the simultaneous activation of agonist and antagonist muscles, caused by the continuous firing of α-motor neurons. An imbalance between noradrenergic and GABA-ergic pathways involved in the control of muscle tone may be at the origin of the disease (5). Most patients respond to pharmacological treatments with high doses of drugs which potentiate the action of GABA, such as benzodiazepines, baclofen or sodium valproate. These pharmacological treatments, however, do not affect the slowly progressive course of the disease. Patients may be eventually bedridden and respiratory failure or sudden death has been reported (7). Neuropathological studies of SMS have been meager, and, when available, have shown no specific abnormalities. In a few cases, however, evidence of a perivascular lymphocytic infiltration has been observed. Because of its rarity, of the subjective nature of the symptoms and the lack of confirmatory laboratory tests, SMS often remained undiagnosed for a long period of time. In many cases, SMS patients are erroneously diagnosed as hysteric (8). Depression, dysphoria, unusual personality traits and paroxysmal fear when crossing a free space unaided are common in SMS patients, even beyond what one may expect as a natural reaction to the difficulties of living with this unusual syndrome (5, 4, 9).

Evidence for an Autoimmune Pathogenesis of Stiff-man Syndrome

Increasing evidence suggests that at least in the majority of cases SMS has an autoimmune pathogenesis (10, 11). Key observations in support of this hypothesis include:

- presence of high titer autoantibodies directed against presynaptic antigens [glutamic acid decarboxylase (GAD), amphiphysin] in the serum (12-16).
- presence of anti-GAD or anti-amphiphysin autoantibodies in the patients' cerebrospinal fluid at concentrations which suggest their production within the central nervous system (12-14).
- frequent occurence of oligoclonal IgG in the cerebrospinal fluid, a general indication of IgG production within the central nervous system (12-14).
- high prevalence of organ specific-autoimmune diseases among SMS patients positive for anti-GAD autoantibodies and their relatives (4, 12, 13).
- presence of breast cancer among SMS patients positive for anti-amphiphysin autoantibodies (paraneoplastic SMS) suggesting an autoimmune paraneoplastic pathogenesis in theses cases (14-16).
- frequent and sometimes dramatic improvement of the symptoms following different immunotherapies including plasmapheresis, steroids, and intravenous immunoglobulins (17-19).

We still do not know, however, the precise role of autoimmunity against GAD or amphiphysin in the pathogenesis of SMS and why these proteins are such prominent autoantigens. Strikingly, both antigens are cytosolic proteins associated with the cytosolic surface of synaptic vesicles (20, 21), the organelles responsible for secretion of non-peptide neurotransmitters, such as GABA. Because of the intracellular localization of GAD and amphiphysin, it is unlikely that anti-GAD or anti-amphiphysin autoantibodies are directly pathogenic. SMS may be caused, for example, by autoantibodies directed against yet to be identified neuronal surface autoantigen(s), including intrinsic membrane protein(s) of synaptic vesicles which interacts with GAD and/or amphiphysin at their cytoplasmic side. Such autoantigen(s) would be accessible to circulating autoantibodies.

This hypothesis is based on the observation that in most autoimmune diseases autoimmunity is directed against a macromolecular complex, rather than against a single antigen.

Classification of SMS Based on Immunological Responses

In our laboratories, we screened the sera of more than 600 neurological patients by immunocytochemistry and western blotting for autoantibodies directed against brain antigens. In a recent survey, 90 of these patients, many of whom at one time were diagnosed as hysteric, fulfilled the stringent clinical criteria for the diagnosis of SMS established by Lorish and coworkers (3). According to the presence of autoantibodies directed against neuronal antigens, these 90 patients can be classified in three distinct groups (Table I).

Group I: (~50% of SMS patients) is characterized by the presence of anti-GAD autoantibodies in the serum and in the cerebrospinal fluid (10, fig. 1). Seventy percent of patients in this group are also affected by other autoimmune diseases (such as hypothyroidism, hyperthyroidism, pernicious anemia, vitiligo), and primarily IDDM (46% of the cases). Some of the patients suffer from depression and dysphoria and paroxysmal fear when crossing a free-space unaided (9). Finally, five (~10%) of these patients suffer from epilepsy.

SMS patients	Autoantibodies	Associated Disorders	Prognosis
46 [M:12; F:34]	Anti-GAD	Autoimmune disorders: 71% Insulin-dependent diabetes mellitus: 46% Epilepsy: 10% Psychiatric disorders: 17%	Chronic; slowly progressive; in some cases partial improvement after immunotheraphy
5 [M:0; F: 5]	Anti-Amphiphysin	Breast cancer : 100%	Partial improvement after cancer removal
39 [M:15; F:24]	Not detected	Autoimmune diseases: 30% Insulin-dependent diabetes mellitus: 8% Epilepsy: 5% Psychiatric disorders: 5%	Chronic; slowly progressive; in some cases partial improvement after immunotheraphy

Table I. Classification of SMS patients according to clinical features and presence of autoantibodies.

Group II: (~5% of SMS patients) is characterized by the presence of autoantibodies directed against amphiphysin (14-16). Amphiphysin is a neuron specific protein which is highly concentrated in nerve terminals (14, 15, 21) and with a putative role in endocytosis for synaptic vesicles (16, 22) None of these five SMS patients have anti-GAD autoantibodies and all suffer from breast cancer. In two of these five patients, the search for an occult breast cancer was

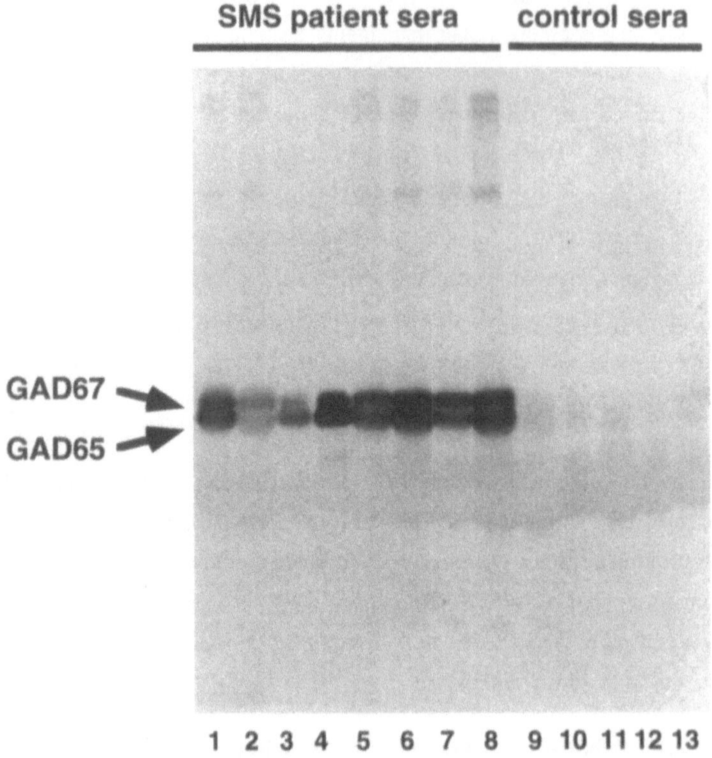

Fig. 1. Presence of anti-GAD autoantibodies in the serum of Stiff-Man syndrome patients. Stiff-Man syndrome patient's and control subject's sera were used for immunoprecipitation from rat brain Triton-X100 extracts. Immunoprecipitates were separated by SDS-gel electrophoresis, transferred on nitrocellulose, and immunoblotted by alkaline phosphatase with an anti-GAD rabbit antiserum. Autoantibodies of SMS patients immunoprecipitated both GAD65 and GAD67 (lanes 1–8), whereas neither GAD65 or GAD67 were detected in immunoprecipitates of control subject's sera (lanes 9–13).

prompted by our previous detection of anti-amphiphysin autoantibodies in the patient serum. In 4 cases (one patient could not undergo surgery) removal of breast cancer resulted in a drastic improvement of the neurological symptoms. This observation suggests that SMS has a paraneoplastic origin in these patients. It also suggests that rigidity is caused by a functional impairment rather than a destructive process of neurons controlling muscle tone.

Group III: (~45% of SMS patients) is characterized by SMS patients with no evident autoantibodies directed against neuronal antigens (14, 23). Despite the lack of autoantibodies detectable by standard assays, some of these patients suffer from autoimmune diseases, psychiatric disorders, and epilepsy. Therefore, it is possible that similar mechanisms may lead to the pathogenesis of SMS in patients of group I and III.

General Properties of GAD, the Dominant Autoantigen of SMS

GAD is a cytosolic protein represented by two isoforms of 65 (GAD65) and 67 (GAD67) kD, respectively (24). GAD65 and GAD67 are the products of two distinct genes (20), each of which has been cloned in several species. Both GAD65 and GAD67, bind piridoxal phosphate (PLP), are enzymatically active (25, 26) and are expressed in GABA-ergic neurons (27-30) and pancreatic

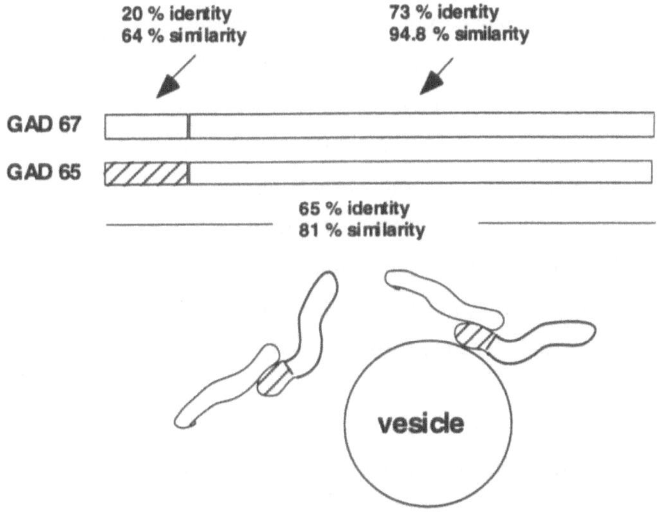

Fig. 2. Schematic representation of GAD67 and GAD65 homologous regions. Percentages of identities and similarities refer to rat GAD 67 and GAD65. Association of GAD67 with membranes involves its interaction with the N-terminal region of GAD65.

ß-cells (31) although their level of expression varies according to the cell type and the species (30, 32, 33).Each isoform is highly conserved during evolution with >95% identity among human, rat and mouse. Overall, GAD65 and GAD67 are very similar to each other (65% identity in humans), but are significantly divergent in their NH2-terminal regions (Fig. 2). The NH2-terminal region of GAD65, but not of GAD67, is palmitoylated and contains information for its association with membranes (34, 35).

SMS Autoantibodies Primarily Recognize the C- and N-terminal Regions of GAD65

With few exceptions anti-GAD autoantibodies of SMS patients can be revealed by immunocytochemistry and western blotting, as well as by immunoprecipitation and ELISA. All SMS patients develop autoantibodies directed against both GAD isoforms, but, preferentially, against GAD65 (23, 36). More specifically, GAD65 is the only isoform recognized by SMS anti-GAD autoantibodies in western blots (36). The dominant epitope of GAD65 recognized by virtually all group I SMS sera is a conformational epitope resistant to SDS denaturation contained within the last 110 amino acids (a.a. 475-485) of GAD65 (epitope SMS 1) (36). Interestingly, this region is the most conserved between the two GAD isoforms, yet the epitope SMS 1 is only present in GAD65. Deletion of a few amino acids at either side of this region of GAD65 results in the nearly complete loss of immunoreactivity. The C-terminal region of GAD65 is also the primary target of anti-GAD65 autoantibodies of IDDM patients (37-39) and contains the early NOD T-cell GAD65 determinants (40). Almost all SMS patients in group I have, in addition, autoantibodies directed against the N-terminal region (amino acids 1-95) of GAD65 (epitope SMS 2) (36, 41). This is the region of GAD65 responsible for its association with membrane compartments (fig. 3). Differently from SMS, IDDM anti-GAD65 autoantibodies do not generally recognize the epitopes SMS 1 or SMS 2 (39). This data suggests that different clinical conditions may correlate with different patterns of humoral autoimmunity (36).

Association of GAD with Synaptic Vesicles and Synaptic-like Microvesicles

Previous immunocytochemical studies with antisera which recognize both GAD isoforms had indicated that GAD is concentrated in nerve terminals of GABA-ergic neurons (27, 42) and in pancreatic ß-cells and in the Golgi complex region of both cells (27, 35, 43). The development of antibodies specifically directed against either GAD65 or GAD67 made it possible to address the cellular and intracellular localization of each isoform. Both GAD isoforms are present in the presynaptic terminals of the majority of GABA-ergic neurons, with a relative enrichment of GAD65 compared to GAD67 (28, 30). Double immunostaining on rat pancreatic sections with anti-glucagon and anti-somatostatin antibodies indicate that GAD67, like GAD65, is only expressed in ß-cells (30). Most ß-cells express high levels of GAD65 and very low levels of GAD67, but a few ß-cells are strongly positive for GAD67 and do not contain detectable levels of GAD65 (30). The physiological significance of the heterogeneous expression of the two GAD isoforms in ß-cells remains to be established.

Immunogold electron microscopy for GAD in rat synaptosomes and pancreatic islets demonstrated that a pool of GAD is localized in close proximity to synaptic vesicles and synaptic-like microvesicles in GABA-ergic neurons and pancreatic ß-cells, respectively (20) (Fig. 2). Synaptic vesicles are concentrated at neuronal presynaptic terminals and are responsible for the storage and secretion of non-peptide neurotransmitter molecules including GABA. Upon depolarization of the nerve terminal, synaptic vesicles fuse with the plasma membrane and release their neurotransmitter content. ß-cell synaptic-like microvesicles, similarly to synaptic vesicles of GABA-ergic neurons, have a GABA-specific transport activity, suggesting that these vesicles may be responsible for the paracrine secretion of GABA from ß-cells. Association of GAD with the cytoplasmic surface of synaptic vesicles and synaptic-like microvesicles may provide a mechanism for the rapid uptake of the newly synthesized GABA into synaptic vesicles (44). The association of GAD, and primarily of GAD65, with these organelles is most likely mediated via its binding to a synaptic vesicle membrane protein. The hypothesis that protein-protein interaction is responsible for the anchoring of GAD65 to synaptic vesicles and synaptic-like microvesicles is supported by the evidence that palmitoylation of GAD65 is not required for its association to membranes (34, 35).

Molecular mechanisms in GAD membrane interaction

The nature of the interaction of GAD with membranes was investigated by transfection of GAD65 and GAD67 in CHO and COS cells (fig. 4) (45). In these cells, which lack synaptic vesicles and synaptic-like microvesicles, GAD65 was found to be selectively targeted to the Golgi complex region. Conversely, GAD67 was evenly distributed in the cell cytosol of transfected fibroblasts, compatible with GAD67 being mostly a soluble protein. Accordingly, upon subcellular fractionation of CHO cells transfected independently with GAD65 and GAD67, approximately 40-50% of GAD65 was recovered in the high speed pellet partitioned in the detergent phase after Triton X-114 extraction, whereas virtually all GAD67 was recovered in the high speed supernatant and in the Triton X-114 aqueous phase. Using a variety of chimeric constructs, we found that a.a. 1-83 of GAD65 contain information required for the targeting of the remaining portion of the GAD molecule (i.e. the region highly similar between GAD65 and GAD67) to the region of the Golgi apparatus (fig. 4) (45). The same GAD65 domain is also sufficient to target to the Golgi complex region a GAD-unrelated protein (ß-galactosidase) (32) (fig.4). It appears, therefore, that a.a. 1-83

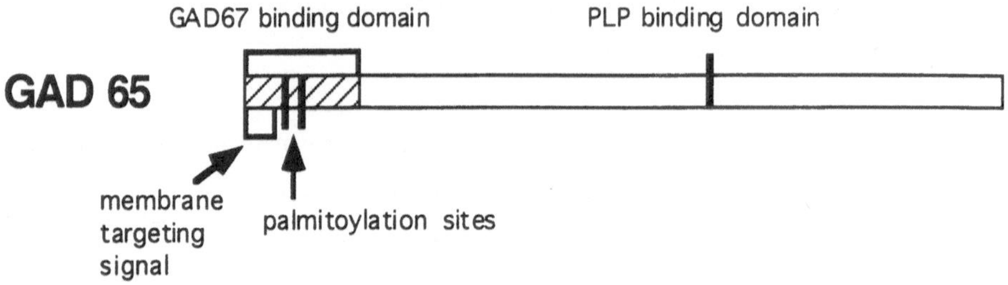

Fig. 3. Schematic representations of the domains of GAD65.

of GAD65 contains information which is necessary and sufficient to target GAD65 and other cytosolic proteins to the Golgi complex area.

It had been proposed that palmitoylation on cysteine residue(s) at the NH2-terminal region of GAD65 is responsible for anchoring GAD65 to membranes (46). Amino acids 1-83 of GAD65 contain six cysteines at position 30, 45, 73, 75, 80, 82. Following transfection of CHO cells with GAD65 constructs in which, individually or simultaneously, each of these six cysteines was replaced by a serine residue, it was found that every Cys-mutated GAD65 protein, including the GAD65 in which all six cysteines had been mutagenized (GAD65S1-6) (fig.4), was still primarily localized in the Golgi complex region (35). Upon subcellular fractionation, GAD65S1-6 was recovered in the high speed pellet, further indicating its ability to interact with subcellular particles. A chimeric GAD65-GAD67 protein in which a.a. 1-29 of GAD67 were replaced by a.a. 1-27 of GAD65 was also targeted to the Golgi complex area (fig. 4). The region of GAD65 corresponding to a.a. 1-27 does not contain any cysteine residues. These findings, which are in agreement with those reported by Shi and coworkers, 1994 (34), exclude the possibility that palmitoylation of GAD65 is directly responsible for the targeting of the protein to membrane organelles. However, both GAD65S1-6 and GAD65(1-27)/GAD67 retained some hydrophobic properties, as demonstrated by their partial recovery in the Triton X-114 detergent phase. These results support the hypothesis that the NH2-terminal domain of GAD65 may undergo a second hydrophobic post-translation modification (46), yet to be characterized. Conversely, replacement of a.a. 1-27 of

GAD65 with a.a 1-29 of GAD67 generated a protein [GAD67(1-29)/GAD65] which, similarly, to GAD67, was homogeneously distributed in the cytosol (fig. 4). The GAD67(1-29)/GAD65 protein was recovered only in the high speed supernatant and in the Triton X-114 aqueous phase. This data suggests that targeting to the Golgi complex region is required for the protein to undergo hydrophobic post-translational modifications, such as palmitoylation. While palmitoylation may stabilize the anchoring of GAD65 to membranes, other mechanisms such as protein-protein interactions are likely to account for the association of GAD65 to specific intracellular compartments (35, 45).

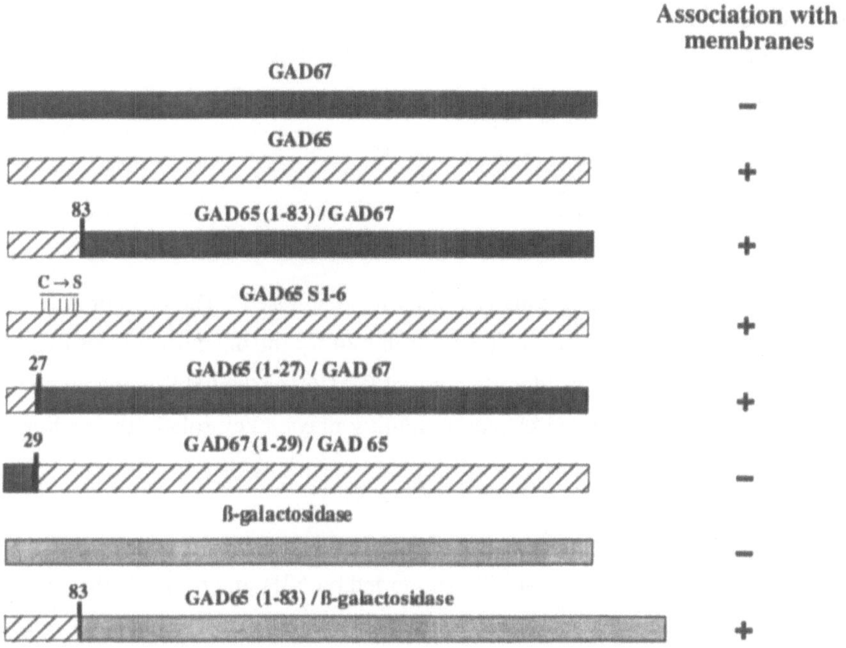

Fig. 4. Ability of various GAD constructs to interact with membranes in transfected fibroblasts.

GAD65 and GAD67 Form a Heterodimeric Complex *In Vivo*.

GAD67 from transfected cells is recovered in the high speed supernatant only and in the Triton X-114 aqueous phase (45, 46). In contrast, a considerable pool of rat brain GAD67 sediments together with GAD65 in pellet fractions (30, 47). Furthermore, immunoprecipitation and affinity purification of brain GAD65 with a monoclonal antibody which selectively recognizes GAD65 results in the co-purification of GAD67, despite the complete lack of cross-reactivity of this antibody with GAD67 (36, 47). It was suggested, therefore, that GAD65 and GAD67 form heterodimers, both as cytosolic proteins as well as in association to membrane compartments (47). Consistent with this hypothesis, a pool of GAD67 colocalizes with GAD65 in the Golgi complex region of neurons and pancreatic ß-cells, both of which express GAD65 and GAD67. Similarly, co-expression of GAD65 and GAD67 in CHO cells results in targeting of GAD67 to the Golgi complex region (30) and co-immunoprecipitation of the two proteins. To establish which domain of GAD65 interacts with GAD67, GAD65(1-83)/ß-galactosidase and GAD67 constructs were co-transfected in CHO cells and their distribution analyzed by immunocytochemistry (30). GAD67 and GAD65(1-83)/ß-galactosidase fusion protein colocalized in the Golgi complex region, indicating that the domain encompassing a.a 1-83 of GAD65 is directly responsible for the association of GAD65 with GAD67 (fig. 2 and 3). Additional data indicates that palmitoylation of GAD65 is not required for its association with GAD67.

Concluding Remarks

SMS is a very disabling disease, which is often misdiagnosed. The detection of anti-GAD autoantibodies in SMS patients has provided a useful tool for the diagnosis of this disease. This discovery, in addition, has led to the identification of GAD65 as the 64kD autoantigen of IDDM and therefore to the discovery that GAD autoimmunity plays a key role in the pathogenesis of IDDM (13, 48).

Acknowledgments. This work was partially supported by NIH grants AI30248, DK43078 and a McKnight Research Project Award to PDC, by a Juvenile Diabetes Foundation International Career Development Award and NIH pilot grant DK-45735 to MS, by a traveling fellowship from the Association Française contre les Myopathies to J-MH and by a U.S. Army Medical Research Postdoctoral Fellowship to CD.

References

1. Moersch FP, Woltman HW. Progressive and fluctuating muscular rigidity and spasm (stiff-man syndrome): report of a case and some observation in 13 other cases. Mayo Clin. Proc. 1956; 31: 421-427.

2. Layzer RB. Stiff-man syndrome: an autoimmune disease? N. Engl. J. Med. 1988; 318: 1060 -1063.

3. Lorish TR, Thorsteinsson G, Howard FM. Stiff-man syndrome updated. Mayo Clin Proc. 1989; 64: 629-636.

4. Blum P, Jankovic J. Stiff-Person syndrome: an autoimmune disease. Mov. Dis. 1991; 6: 12-20.

5. McEvoy KM.. Stiff-Man syndrome. Mayo Clin. Proc. 1991; 66: 300-304.

6. Meinck HM, Ricker K, Conrad B. The stiff-man syndrome: new pathophysiological aspects from abnormal exteroceptive reflexes and ther response to clorimipramine clonidine and tizanidine. J. Neurol. Neurosurg. Psychiatry 1984; 47: 280-287.

7. Mitsumoto H, Schwartzman MJ, Estes ML, Chou SM, La Franchise EF, De Camilli P, and Solimena M. Sudden death and paroxysmal autonomic dysfunction in stiff-man syndrome. J. Neurol. 1991; 238: 91-96.

8. Tarsey D, and Miyawaki EK. Stiff-man syndrome. Report of a case. Arch. Int. Med. 1994; 154: 1285-1288.

9. Meinck HM, Ricker K, Hulser PJ, Schmid E, Pfeiffer J, and Solimena M. Stiff-man syndrome: clinical and laboratory findings in eight patients. J. Neurol. 1994; 241: 157-166.

10. Solimena M, Butler M, and De Camilli P. GAD, diabetes and stiff-man syndrome: some progress and more questions. J. Endocrin. Invest. 1994a; 17: 509-520.

11. McEvoy KM and Lennon VA. 1994. Stiff-man syndrome: clinical aspects and anti-islet cell antibodies as a disease marker. In: Motor Unit Hyperactivity States. RB Layzer, editor. New York: Raven Press 1994: 45-51.

12. Solimena M, Folli F, Denis-Donini S, Comi GC, Pozza G, De Camilli P, and Vicari AM. Autoantibodies to glutamic acid decarboxylase in a patient with Stiff-Man syndrome epilepsy and type I diabetes mellitus. N. Engl. J. Med. 1988; 318: 1012-1020.

13. Solimena M, Folli F, Aparisi R, Pozza G, and De Camilli P. Autoantibodies to GABAergic neurons and pancreatic beta cells in Stiff-Man syndrome. N. Engl. J. Med. 1990; 322: 1555-1560.

14. Folli F, Solimena M, Cofiell R, Austoni M, Tallini G, Fassetta G, Bates D, Cartlidge N, Bottazzo GF, Piccolo G, and De Camilli P. Autoantibodies to a 128 kd synaptic protein in Stiff-Man syndrome with breast cancer. N. Engl. J. Med. 1993; 328: 546-551.

15. De Camilli P,Thomas A, Cofiell R, Folli F, Lichte B, Piccolo G, Meinck H-M, Austoni M, Fassetta G, Bottazzo GF, Bates D, Cartlidge N, Solimena M, and Kilimann MW. The synaptic vesicle-associated protein amphiphysin is the autoantigen of Stiff-Man syndrome with breast cancer. J. Exp. Med. 1993; 178: 2219-2223.

16. David C, Solimena M, and De Camilli P. Autoimmunity to Stiff-man syndrome with breast cancer is targeted to the C-terminal region of human amphiphysin, a protein homologous to the yeast protein, Rvs167 and Rvs161. FEBS Letter 1994; 351: 73-79.

17. Vicari AM, Folli F, Pozza G, Comi GC, Comola M, Canal N, Besana C, Borri A, Tresoldi M, Solimena M, and De Camilli P. Plasmapheresis in the treatment of stiff-man syndrome. N. Engl. J. Med. 1989; 320 :1499.

18. Piccolo G, Cosi V, Zandrini C, Moglia A. Steroid-responsive and dependent stiff-man syndrome: a clinical and electrophysiological study of two cases. Ital. J. Neurol Sci. 1988; 9 : 559-66.

19. Karlson EW, Sudarsky L, Ruderman E, Pierson S, Scott M, and Helfgott SM. Treatment of stiff-man syndrome with intravenous immune globulin. Arthritis Rheum 1994; 37: 915-8.

20. Reetz A, Solimena M, Matteoli M, Folli F, Takei K, and De Camilli P. GABA and pancreatic ß-cells: colocalization of glutamic acid decarboxylase (GAD) and GABA with synaptic-like microvesicles suggests their role in GABA storage and secretion. EMBO J. 1991; 10: 1275-1284.

21. Lichte B, Veh RW, Meyer HE, and Kilimann MW. Amphiphysin, a novel protein associated with synaptic vesicles. EMBO J. 1992; 11: 2521-2530.

22. David, C., McPherson, P.S., Cho, Y., Solimena, M., and De Camilli, P. Amphiphysin, a nerve terminal protein similar to yeast RVS161 and RVS167, binds dynamin and P145 via its SH3 domain. 1994. 34th meeting of the American Society for Cell Biology

23. Solimena M, and De Camilli P. Autoimmunity to glutamic acid decarboxylase (GAD) in Stiff-Man syndrome and insulin-dependent diabetes mellitus. TINS 1991; 14: 452-457.

24. Erlander MG, Tillakaratne NJK, Feldblum S, Patel N, and Tobin AJ. Two genes encode distinct glutamate decarboxylases. Neuron 1991; 7: 91-100.

25. Erlander MG, and Tobin AJ. The structural and functional heterogeneity of glutamic acid decarboxylase: a review. Neurochem. Res. 1991; 16: 215-226.

26. Martin DL, Martin SB, Wu SJ and Espina N. Regulatory properties of brain glutamate decarboxylase (GAD): the apoenzyme of GAD is present principally as the smaller of two molecular forms of GAD in brain. J. Neurosci. 1991, 11: 2725-2731.

27. Mugnaini E, and Oertel WH. An atlas of the distribution of GABAergic neurons and terminals in the rat CNS as revealed by GAD immunohistochemistry. In: Handbook of Chemical Neuroanatomy Vol. 4: GABA and Neuropeptides in the CNS Part I. A. Bjorklund and T. Hökfelt, editors. Amsterdam: Elsevier Science Publishers B.V. 1985: 436-608.

28. Kaufman DL, Houser CR, and Tobin AJ. Two forms of the GABA synthetic enzume glutamate decarboxylase have distinct intraneuronal distributions and cofactor interactions. J. Neurochem. 1991; 56: 720-723.

29. Esclapez M, Tillakaratne NJK, Kaufman DL, Tobin AJ, and Houser CR. Comparative localization of two forms of glutamic acid decarboxylase and their mRNAs in rat brain supports the concept of functional differences between the forms. J. Neurosci 1994; 14: 1834-1855.

30. Dirkx R., Thomas A., Li L., Lernmark Å., Sherwin R., De Camilli P., and Solimena M. Targeting of GAD67 to membranes via heterodimeric interaction with the N-terminal region of GAD65. J. Biol. Chem. 1994; 270: 2241-2246.

31. Vincent SR, Hökfelt T, Wu JY, Elde RP, Morgan LM, and Kimmel JR. Immunohistochemical studies of the GABA system in the pancreas. Neuroendocrin. 1983; 36: 197-204.

32. Faulkner-Jones BE, Cram DS, Kun J, and Harrison LC. Localization and quantitation of expression of two glutamate decarboxylase genes in pancreatic ß-cells and other peripheral tissues of mouse and rat. Endocrinol. 1993; 133: 2962-2972.

33. Kim J, Richter W, Aanstoot H-J, Shi Y, Fu Q, Rajotte R, Warnock G, and Bækkeskov S. Differential expression of GAD65 and GAD67 in human rat and mouse pancreatic islets. Diabetes 1993; 42: 1799-1808.

34. Shi Y, Veit B, and Bækkeskov S. Amino acid residues 24-31 but not palmitoylation of cysteins 30 and 45 are required for membrane anchoring of glutamic acid decarboxylase GAD65. J. Cell Biol. 1994;124: 927-934.

35. Solimena M, Dirkx R, Radzynski M, Mundigl O, and De Camilli P. A signal located within amino acids 1-27 of GAD65 is required for its targeting to the Golgi complex region. J. Cell Biol. 1994b, 126: 331-41.

36. Butler M, Solimena M, Dirkx R, Hayday A, and De Camilli P. Identification of a dominant epitope of glutamic acid decarboxylase (GAD65) recognized by autoantibodies in Stiff-Man syndrome. J. Exp. Med. 1993; 178: 2097-2106.

37. Kaufman DL, Erlander MG, Claire-Salzler M, Atkinson MA, Maclaren NK, and Tobin AJ. Autoimmunity to two forms of glutamate decarboxylase in insulin dependent diabetes mellitus. J. Clin. Invest. 1992; 89: 283-292.

38. Mauch L, Seissler J, Haubruck H, Cook NJ, Abney CC, Berthold H, Wirbelauer C, Liedvogel B, Scherbaum WA, and Northemann W. Baculovirus-mediated expression of human 65 kDa and 67 kDa glutamic acid decarboxylases in SF9 insect cells and their relevance in diagnosis of insulin-dependent diabetes mellitus. J. Biochem 1993; 113: 699-704.

39. Richter W, Shi Y, and Bækkeskov S. Autoreactive epitopes defined by diabetes-association human monoclonal antibodies are localizaed in the middle and C-terminal domains of the smaller form of glutamate decarboxylase. Proc. Natl. Acad. Sci. USA 1993; 90: 2832-2836.

40. Kaufman DL, Clare-Salzer M, Tian J, Forsthuber T, Ting GSP, Robinson P, Atkinson MA, Sercarz EE, Tobin AJ, and Lehmann PV. Spontaneous loss of T-cell tolerance to glutamic acid decarboxylase in murine insulin-dependent diabetes. Nature 1993; 366: 69-72.

41. Kim J, Namchuk M, Bugawan T, Fu Q, Jaffe M, Shi Y, Aanstoot HJ, Turck CW, Erlich H, Lennon V, and Baekkeskov S. Higher autoantibody levels and recognition of a linear NH2-terminal epitope in the autoantigen GAD65, distinguish stiff-man syndrome from insulin-dependent diabetes mellitus. J. Exp. Med. 1994; 180:595-606.

42. McLaughlin BJ, Barber R, Saito K, Roberts E, Wu JY. Immunocytochemical localization of glutamate decarboxylase in rat spinal cord. J. Comp. Neurol. 1975; 164: 305-321.

43. Sorenson RL, Garry DG, and Brelje TC. Structural and functional considerations of GABA in islets of Langerhans ß-cells and nerves. Diabetes 1991; 41: 1365-1374.

44. Thomas-Reetz A, Hell JW, During MJ, Walch-Solimena C, Jahn R, and De Camilli P. A γ-aminobutyric acid transporter driven by a proton pump is present in synaptic-like microvesicles of pancreatic ß cells. Proc. Natl. Acad. Sci. USA 1993; 90: 5317-5321.

GABA: Receptors, Transporters and Metabolism
ed. by C. Tanaka & N.G. Bowery

ANTI-GAD ANTIBODY IN VARIOUS NEUROLOGICAL DISEASES

Shigenobu Nakamura, Hidekazu Kamei and Yasuyo Mimori

Third Department of Internal Medicine, Hiroshima University School of Medicine, Hiroshima, Japan

Summary: Anti-GAD antibody was measured in the cerebrospinal fluid (CSF) of 84 neurological patients and in the serum of 2258 patients with insulin-dependent diabetes mellitus (IDDM) or other diseases and 461 normal subjects. Anti-GAD antibody was detected in the CSF of a patient with epilepsy and cerebellar atrophy who was administered phenytoin for more than 20 years. Anti-GAD antibody was detected in the serum of 36.6% patients with IDDM.

Introduction

The rate limiting step in the synthesis of γ-aminobutyric acid(GABA) is the decarboxylation of glutamate, a reaction catalyzed by glutamate decarboxylase (GAD). Two GADs, 65-kDa and 67-kDa have been reported to be each encoded by a single gene in human tissues (1, 2). Both types of GAD are expressed in rat brain, rat islets, dog brain, monkey brain and the human brain (1). An exclusive expression of GAD 65-kDa has been reported at both mRNA and protein level in human islet (1). Antibodies against 64-kDa has been identified in patients with stiff-man syndrome, epilepsy, and insulin-dependent diabetes mellitus (3-5). We measure anti-GAD titre in the cerebrospinal fluid (CSF) of patients with various neurological diseases and in the serum of patients with insulin-dependent diabetes mellitus (IDDM).

Materials and Methods

Anti-GAD antibody was measured with LIP anti-GAD Hoechst kit developed by Yamaguchi et al. (6). Twenty microliters of serum or cerebrospinal fluid was mixed with 100 microliters of 64 kDa GAD purified from swine brain which was labeled with 125 iodine isotope. The mixture was kept for 2 hours at room temperature. The incubation mixture was mixed with 1 ml of goat anti-human IgG antibody and incubated at room temperature for 30 minutes, followed by centrifugation at 3,000rpm for 30 minutes. The resulting precipitate was separated from the supernatant by a careful aspiration to remove unreacted tracer. The radioactivity in the precipitate was counted with a gamma counter.

We examined anti-GAD titre in the cerebrospinal fluid of patients with various neurological diseases. Those neurological diseases consisted mainly of diseases caused by autoimmune mechanisms (Table 1).

Table 1. Subjects examined

Multiple Sclerosis 33 cases (male 8, female 25) age: 37.0 ± 14.3 years old
Mixed Connective Tissue Disease with Neurological Manifestations 7 cases (male 0, female 7) age: 56.4 ± 19.0 years old
Rheumatoid Arthritis with Neurological Manifestations 3 cases (male 1, female 2) age: 70.0 ± 7.0 years old
Systemic Lupus Erythematosus with Neurological Manifestations 2 cases (male 0, female 2) age: 41.5 years old
Sarcoidosis 1 case (female) 54 years old
NeuroBehçet 3 cases (male 0, female 3) age: 21.7 ± 6.7 years old
Myoclonus epilepsy 5 cases (male 1, female 5) age: 37.4 ± 22.3 years old
Epilepsy 2 cases (male 2, female 0) age: 33 years old
Dystonia 8 cases (male 2, female 5) age: 40.4 Å} 13.4 years old
Huntington's Disease 4 cases (male 2, female 2) age: 55.0 ± 4.2 years old
Progressive Supranuclear Palsy 1 case (male) 48 years old
Pure Akinesia 1 case (male) 66 years old
Wilson's Disease 1 case (female) 16 years old
Pick's Disease 1 case (male) 45 years old
Myotonic Dystrophy 1 case (male) 30 years old
Cervical Spondylosis 3 cases (male 2, female 1) age: 60.0 ± 15.7 years old
Multifocal Motor Neuropathy 2 cases (male 2) age: 49 years old
Hashimoto's Thyroiditis with Neurological Manifestations 2 cases (male 1, female 1) age: 48.5 years old
Harada's Disease with Neurological Manifestations 2 cases (male 2) age: 31.5 years old
Sleep apnea 1 case (male) age: 58 years old
X-Linked Recessive Spinobulbar Muscular Atrophy 1 case (male) 31 years old

Patients with Huntington's disease have been reported to show a decrease in GAD activity in their striatum (7). So, we detected anti-GAD antibody in the cerebrospinal fluid of patients with Huntington's disease. A case of progressive supranuclear palsy, pure akinesia, Wison's disease, Pick's disease or myotonic dystrophy was also assessed.

The collaboration working group examined the anti-GAD antibody in the serum of 1071 patients with insulin-dependent diabetes mellitus. among 1071 cases 164 patients were within 1 year after onset and 807 cases were over 1 year after the onset of diabetes. Nine hundred twenty six patients with non-insulin dependent diabetes mellitus were examined for anti-GAD antibody. We detected anti-GAD antibody in the serum of autoimmune diseases and other diseases. We also measured anti-GAD antibody in the serum of 461 normal subjects.

Results

Totally, 84 cerebrospinal fluid samples were examined for anti-GAD antibody. Among them, anti-GAD antibody was detected in the CSF of only one patient with myclonus epilepsy and cerebellar atrophy who had been administered phenytoin for more than 20 years.

Report of a case

We have studied a 32-year-old woman. Her chief complaint was convulsion, gait disturbance and dementia. As a family history, her mother suffered from convulsive seizure, but no consanguinity. She had measles at 2 years old and mumps at 3 years old. The first tonic-clonic convulsion occurred at age 7, in 1970. Soon she visited a pediatrician and was diagnosed as epilepsy and was administered anticonvulsant (phenytoin). But epileptic fits occurred occasionally and then increased in frequency at age 8. So her dose of anticovulsant increased to prevent attacks thereafter. At age 15, she stopped medication according to her friend's suggestion. One week after, however, epileptic fits occurred quite frequently. So, she resumed anticonvulsants and since then, she was free from convulsive seizure until age 18. But her school record became poor. At her age 19, she felt weakness and tremor in her upper extremities. She also recognized drunken gait and an attack occurred once a year. The head CT-scan taken in 1983 revealed minimal cerebellar cortical atrophy with a slight atrophy at brain stem. Her physical conditions were getting worse slowly. At her age 28, she was admitted to our hospital for progressive gait disturbance and epileptic attack.

Her consciousness was alert, but her intelligence score by Dr. Hasegawa was 18 points in 30 points at full mark. She did not have any meningeal irritation, and her cranial nerve function was normal. Muscle manual test revealed normal. But her deep tendon reflexes were slightly reduced. She showed a dysmetria and decomposition in her four extremities. Tremurous involuntary movements were found in her all limbs. Her sensory functions were normal. She had a mild orthostatic hypotension, suggesting dysautonomia.

Her laboratory findings disclosed that glucose tolerance test was within normal range. Serum lactate and pyruvate concentrations were normal either at rest or after aerobic exercise. Her cerebrospinal fluid obtained by lumbar puncture was clear and colorless. The initial pressure was 70 mm of water and showed an ample rise by pressing her jugular vein. Cerebrospinal fluid did not contain cells and the concentration of glucose and protein was 52 and 77 mg/dl, respectively. The anti-GAD antibody was 4.7 units/ml and elevated. Serum anti-GAD antibody titer was 2,861 unit/ml and higher, when compared with that of normal controls which is below 4 units /ml. Her surface electromyogram showed brief synchronous discharges. Electroencephalogram showed spiky paroxysmal discharges and photomyoclonic response as shown in the next slide.

Her electroencephalogram disclosed spike and sharp waves diffusely in almost all cortical leads (Fig. 1).

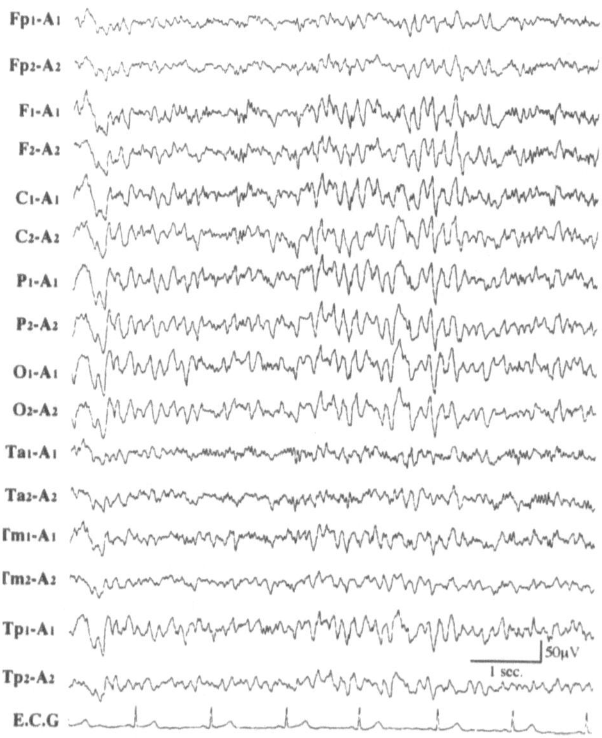

Figure 1. The electroencephalogram of 32-year-old female with a high anti-GAD titre in her cerebrospinal fluid.

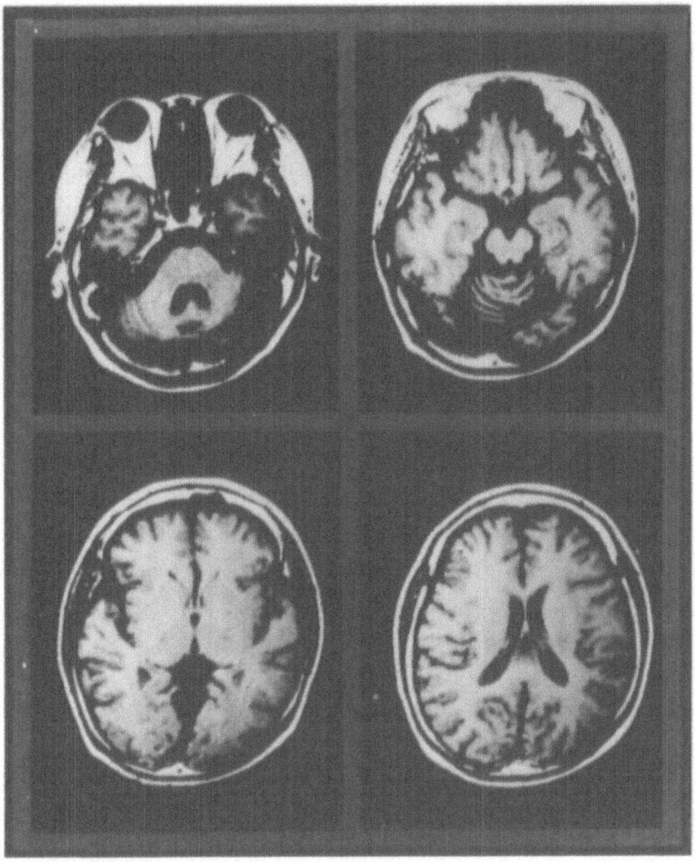

Figure 2. The magnetic resonance imaging of 32-year-old female with a high anti-GAD titre in her cerebrospinal fluid.

Magnetic resonance imaging revealed a remarkable atrophy in cerebellar cortex (Fig. 2). The forth ventricle was enlarged and the brain stem was slightly atrophied. However, her cerebral cortex was well preserved. The grade of cerebellar cortical atrophy demonstrated in 1993 was much more remarkable than that in 1983. Those observations suggest that the patient with myoclonus epilepsy accompanied by cerebellar atrophy exhibits a high anti-GAD antibody both in the serum and cerebrospinal fluid.

Serum anti-GAD antibody in IDDM patients

The anti-GAD antibody titer was measured in the serum of Japanese patients with insulin-dependent diabetes mellitus. More than 30% of IDDM patients showed higher anti-GAD antibody titer than cut off line, 4 units/ml. Female patients represented with white circles seem to show slightly higher anti-GAD titer.

The prevalence of high titres of autoantibody to GAD was 36.6% in total IDDM patients. Higher incidence was observed in female IDDM patients than male patients. The prevalence seems higher in IDDM patients who have suffered the disease less than 1 year than in those more than 1 year. The prevalence of anti-GAD antibody was higher in younger onset (< 26 years)of the disease than later onset.

The incidence of anti-GAD antibody was much lower in non-insulin-dependent diabetes mellitus NIDDM) than insulin-dependent diabetes mellitus (IDDM). The prevalence of anti-GAD antibody was less than 4% in patients with NIDDM. Female patients seem to show higher prevalence than male patients and non-obese NIDDM patients seem to contain anti-GAD antibody more frequently than obese patients. However, the prevalence was much lower, compared with IDDM.

The prevalence of autoantibody to GAD is low in general autoimmune diseases and less than 5%. Neurological diseases seem to show a higher prevalence of anti-GAD antibody than other diseases, although the total numbers of the patient are too small to deduce the conclusion.

Discussion

Adult-onset cerebellar atrophy is caused by various genetic and acquired diseases. Genetic origins involving autosomal recessive inheritance (8), fresh dominant mutation (9), or mitochondrial inheritance are suggested. The parenchymal cerebellar cortical atrophy is caused by acquired factors, such as alcoholism, hypothyroidism, toxicity, long-term anticonvulasant therapy, or paraneoplastic process with major autoimmune participation (10). However, there is no explanation of the degenerative process that causes the sporadic pure cortical atrophy as described by Marie et al. (11).

Honnorat and coworkers (12) have recently reported a paper on autoantibodies to GAD in a patient with cerebellar cortical atrophy, peripheral neuropathy, and slow eye movements. In the paper, they showed that serum and cerebrospinal fluid samples taken from one patient with sporadic cortical cerebellar atrophy associated with peripheral neuropathy and slow eye movements contained anti-GAD autoantibodies.

Decreased GABAergic activity is implicated as a cause of epilepsy (13). Decreased GAD activity in the brain of experimental animals produce a model of epilepsy. Moreover, anticonvulsant, phenytoin has been known to produce a cerebellar atrophy associated with ataxia, dysarthria, exaggerated deep tendon reflexes (14). The present case had led a normal physical life at least at her age 18, suggesting that the cerebellar atrophy might be an acquired one and caused through an autoimmune mechanism, probably related with phenytoin, distinct from the case reported by Honnorat et al. (12). Both of these cases suggest a participation of autoimmunity in the pathogenesis of a group of sporadic cerebellar cortical atrophy and the involvement of the cerebellar GABAergic system in these diseases.

The autoimmune disease IDDM is thought to result from T-cell-mediated destruction of pancreatic-islet beta cells (15). A 64 kDa β-cell autoantigen is a target of autoantibodies in IDDM. The 64 kDa autoantibodies are present in ≥80% of newly diagnosed IDDM patients and have been detected up to several years before clinical onset of IDDM concomitant with a gradual loss of β-cell (16).

Several patients with a neurological disease called stiff-man syndrome have autoantibodies to GABAergic neurons, especially GAD (3). A significant fraction of patients with stiff-man syndrome have IDDM (17). The pancreatic islet β-cell 64 kDa autoantigen has been identified as GAD (18). Thus, autoantibodies to the 64 kDa GAD are the earliest and most reliable predictive markers of IDDM in humans. However, the prevalence of anti-GAD antibody was much less in Japanese patients with IDDM. Probably, the difference might be attributable to the genetic factor such as HLA.

In addition, a case with cerebellar atrophy showed a high anti-GAD antibody titre, suggesting an autoimmune mechanism for cerebellar cortical atrophy. Measures that could divert GAD immunity away from a cellular response might retard cerebellar cortical atrophy as well as progressive β-cell destruction.

Acknowledgments

We thank Dr. Yamaguchi A in Hoechst Japan, Pharma Research and Development Division, Drug Development Research Laboratories Analytical Research, Biochemistry for his technical assistance. We also thank Drs Matsuba I, Turuoka A, and Ikeda Y in The Jikei University School of Medicine, Tokyo), Toyoda T(Tohoku University, School of Medicine, Sendai, Dr. Nagataki S in Nagasaki University, School of Medicine, Nagasaki, and Dr. Ishiki G in Osaka City University Medical School, Osaka, for collaborative study on anti-GAD antibody in IDDM patients, and Drs Nakayama T, Imon Y, Katayama S, Yamaguchi S, and Kohriyama T in Hiroshima University School of Medicine, Hiroshima for clinical studies on neurological patients.

References

1. Karlsen AE, Hagopian WA, Grubin CE, Dube S, Disteche CM, Adler DA, Barmeier H, Mathewes S, Grant FJ, Foster D, Lernmark A. Cloning and primary structure of a human islet isoform of glutamic acid decarboxylase from chromosome 10. Proc Natl Acad Sci USA 1991; 88:8337-8341.

2. Bu DF, Erlander MG, Hitz BC, Tillakaratne NJK, Kaufman DL, Wagener-McPherson CB, Evans GA, Tobin AJ. Two human glutamate decarboxylases, 65-kDa GAD, are each encoded by a single gene. Proc Natl Acad Sci USA 1992; 89:2115-2119.

3. Solimena M, Folli F, Denis-donini S, Comi GC, Pozza G, Gamilli PDe, Vicari AM. Autoantibodies to glutamic acid decarboxylase in a patient with stiff-man syndrome, epilepsy, and type 1 diabetes mellitus. New Engl J Med 1988; 318:1012-1020.

4. DeAizpurua HJ, Wilson Y, Harrison LC. Glutamic acid decarboxylase (GAD) autoantibodies in pre-clinical insulin-dependent diabetes. Proc Natl Acad Sci USA 1992; 89:9841-9845.

5. Kaufman DL, Erlander MG, Clare-Salzer M, Atkinson MA, Maclaren NK, Tobin AJ. Autoimmunity to two forms of glutamate decarboxylase in insulin-dependent diabetes mellitus. J Clin Invest 1992; 89:283-292.

6. Yamaguchi A, Ogata K, Kubo H, Ohta K, Mizushima Y, Watanabe H, Ishige H, Yamane R, Kawasaki E, Nagataki S, Tsuruoka A, Matsuba I, Ikeda Y. Performance and characteristic study of "RIP Anti-GAD Hoechst" RIA kit in detection of anti-GAD antibodies. Igaku to Yakugaku 1994; 31:419-431.

7. Bird ED, Iversen LL. Huntington's chorea. Post-mortem measurement of glutamic acid decarboxylase, choline acetyltransferase and dopamine in basal ganglia. Brain 1974; 97:457-472.

8. Kawakami H, Maruyama H, Nakamura S, Kawaguchi Y, Kakizuka A, Doryu M, Sobue G. Unique features of the CAG repeats in Machado-Joseph disease. Nature Genet 1995; 9: 344-345.

9. Harding AE. 'Idiopathic' late onset cerebellar ataxia: a clinical and genetic study of 36 cases. J Neurol Sci 1981; 51: 259-271.

10. Posner JB. Pathogenesis of central nervous system paraneoplastic syndromes. Rev Neurol (Paris) 1992; 148: 502-512.

11. Marie P, Foix C, Alajouanine T. De l'atrophie céréblleuse tardive à prédominance corticale. Rev Neurol (Paris) 1922; 2: 849-885.

12. Honorat J, Trouillas P, Thivolet C, Aguera M, Belin M-F. Autoantibodies to glutamate decarboxylase in a patient with cerebellar cortical atrophy, peripheral neuropathy, and slow eye movements. Arch Neurol 1995;52: 462-468.

13. Meldrum BS. The epilepsies. In: Neurotransmitters. Webster RA, Jordan CC, editors. Oxford: Blackwell Scientific Pub, 1989: 301-335.

14. Porter IH. Diphenylhytantoin toxicity. In: Handbook of Clinical Neurology Vol 42. Vinken PJ, Bruyn GW, editors. Amsterdam: Northholand Publisher, 1981: 641-642.

15. Harrison LC, Honeyman MC, DeAizpurua HJ, Schmidi RS, Colman PG, Tait BD, Cram DS. Inverse relation between humoral and cellular immunity to glutamic acid decarboxylase in subjects at risk of insulin-dependent diabetes. Lancet 1993; 341: 1365-1369.

16. Atkinson MR, Maclaren NK, Scharp DW, Lacy PE, Riley WJ. 64,000 Mr autoantibodies as predictors of insulin-dependent diabetes. Lancet 1990; 335:1357-1360.

17. Solimena M, Folli F, Aparisi R, Pozza G, DeCamilli P. Autoantibodies to GABA-ergic neurons and pancreatic beta cells in stiff-man syndrome. N Engl J Med 1990;322: 1555-1560.

18. Bekkeskov S, Aanstoot H-J, Christgau S, Reetz A, Solimena M, Cascalho M, Folli F, Richter-Olesen H, De Camilli P. Identification of the 64K autoantigen in insulin-dependent diabetes as the GABA-synthesizing enzyme glutamic acid decarboxylase. Nature 1990; 347: 151-156

GABA: Receptors, Transporters and Metabolism
ed. by C. Tanaka & N.G. Bowery

GABA TRANSPORTERS: STRUCTURE, FUNCTION AND MECHANISM

Baruch I. Kanner

Department of Biochemistry, Hadassah Medical School, The Hebrew University, P.O.Box 12272, Jerusalem 91120, Israel.

Summary: The removal of neurotransmitters by their transporters — located in the plasma membranes of nerve terminals and glial cells — plays an important role in the termination of synaptic transmission. The GABA transporter GAT-1 is the first neurotransmitter transporter to be purified, reconstituted and cloned. It belongs to a large superfamily of neurotransmitter transporters. Its properties and its relation to the other family members will be discussed.

Introduction

The reuptake of neurotransmitters from the synaptic cleft by high-affinity transport appears to play an important role in the overall process of synaptic transmission, namely in its termination [1,2]. The process is catalyzed by sodium-coupled neurotransmitter transport systems [reviewed in 3–5] located in plasma membranes of nerve endings and glial cells. These transport systems have been investigated in detail by using plasma membranes obtained upon osmotic shock of synaptosomes. It appears that these transporters are coupled not only to sodium, but also to additional ions like potassium or chloride.

The most abundant and well characterized of these uptake systems in rat brain are those for GABA and L-glutamate [3–5], two major neurotransmitters in the central nervous system. Multiple GABA transporter species have been detected in membrane vesicles and reconstituted preparations from rat brain [6,7], suggesting diversity of these systems. The application of molecular cloning techniques to this area has revealed that this diversity is larger than expected (see below). In this review emphasis will be laid on the GABA transporter and those related to it. The properties of the L-glutamate transporter which is structurally and functionally different, have been reviewed recently [8].

Mechanistic studies

GABA is accumulated by electrogenic co-transport with sodium and chloride. The electrogenicity of the process has recently been shown directly [9,10]. We have been able to demonstrate directly that both sodium as well as chloride ions are cotransported with GABA by the transporter. This has been accomplished using a partly purified transporter preparation which was reconstituted into liposomes and the use of Dowex columns to terminate the reactions. These proteoliposomes catalysed GABA- and chloride-dependent $^{22}[Na^+]$ transport, as well as GABA- and sodium-dependent $^{36}[Cl^-]$ translocation [11]. Using this system the stoichiometry has also been determined kinetically, i.e., by comparing the initial rate of the fluxes of $[^3H]$-GABA, $^{22}[Na^+]$ and $^{36}[Cl^-]$. The results [11] are similar to those found using the thermodynamic method [12], yielding an apparent stoichiometry of 2.5 Na^+: 1 Cl^-:1 GABA. This is in harmony with the predicted restrictions; thus if GABA is translocated in the zwitterionic form - the predominant one at physiological pH - an electrogenic cotransport of the three species requires a stoichiometry of nNa^+:mCl^-: GABA with $n > m$. Many other neurotransmitter transporters, including those for norepinephrine, dopamine, serotonin, choline and glycine, require chloride in addition to sodium for optimal activity [13].

Reconstitution and purification

Using the reconstitution methodology which enables to reconstitute many samples simultaneously and rapidly [14], one of the subtypes of the GABA transporter [15] has been purified to an apparent homogeneity. It is a glycoprotein and has an apparent molecular weight of 70-80 kDa. The transporter retains all the properties observed in membrane vesicles.

Biochemical characterization of the GABA transporter

The effect of proteolysis on the function of the transporter was examined. It was purified using all steps except for the lectin chromatography [15]. After papain treatment and lectin chromatography, GABA transport activity was eluted with N-acetyl glucosamine. The characteristics of transport were the same as that of the pure transporter [16].

In order to define which regions of the transporter were cleaved off, antibodies were raised against synthetic peptides corresponding to several regions of the rat brain GABA transporter. According to our model (see below), this glycoprotein has 12 transmembrane α-helices with both amino and carboxyl terminii located in the cytoplasm. The antibodies recognized the intact transporter on Western blots. The papainized transporter runs on sodium dodecyl sulfate-polyacrylamide gels as a broad band with an apparent molecular mass between about 58 and 68 kDa as compared to 80 kDa for the untreated transporter. The transporter fragment was recognized by all the antibodies, except for that raised against the amino terminus. Pronase cleaves the transporter to a relatively sharp 60 kDa band, which reacts with the antibodies against the internal loops but not with either the amino- or the carboxyl-terminii. This pronase-treated transporter, upon isolation by lectin chromatography, was reconstituted. It exhibits full GABA transport activity. This activity exhibits the same features as the intact system including an absolute dependence on sodium and chloride as well as electrogenicity. Thus the amino- and carboxyl-terminal parts of the transporter are not required for functionality [17].

Fragments of the (Na^+ + Cl^-)-coupled $GABA_A$ transporter were produced by proteolysis of membrane vesicles and reconstituted preparations from rat brain [18]. The former were digested with pronase, the latter with trypsin. Fragments with different apparent molecular masses were recognized by sequence directed antibodies raised against this transporter. When GABA was present in the digestion medium the generation of these fragments was almost entirely blocked [18]. At the same time, the neurotransmitter largely prevented the loss of activity caused by the protease. The effect was specific for GABA; protection was not afforded by other neurotransmitters. It was only observed when the two cosubstrates, sodium and chloride, were present on the same side of the membrane as GABA [18]. The results indicate that the transporter may exist in two conformations. In the absence of one or more of the substrates, multiple sites located throughout the transporter are accessible to the proteases. In the presence of all three substrates — conditions favoring the formation of the translocation complex — the conformation is changed such that these sites become inaccessible to protease action.

A new superfamily of Na-dependent neurotransmitter transporters

Partial sequencing of the purified $GABA_A$ transporter allowed the cloning of the first member of the new family of Na-dependent neurotransmitter transporters [19]. After expression cloning of the noradrenaline transporter [20], it became clear that it had significant homology with the $GABA_A$ transporter. The use of functional c-DNA expression assays and amplification of related sequences by polymerase chain reaction (PCR) resulted in the cloning of additional transporters which belong to this family, such as the dopamine [21–23] and serotonin [24,25] transporters, additional GABA transporters [26–29], transporters of glycine [30–32], proline [33], taurine [34,35], betaine [36] and two 'orphan' transporters, whose substrates are still unknown [37,38]. In addition, another family member which was originally thought to be a choline transporter [39], probably is in fact a creatine transporter [40]. A novel glycine transporter cDNA encoding for a 799 amino acid protein has recently been isolated. This is significantly longer than most members of the superfamily. If we take into account that part of the mass of these transporters is constituted by sugar, it could encode for the 100 kDa glycine transporter which was purified and reconstituted [41].

Shared features of family members

The deduced amino acid sequences of these proteins reveal 30–65% identity between different members of the family. Based on these differences in homology the family can be divided into four subgroups: **A.** transporters of biogenic amines (noradrenaline, dopamine and serotonin); **B.** various GABA transporters as well as transporters of taurine and creatine; **C.** transporters of proline and glycine; and **D.** 'orphan' transporters. These proteins share some features of a common secondary structure. Each transporter is composed of 12 hydrophobic putative transmembrane α-helices. The lack of a signal peptide suggests that both amino- and carboxy-terminii face the cytoplasm. These regions contain putative phosphorylation sites, that may be involved in regulation of the transport process. The second extracellular loop between helices 3 and 4 is the largest, and it contains putative glycosylation sites.

Alignment of the deduced amino acid sequences of 13 different members of this superfamily, whose substrates are known (subgroups A-C) revealed that some segments within these proteins share a higher degree of homology than others. The most highly conserved regions (>50% homology) are helix 1 together with the extracellular loop connecting it with helix 2, and helix 5 together with a short intracellular loop connecting it with helix 4 and a larger extracellular loop connecting it with helix 6. These domains may be involved in stabilizing a tertiary structure, that is essential for the function of all these transporters. Alternatively, they may be related to a common function of these transporters, such as the translocation of sodium ions. The region stretching from helix 9 on is far less conserved than the segment containing the first 8 helices. Possibly, this domain contains some residues that are involved in translocating the different substrates. The least conserved segments are the amino- and carboxy-terminii. As was mentioned above, these areas may be involved in regulation of the transport process. The 'orphan' transporters differ from all other members of the family in three regions. They contain much larger extracellular loops between helices 7-8 and helices 11-12.

Structure-function relationships

It has been shown previously that parts of amino- and carboxyl-terminii of the $GABA_A$ transporter are not required for function [17]. In order to define these domains, a series of deletion mutants was studied in the GABA transporter [42]. Transporters truncated at either end until just a few amino-acids distance from the beginning of helix 1 and the end of helix 12, retained their ability to catalyze sodium and chloride-dependent GABA transport. These deleted segments did not contain any residues conserved among the different members of the superfamily. Once the truncated segment included part of these conserved residues, the transporter's activity was severely reduced. However, the functional damage was not due to impaired turnover or impaired targeting of the truncated proteins [42].

The substrate translocation performed by the various members of the superfamily is sodium- and usually chloride-dependent. In addition, some of the substrates contain charged groups as well. Therefore, charged amino acids in the membrane domain of the transporters may be essential for their normal function. This was tested using the GABA transporter [43]. Out of 5 charged amino acids within its membrane domain only one, $arginine_{69}$ in helix 1, is absolutely essential for activity. It is not merely the positive charge which is important, as even its substitution to other positively charged amino acids does not restore activity. The functional damage is not due to impaired turnover or impaired targeting of the mutated protein. The three other positively charged amino acids and the only negatively charged one are not critical [43].

The transporters of biogenic amines contain an additional negatively charged residue in helix 1. Replacement of $aspartate_{79}$ in the dopamine transporter with alanine, glycine or glutamate significantly reduced the uptake of dopamine, MPP^+ (parkinsonism inducing neurotoxin), and CFT (cocaine analogue) without affecting Bmax. Apparently, $aspartate_{79}$ in helix 1 interacts with dopamine's amino group during the transport process. $Serine_{356}$ and $serine_{359}$ in helix 7 are also involved in dopamine binding and translocation, perhaps by interacting with the hydroxyls on the catechol [44].

Studies of other proteins indicate that in addition to charged amino acids, aromatic amino acids containing π-electrons are also involved in maintaining the structure and function of these proteins [45]. Therefore, tryptophan residues in the membrane domain of the GABA transporter were mutated into serine as well as leucine [46]. Mutations at the 68 and 222 position (in helix 1 and helix 4, respectively) led to a decrease of over 90% of the GABA uptake. The available data indicate the importance of helices 1, 4 and 7 in the function of the different transporters, but the nature of their contribution to the overall translocation process remains unclear.

We have explored the role of the hydrophillic loops connecting the putative transmembrane domains. Deletions of randomly picked non-conserved single amino acids in the loops connecting helices 7 and 8 or 8 and 9 result in inactive transport upon expression in HeLa cells. However, transporters where these amino acids are replaced with glycine retain significant activity. The expression levels of the inactive mutant transporters was similar to that of the wild-type, but one of these, ΔVal-348, appears to be defectively targetted to the plasma membrane. Our data are compatible with the idea that a minimal length of the loops is required, presumably to enable the transmembrane domains to interact optimally with each other [47].

Conclusions

Recent breakthroughs including the purification of some of the sodium-coupled neurotransmitter transporters, followed by the cloning of their cDNAs, have considerably improved our understanding of the structure of these transporters. Studies using site-directed mutagenesis revealed the importance of specific residues in the function of these transporters. Additional mutations and further functional characterization of all the mutated transporters should help to understand the functional contribution of different segments of these proteins to the overall transport process. Applying independent structural approaches will complement and extend our insight on the structure and function of these transporters.

References

1. Iversen LL: Role of transmitter uptake mechanism in synaptic neurotransmission. Br J Pharmacol 1971;41:571–591.
2. Kuhar JM: Neurotransmitter uptake: a tool in identifying neurotransmitter-specific pathways. Life Sci 1973;13:1623–1634.
3. Kanner BI: Bioenergetics of neurotransmitter transport. Biochim Biophys Acta 1983;726:293–316.
4. Kanner BI: Ion-coupled neurotransmitter transporter. Curr Opin Cell Biol 1989;1:735–738.
5. Kanner BI, Schuldiner S: Mechanism of transport and storage of neurotransmitters. CRC Crit Rev Biochem 1987;22:1–39.
6. Mabjeesh NJ, Kanner BI: Low affinity γ-aminobutyric acid transport in rat brain. Biochemistry 1989;28:7694–7699.
7. Kanner BI, Bendahan A: Two pharmacologically distinct sodium- and chloride-coupled high-affinity γ-aminobutyric acid transporters are present in plasma membrane vesicles and reconstituted preparations from rat brain. Proc Natl Acad Sci USA 1990;87:2550–2554.
8. Kanner BI: Glutamate transporters from brain: a novel neurotransmitter transporter family. FEBS Lett 1993;325:95–99.
9. Mager S, Naeve J, Quick M, Labarca C, Davidson N, Lester HA: Steady states, charge movements and rates for a cloned GABA transporter expressed in *Xenopus* oocytes. Neuron 1993;10:177-188.
10. Kavanaugh MP, Arriza JL, North RA, Amara SG: Electrogenic uptake of γ-aminobutyric acid by a cloned transporter expressed in oocytes. J Biol Chem 1992;267:22007-22009.
11. Keynan S, Kanner BI: γ-Aminobutyric acid transport in reconstituted preparations from rat brain: coupled sodium and chloride fluxes. Biochemistry 1988;27:12–17.
12. Radian R, Kanner BI: Stoichiometry of sodium- and chloride-coupled γ-aminobutyric acid transport by synaptic plasma membrane vesicles isolated from rat brain. Biochemistry 1983;22:1236–1241.
13. Kuhar MJ, Zarbin MA: Synaptosomal transport: a chloride dependence for choline, GABA, glycine and several other compounds. J Neurochem 1978;31:251-256.
14. Radian R, Kanner, BI: Reconstitution and purification of the sodium- and chloride-coupled γ-aminobutyric acid transporter from rat brain. J Biol Chem 1985;260:11859–11865.
15. Radian R, Bendahan A, Kanner BI: Purification and identification of the functonal sodium- and chloride-coupled γ-aminobutyric acid transport glycoprotein from rat brain. J Biol Chem 1986;261:15437–15441.
16. Kanner BI, Keynan S, Radian R: Structural and functional studies on the sodium- and chloride-coupled γ-aminobutyric acid transporter. Deglycosylation and limited proteolysis. Biochemistry 1989;28:3722–3727.
17. Mabjeesh NJ, Kanner, BI: Neither amino nor carboxyl termini are required for function of the sodium- and chloride-coupled γ-aminobutyric acid transporter from rat brain. J Biol Chem 1992;267:2563–2568.

18. Mabjeesh NJ, Kanner BI: The substrates of a sodium- and chloride-coupled γ-aminobutyric acid transporter protect multiple sites throughout the protein against proteolytic cleavage. Biochemistry 1993;32:8540–8546.
19. Guastella J, Nelson N, Nelson H, Czyzyk L, Keynan S, Miedel MC, Davidson NC, Lester HA, Kanner BI: Cloning and expression of a rat brain GABA transporter. Science 1990;249:1303–1306.
20. Pacholczyk T, Blakely RD, Amara SG: Expression cloning of a cocaine and antidepressant-sensitive human noradrenaline transporter. Nature 1991;350:350–354.
21. Shimada S, Kitayama S, Lin CL, Patel A, Nanthakumar E, Gregor P, Kuhar M, Uhl G: Cloning and expression of a cocaine-sensitive dopamine transporter complementary DNA. Science 1991;254:576–578.
22. Kilty JE, Lorang D, Amara SG: Cloning and expression of a cocaine-sensitive rat dopamine transporter. Science 1991;254:578–579.
23. Usdin TB, Mezey E, Chen C, Brownstein MJ, Hoffman BJ: Cloning of the cocaine sensitive bovine dopamine transporter. Proc Natl Acad Sci, USA 1991;88:11168–11171.
24. Hoffman BJ, Mezey E, Brownstein MJ: Cloning of a serotonin transporter affected by antidepressants. Science 1991;254:579–580.
25. Blakely RD, Benson HE, Fremeau RT Jr, Caron MG, Peek MM, Prince HK, Bradley CC: Cloning and expression of a functional serotonin transporter from rat brain. Nature 1991;353:66–70.
26. Clark JA, Deutch AY, Gallipoli PZ, Amara SG: Functional expression and CNS distribution of a β-alanine sensitive neuronal GABA transporter. Neuron 1992;9:337–348.
27. Borden LA, Smith KE, Hartig PR, Branchek TA, Weinshank RL: Molecular heterogeneity of the GABA transport system. J Biol Chem 1992;267:21098–21104.
28. Lopez-Corcuera B, Liu QR, Mandiyan S, Nelson H, Nelson N: Expression of a mouse brain cDNA encoding novel γ-aminobutyric acid transporter. J Biol Chem 1992;267:17491–17493.
29. Liu QR, Lopez-Corcuera B, Mandiyan S, Nelson H, Nelson N: Molecular characterization of four pharmacologically distinct γ-aminobutyric acid transporters in mouse brain. J Biol Chem 1993;268:2104–2112.
30. Smith KE, Borden LA, Hartig PA, Branchek T, Weinshank RL: Cloning and expression of a glycine transporter reveal colocalization with NMDA receptors. Neuron 1992;8:927–935.
31. Liu QR, Nelson H, Mandiyan S, Lopez-Corcuera B, Nelson N: Cloning and expression of a glycine transproter from mouse brain. FEBS Lett 1992;305:110–114.
32. Guastella J, Brecha N, Weigmann C, Lester HA: Cloning, expression and localization of a rat brain high affinity glycine transporter. Proc Natl Acad Sci, USA 1992;89:7189–7193.
33. Fremeau RT Jr, Caron MG, Blakely RD: Molecular cloning and expression of a high affinity l-proline transporter expressed in putative glutamatergic pathways of rat brain. Neuron 1992;8:915–926.

34. Uchida S, Kwon HM, Yamauchi A, Preston AS, Marumo F, Handler JS: Molecular cloning of the cDNA for an MDCK cell Na^+- and Cl^--dependent taurine transporter that is regulated by hypertonicity. Proc Natl Acad Sci, USA 1992;89:8230–8234.
35. Liu QR, Lopez-Corcuera B, Nelson H, Mandiyan S, Nelson N: Cloning and expression of a cDNA encoding the transporter of taurine and β-alanine in mouse brain. Proc Natl Acad Sci, USA 1992;89:12145–12149.
36. Yamauchi A, Uchida S, Kwon HM, Preston AS, Robey RB, Garcia-Perez A, Burg MB, Handler JS: Cloning of a Na^+ and Cl^- dependent betaine transporter that is regulated by hypertonicity. J Biol Chem 1992;267:649–652.
37. Uhl GR, Kitayama S, Gregor P, Nanthakumer E, Persico A, Shimada S: Neurotransmitter transporter family cDNAs, in a rat midbrain library: 'orphan transporters' suggest sizable structural variations. Molec Brain Res 1992;16:353–359.
38. Liu QR, Mandiyan S, Lopez-Corcuera B, Nelson H, Nelson N: A rat brain cDNA encoding the neurotransmitter transporter with an unusual structure. FEBS Lett 1993;315:114–118.
39. Mayser W, Schloss P, Betz H: Primary structure and functional expression of a choline transporter expressed in the rat nervous system. FEBS Lett 1992;305:31–36.
40. Guimbal C, Kilimann MW: A Na^+ dependent creatine transporter in rabbit brain, muscle, heart and kidney. cDNA cloning and functional expression. J Biol Chem 1993;268:8418–8421.
41. Lopez-Corcuera B, Vazquez J, Aragon C: Purification of the sodium- and chloride-coupled glycine transporter from central nervous system. J Biol Chem 1991;266:24809-24814.
42. Bendahan A, Kanner BI: Identification of domains of a cloned rat brain GABA transporter which are not required for its functional expression. FEBS Lett 1993;318:41–44.
43. Pantanowitz S, Bendahan A, Kanner BI: Only one of the charged amino acids located in the transmembrane α helices of the γ-aminobutyric acid transporter (subtype A) is essential for its activity. J Biol Chem 1993;268:3222–3225.
44. Kitayama S, Shimada S, Xu H, Markham L, Donovan DM, Uhl GR: Dopamine transporter site-directed mutations differentially alter substrate transport and cocaine binding. Proc Natl Acad Sci, USA 1992;89:7782–7785.
45. Sussman JL, Silman I: Acetylcholinesterase: structure and use as a model for specific cation-protein interactions. Curr Opin Struc Biol 1992;2:721–729.
46. Kleinberger-Doron N, Kanner BI: Identification of tryptophan residues critical for the function and targeting of the γ-aminobutyric acid transporter (subtype A). J Biol Chem 1994;269:3063–3067.
47. Kanner BI, Bendahan A, Pantanowitz S, Su H: The number of amino acid residues in hydrophillic loops connecting transmmebrane domains of the GABA transporter GAT-1 is critical for its function. FEBS Lett 1994;356:191–194.

GABA: Receptors, Transporters and Metabolism
ed. by C. Tanaka & N.G. Bowery

HETEROGENEITY OF BRAIN GABA TRANSPORTERS

Kelli E. Smith, Eric L. Gustafson, Laurence A. Borden, T.G. Murali Dhar, Margaret M. Durkin, Pierre J.-J. Vaysse, Theresa A. Branchek, Charles Gluchowski, and Richard L. Weinshank

Synaptic Pharmaceutical Corporation
215 College Road, Paramus, New Jersey, USA

Summary: Molecular cloning has revealed a surprising degree of heterogeneity in GABA transport systems. To investigate the potential of GABA transporters as therapeutic targets, we have applied a combined approach including cDNA cloning, pharmacological characterization, mRNA localization, and synthesis of novel GABA uptake inhibitors. Four high-affinity GABA transporters are present in brain and each has a unique pattern of mRNA distribution, with multiple transporters contributing to both neuronal and glial GABA uptake. We have designed and synthesized a novel GABA uptake inhibitor, (S)-SNAP-5114, with moderate affinity and selectivity for human GAT-3. The identification of selective inhibitors will help to elucidate the function and therapeutic potential of GABA transporters in the brain.

INTRODUCTION

Four transporters exhibiting high affinity for GABA are present in brain and may contribute to the modulation of GABAergic transmission. Termed GAT-1, GAT-2 , GAT-3, and BGT-1, the cloned GABA transporters belong to the family of Na^+ and Cl^--dependent neurotransmitter transporters mediating the uptake of inhibitory amino acid and biogenic amine transmitters into presynaptic terminals and surrounding glial cells. Since inhibition of uptake by pharmacologic agents increases the levels of neurotransmitter in the synapse, and thus enhances synaptic transmission, neurotransmitter transporters provide important targets for therapeutic intervention. Specifically, inhibition of GABA uptake provides a novel therapeutic approach to enhance inhibitory GABAergic transmission in the central nervous system.

To evaluate the therapeutic potential of the multiple GABA transporters in the brain, we have carried out a comprehensive program including molecular cloning, pharmacological characterization, localization of transporter mRNAs, and synthesis of novel GABA uptake inhibitors. Since considerable differences in distribution and pharmacology can exist between species homologues of the same gene product, we have isolated clones encoding human transporters for use in our drug discovery program. We have also reexamined the question of neuronal versus glial GABA transport using primary cultures from rat brain. Further, we have characterized the specificity of previously known GABA uptake inhibitors using the cloned transporters, and synthesized compounds with novel patterns of selectivity to serve as design leads for GABA uptake inhibitors with higher affinity and selectivity.

CLONING OF RAT AND HUMAN GABA TRANSPORTERS

Cloning of GABA transporters has revealed considerable diversity in this system. In addition to the original GABA transporter clone obtained from rat brain (GAT-1; 1), we have identified two additional rat brain clones encoding high-affinity GABA transporters which we term GAT-2 and GAT-3 (2). A clone identical to GAT-3 was described by Clark et al. (3) and referred to as GAT-B. An additional member of the GABA transporter family was cloned from dog kidney and transports both GABA and the osmolyte betaine, and was thus referred to as BGT-1 (4). Clones for all four GABA transporters have been isolated from mouse (5) using a different nomenclature: mouse GAT1 is the homologue of rat GAT-1; mouse GAT2 is the homologue of BGT-1; mouse GAT3 is the homologue of rat GAT-2; mouse GAT4 is the homologue of rat GAT-3. We have also cloned a taurine transporter (TAUT) from rat brain (6) which is sensitive to inhibition by β-alanine and transports GABA with low affinity, and thus should be considered as a potential contributor to brain GABA transport.

Because of possible species differences in transporter pharmacology, we have cloned the human homologues of GABA transporters for use in our drug design program. The isolation from human brain of a BGT-1 cDNA that exhibits high affinity for GABA (7, 8) suggests a role for this transporter in regulating GABAergic transmission in the human brain. Both the pharmacological profile and primary sequence (Table 1) of hBGT-1 diverge somewhat from those of dog (4) and mouse (9) BGT-1. Similarly, the partial amino acid sequence of human GAT-2 (10) exhibits only ~90% identity with rat (2) and mouse (5) clones. In contrast, human GAT-3 (11) exhibits a greater degree of similarity in primary sequence (Table 1) and pharmacological properties to its rat (2, 3) and mouse (5) homologues, as does human GAT-1 (1, 5, 12). The degree of species variations in transporter primary sequence, distribution, and pharmacological profiles is unpredictable, underscoring the importance of using human clones in the development of therapeutic agents acting at these sites.

Table 1. Transporter Profiles: High and Low Affinity GABA Transporters

	Amino Acid Identities	Pharmacology of Cloned Transporters	Regional mRNA Distribution	Brain mRNA Distribution
GAT-1	Human vs. rat: 97% Human vs. mouse: 96% rGAT-1 vs. rGAT-2: 53%	High-affinity for GABA, (±)-Nipecotic Acid, ACHC, NNC-711 --- Low-affinity for β-Alanine	CNS only (not detected in peripheral tissues)	Widely distributed, generally parallels distribution of GABAergic neurons
GAT-2	Human[a] vs. rat: 90% Human[a] vs. mouse: 91% rGAT-2 vs. rGAT-3: 67%	High affinity for GABA, β-Alanine --- Low-affinity for ACHC, NNC-711	CNS and some peripheral tissues	Confined to leptomeninges (pia-arachnoid)
GAT-3	Human vs. rat: 95% Human vs. mouse: 95% rGAT-3 vs. rGAT-1: 52%	High affinity for GABA, β-Alanine --- Low-affinity for ACHC, NNC-711	CNS only (not detected in peripheral tissues)	Predominantly subcortical, possibly glial
BGT-1	Human vs. rat[b]: 86% Human vs. mouse: 87% hBGT-1 vs. rGAT-2: 70%	High-affinity for GABA --- Low-affinity for ACHC, (±)-Nipecotic Acid, NNC-711, β-Alanine	CNS and some peripheral tissues	Evenly distributed throughout brain
TAUT	Rat vs. human: 94% Rat vs. mouse: 95% rTAUT vs. rGAT-3: 60%	Low-affinity for GABA, ACHC, (±)-Nipecotic Acid --- High-affinity for β-Alanine, taurine	CNS and all peripheral tissues examined	N.D.

a Partial human cDNA sequence
b Partial rat sequence generated by PCR

DISTRIBUTION OF GABA TRANSPORTER mRNAS

Brain vs. peripheral distribution

We examined the distribution patterns of GAT-1, GAT-2, GAT-3, BGT-1, and TAUT mRNAs in various rat and human tissues by reverse transcription/polymerase chain reaction (RT-PCR) and Northern blot analysis (2, 6, 13). Together these data indicate that the expression of GAT-1 and GAT-3 mRNA is restricted to the CNS, while GAT-2, BGT-1, and TAUT mRNAs are expressed in at least some peripheral tissues as well as in the brain (Table 1). This raises the possibility that the same transporter may have different functions depending on the tissue in which it is expressed, and upon the complement and concentration of endogenous substrates expressed in each tissue. Interestingly, the high-affinity GABA transporters whose localization is restricted to CNS (GAT-1 and GAT-3) show less divergence between species than the transporters with wider distribution patterns.

Regional brain distribution of GABA transporter mRNAs: in situ hybridization studies

Localization of the messenger RNAs encoding the four high-affinity GABA transporters, termed GAT-1, GAT-2, GAT-3, and BGT-1, has been carried out in rat (GAT-1, GAT-2, GAT-3) and mouse (BGT-1) brain using radiolabeled oligonucleotide probes and in situ hybridization histochemistry (Figure 1, 7, 14). Hybridization signals for GAT-1 mRNA are observed over many regions of the rat brain, including the retina, olfactory bulb, neocortex, ventral pallidum, hippocampus, and cerebellum (Figure 1). At the microscopic level, this signal appears to be restricted to neuronal profiles, and the overall distribution of GAT-1 mRNA closely parallels that seen in other studies with antibodies to GABA. Areas containing hybridization signals for GAT-3 mRNA include the retina, olfactory bulb, subfornical organ, hypothalamus, midline thalamus, and brainstem (Figure 1). In some regions, the hybridization signal for GAT-3 seems to be preferentially distributed over glial cells, although hybridization signals are also observed over neurons, particularly in the retina and olfactory bulb. Notably, hybridization signal for GAT-3 mRNA is absent from the neocortex and cerebellar cortex, and is very weak in the hippocampus. In contrast to the parenchymal localization obtained for GAT-1 and GAT-3 mRNAs, hybridization signals for GAT-2 mRNA are found only over the leptomeninges (pia and arachnoid). BGT-1 mRNA localization reveals a broad distribution in mouse brain with an intense hybridization signal over the lateral septum. Northern blot analysis of BGT-1 mRNA in human brain reveals approximately equal levels of hBGT-1 transcript in all regions examined (8).

In summary, the distribution of GAT-1 mRNA seems to parallel that of GABA itself, suggesting that GAT-1 serves to terminate GABAergic transmission at the synapse. The localization of GAT-3 to a subpopulation of glial cells suggests that in some brain areas, GAT-3 may take up GABA that has diffused from the synapse. The restricted expression of GAT-2 in leptomeninges suggests that GAT-2 could contribute to the regulation of CSF levels of GABA. Finally, the broad distribution of BGT-1 mRNA throughout the brain is consistent with both neuronal and glial expression; it remains speculative whether BGT-1 in the brain participates in osmoregulation as it does in the kidney. The differential distribution of the four GABA transporters described here suggests that while each may play a role in GABA uptake, they do so via distinct cellular populations.

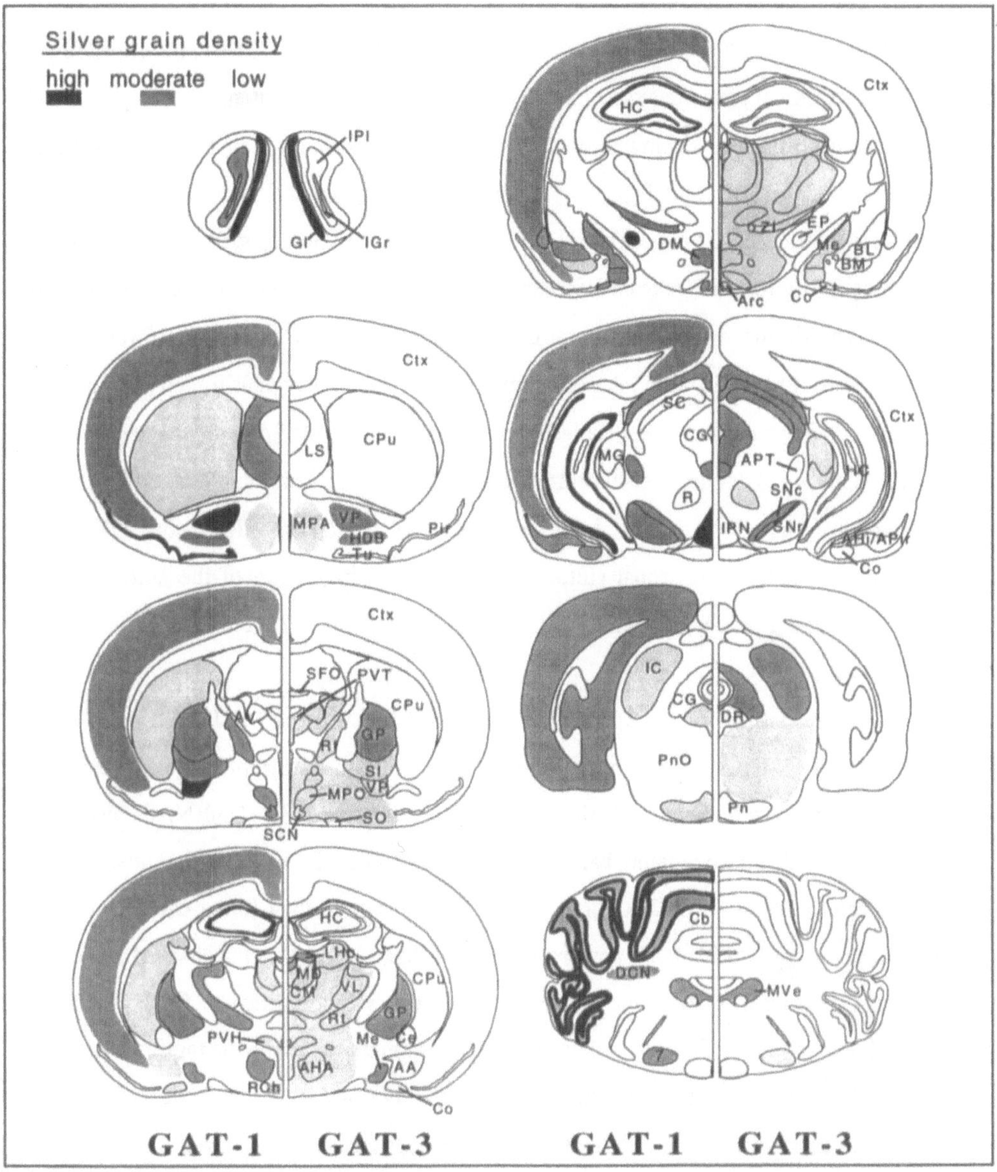

Figure 1: Schematic diagrams of coronal sections through the rat brain showing the distribution of the GABA transporters GAT-1 (left half of each section) and GAT-3 (right half). The shading intensity indicates the relative density of silver grains over a given brain region. Abbreviations: 7, facial n.; AA, anterior amygdala; AHA, anterior hypothalamic area; AHi, amygdalohippocampal area; APir, amygdalo-piriform n.; APT, anterior pretectal n.; Arc, arcuate n.; AV, anteroventral thalamic n.; BL, basolateral n. amygdala; BM, basomedial n. amygdala; Cb, cerebellum; Ce, central n. amygdala; CG, central gray; CM, centromedial n. thalamus; Co, cortical n. amygdala; CPu, caudate-putamen; Ctx, cortex; DCN, deep cerebellar n.; DM, dorsomedial n. hypothalamus; DR, dorsal raphe; EP, entopeduncular n.; Gl, glomerular layer olfactory bulb; GP, globus pallidus; HC, hippocampus; HDB, horizontal n. diagonal band; IC, inferior colliculus; IPN, interpeduncular n.; IGr, internal granule layer olfactory bulb; IPl, inner plexiform layer olfactory bulb; LHb, lateral habenula; LS, lateral septum; MD, mediodorsal n. thalamus; Me, medial n. amygdala; MG, medial geniculate; MPA, medial preoptic area; MPO, medial preoptic n.; MVe, medial vestibular n.; Pir, piriform cortex; Pn, pontine n.; PnO, pontine reticular n.; PVH, paraventricular n. hypothalamus; PVT, paraventricular n. thalamus; R, red n.; RCh, retrochiasmatic area; Rt, reticular n. thalamus; SC, superior colliculus; SCN, suprachiasmatic n. hypothalamus; SFO, subfornical organ; SI, substantia innominata; SNc, substantia nigra, compacta; SNr, substantia nigra reticulata; SO, supraoptic n.; Tu, olfactory tubercle; VL, ventrolateral n. thalamus; VP, ventral pallidum; ZI, zona incerta.

Neuronal vs. glial localization of GABA transporter mRNAs

Previous studies suggested the existence of distinct neuronal and glial GABA transporters with different pharmacological properties (see 15). We employed Northern blot analysis to examine the distribution of transporter mRNAs in neuronal and glial cultures from rat brain (13). cDNA probes for GAT-1, GAT-2, GAT-3, and TAUT were labeled to similar specific activities to allow direct comparison; a fragment of rBGT-1 generated by PCR was labeled at lower specific activity. Northern blots containing 25 μg total RNA from cultured neurons, Type 1 astrocytes, or mixed O-2A/Type 2 astrocytes were hybridized with transporter cDNA probes, then reprobed for 1B15 (cyclophilin; 16) for the purpose of normalization. Autoradiograms were scanned by densitometry to quantitate mRNA levels. The normalized data (Table 2) reveal that GAT-1 mRNA is abundant in neurons, whereas GAT-2 mRNA is not detected in neuronal cultures by Northern blot analysis. GAT-3 mRNA is present in neurons but the levels are about one third those of GAT-1. mRNA for BGT-1 is not detected by densitometry in neuronal cultures, though visual inspection of the autoradiograms sometimes reveals the presence of a faint band. Levels of TAUT mRNA in neurons are about 1.5-fold higher than those of GAT-1.

In Type 1 astrocyte cultures (Table 2), mRNAs for GAT-1, GAT-2, and GAT-3 are not detected, while BGT-1 and TAUT mRNAs are both present. Levels of transporter mRNA expression in O-2A/Type 2 astrocytes are more variable than in the other cell types. GAT-1 mRNA is observed in these cultures at levels approximately half those in neurons. GAT-2 mRNA is the most abundant GABA transporter in O-2A/Type 2 cultures, with levels approximately 4-fold greater than GAT-1 mRNA. Levels of GAT-3 mRNA are approximately half those of GAT-1. mRNAs for BGT-1 and TAUT are also present in O-2A/Type 2 astrocytes at levels similar to that of GAT-1. Thus, O-2A/Type 2 cultures contain all four high-affinity GABA transporters as well as the taurine transporter.

In summary, GAT-1 mRNA is present in both neurons and O-2A/Type 2 astrocytes, and appears to account for the majority of GABA transport in both cell types (Table 2, 13). GAT-3 mRNA is also observed in neurons and O-2A/Type 2 astrocytes, and like GAT-1 appears to be CNS-specific. Despite these similarities, their distribution patterns within the brain are distinct, suggesting different functional roles for each transporter. The abundance of GAT-2 mRNA in O-2A/Type 2 astrocytes, which exhibit high levels of GABA transport, is surprising compared with its restricted leptomeningeal localization in the adult rat brain. Interestingly, it has been suggested that O-2A/Type

Table 2. Quantitation of GABA Transporter mRNA Localization

Cell Culture/ Clone	Neurons	Type 1 Astrocytes	O-2A/Type 2 Astrocytes
GAT-1	++	-	+
GAT-2	-	-	++
GAT-3	+	-	+
BGT-1	(+/-)	+	+
TAUT	++	+++	+

Quantitative Northern blot results are expressed as ratios of transporter mRNA / cyclophilin and converted to +'s according to the following key: + : < 1.0, ++ :1.0- 2.0, +++ : > 2.0

2 astrocytes may represent a subpopulation of glial cells which proliferates in response to injury, but is not present in normal brain (17). In contrast with neurons and O-2A/Type 2 astrocytes, rat Type 1 astrocytes in culture exhibit very low levels of GABA transport, and this activity has low affinity for GABA but high affinity for taurine (13). TAUT and BGT-1 mRNAs are present in Type 1 astrocyte cultures, whereas those for GAT-1, GAT-2, and GAT-3 are not detected. The presence of GAT-1, GAT-3, and BGT-1 in both neurons and glial cells reveals that none of the GABA transporters has a uniquely neuronal distribution. In contrast, GAT-2 is the only GABA transporter whose expression is restricted to nonneuronal cells. Similarly, BGT-1 is the only high-affinity GABA transporter mRNA present in Type 1 astrocytes. Pharmacologic data in cultured neurons and glial cells can thus be explained by the presence of various combinations of the cloned GABA and taurine transporters.

DISCOVERY OF NOVEL GABA TRANSPORT INHIBITORS

Selectivity of known GABA transport inhibitors

Pharmacological inhibition of transport provides a mechanism for increasing GABAergic transmission, which may be useful in the treatment of various neuropsychiatric disorders. Recently, a number of lipophilic GABA transport inhibitors have been designed and synthesized, which are capable of crossing the blood brain barrier and which display anticonvulsive activity. We have now determined the potency of four of these compounds, SK&F 89976-A (N-(4,4-diphenyl-3-butenyl)-3-piperidinecarboxylic acid), Tiagabine ((R)-1-[4,4-Bis(3-methyl-2-thienyl)-3-butenyl]-3-piperidencarboxylic acid), CI-966 ([1-[2-[bis 4-(trifluoromethyl)phenyl]methoxy]ethyl]-1,2,5,6-

tetrahydro-3-pyridinecarboxylic acid), and NNC-711 (1-(2-(((diphenylmethylene)amino)oxy)ethyl)-1,2,4,6-tetrahydro-3-pyridinecarboxylic acid hydrochloride), at each of the four cloned GABA transporters, and find them to be highly selective for GAT-1 (Table 3, 18). These data suggest that the anticonvulsant activity of compounds such as Tiagabine (19) is mediated via inhibition of uptake by GAT-1.

Table 3. Affinities of Transport Inhibitors at Cloned GABA Transporters

		IC50, μM		
Clone / Compound	hGAT-1	rGAT-2	hGAT-3	hBGT-1
GABA	5 ± 1	5 ± 2	7 ± 1	36 ± 3
(±)- Nipecotic Acid	8 ± 0.4	39 ± 4	106 ± 13	2370 ± 617
β-Alanine	5690 ± 1890	19 ± 7	58 ± 3	1320 ± 224
Tiagabine	0.07 ± 0.007	1410 ± 250	91[illegible] ± 193	1670 ± 722
NNC-711	0.04 ± 0.01	171 ± 53	1700 ± 252	622 ± 265
CI-966	0.26 ± 0.05	297 ± 34	333 ± 76	300 ± 10
SK&F 89976-A	0.13 ± 0.01	550 ± 225	944 ± 259	7210 ± 3630
EGYT-3886	26 ± 5	30 ± 4	46 ± 1	39 ± 6
SNAP-5294	133 ± 50	51 ± 4	142 ± 21	(27 ± 10%)
(S)-SNAP-5114	388 ± 92	21 ± 3	5 ± 1	140 ± 30

Data show the IC_{50} values for inhibition of [^{3}H] GABA uptake and represent means ± SEM of at least 3 experiments.

Design and synthesis of novel transport inhibitors

Unlike GAT-1, little is known regarding the functional role(s) of GAT-2, GAT-3, and BGT-1, due in part to the lack of selective inhibitors. EGYT-3886 is a GABA transport inhibitor (20) with a chemical structure different from that of classical inhibitors such as nipecotic acid and guvacine. In contrast with the GAT-1-selective compounds described above, EGYT-3886 lacks selectivity but has modest affinity at all four cloned GABA transporters (Table 3). In order to identify novel GABA transport inhibitors we designed and synthesized a series of triarylnipecotic acid derivatives using EGYT-3886 as a lead compound (21). From this series we identified the lipophilic trimethoxytrityl derivative (S)-SNAP-5114 (Figure 2, Compound 4S in 21) and the cyclic fluorene analog SNAP-5294 (Figure 2, Compound 14 in 21) which are most potent at GAT-3 and GAT-2, respectively (Table 3). (S)-SNAP-5114 displays an IC_{50} at human GAT-3 of 5 μM, and is thus as potent as GABA itself (11). (S)-SNAP-5114 is also relatively active at GAT-2 (IC_{50} ~21 μM), but exhibits

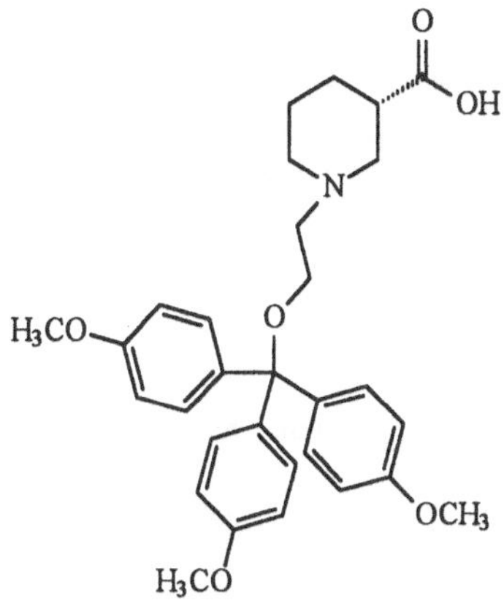

(S)-SNAP 5114

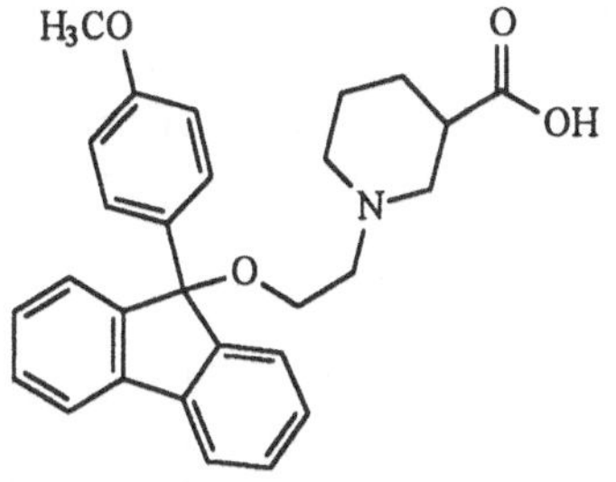

SNAP 5294

Figure 2. Chemical Structures for (S)-5114 and SNAP 5294

greater than 25-fold selectivity for GAT-3 over GAT-1 and BGT-1. Further refinements in the structure of (S)-SNAP-5114 may lead to compounds with increased affinity and selectivity. SNAP-5294 displays moderate affinity for GAT-2 (IC_{50} ~51 μM) and is 3-fold selective for this site as compared to GAT-1 and GAT-3. This compound thus represents a design lead for lipophilic GAT-2 selective GABA transport inhibitors. In conclusion, we have identified two GABA uptake inhibitors, (S)-SNAP-5114 and SNAP-5294 with novel pharmacological profiles at the cloned GABA transporters. These compounds may be useful tools in evaluating the roles of GAT-2 and GAT-3 in neural function.

CONCLUSIONS

The presence of four high-affinity GABA transporters in the brain with unique regional and cellular patterns of distribution suggests that each plays a role in the regulation of GABAergic functions in the CNS. Further, it suggests that each site may represent a distinct human therapeutic target. While GAT-1 selective uptake inhibitors have been shown to exhibit anticonvulsant activity, the functional consequences of modulating transport at GAT-2, GAT-3, and BGT-1 are not yet known. The development of selective inhibitors acting at these sites will help to elucidate the function and therapeutic potential of GABA transporters in the brain.

Acknowledgments

We would like to acknowledge the excellent technical assistance of Mr. Steven Fried, Mr. Sherif Daouti, Mr. Sanjay Patel, Ms. Antoinette Illy, and Yijin She. Ms. Tracy Johnson, and Ms. Janet Lee. We also acknowledge the work of Mr. Sriram Tyagarajan in the synthesis of GABA uptake inhibitors. We thank Mr. George Moralishvilli and Mr. Ernest Lilley for assistance with illustrations. We also thank Dr. Paul Hartig for helpful comments and support throughout the course of this work.

REFERENCES

1. Guastella, J., Nelson, N., Nelson, H., Czyzyk, L., Keynan, S., Miedel, M.C., Davidson, N., Lester, H.A., and Kanner, B.I. (1990) *Science* **249,** 1303-1306.
2. Borden, L.A., Smith, K.E., Hartig, P.R., Branchek, T.A., and Weinshank, R.L. (1992). *J. Biol. Chem.* **267(29),** 21098-21104.
3. Clark, J.A., Deutch, A.Y., Gallipoli, P.Z., and Amara, S.G. (1992) *Neuron* **9**, 337-348.
4. Yamauchi, A., Uchida, S., Kwon, H.M., Preston, A.S., Robey, R.B., Garcia-Perez, A., Burg, M.B., and Handler, J.S. (1992) *J. Biol. Chem.* **267(1)**, 649-652.
5. Liu, Q.-R., Lopez-Corcuera, B., Mandiyan, S., Nelson, H., and Nelson, N. (1993). *J. Biol. Chem.* **268,** 2106-2112.
6. Smith, K.E., Borden, L.A., Wang, C.-H.D., Hartig, P.R., Branchek, T.A., and Weinshank, R.L. (1992) *Mol. Pharmacol.* **42,** 563-569.
7. Smith, K.E., L.A. Borden, E.L. Gustafson, T.A. Branchek, and R.L. Weinshank (1994) *Soc. Neurosci. Abstr.* **20**, 918.
8. Borden, L.A., Smith, K.E., Gustafson, E.L. Branchek, T.A. and Weinshank, R.L. (1995) *J. Neurochem.* **64**, 977-984.
9. Lopez-Corcuera, B., Liu, Q.-R., Mandiyan, S., Nelson, H., and Nelson, N. (1992) *J. Biol. Chem.* **267**, 17491-17493.
10. Smith, K.E., Borden, L.A., Hartig, P., and Weinshank, R.L. (1993) PCT International Patent Application PCT/US93/01959, International Patent No. WO 93/18143, 16 Sep 1993.
11. Borden, L.A., Dhar, T.G.M., Smith, K.E., Branchek, T.A., Gluchowski, C. and Weinshank, R.L. (1994).*Receptors and Channels* **2**, 207-213.
12. Nelson, H., Mandiyan, S., and Nelson, N. (1990) *FEBS* **269(1),** 181-184.
13. Borden, L.A., Smith, K.E., Vaysse, P.J.-J., Gustafson, E.L., Weinshank, R.L. and Branchek, T.A., *Receptors and Channels* **3**, 129-146.
14. Durkin, M.M., Smith, K.E., Borden, L.A., Weinshank, R.L., Branchek, T.A., and Gustafson, E.L. *Mol. Brain Research* 33, 7-21.
15. Krogsgaard-Larsson, P., Falch, E., Larsson, O.M., and Schousboe, A. (1987) *Epilepsy Res.* 1, 77-93.
16. Danielson, P.E., Forss-Pettr, S., Brow, M.A., Calavetta, L., Douglas, J., Milner, R.J., and Sutcliffe, J.G. (1988) *DNA* **7**, 261-267.
17. Marriott, D.R. and Wilkin, G.P. (1993) *J. Neurochem.* **61**,826-834.
18. Borden, L.A., Dhar, T.G.M., Smith, K.E., Weinshank, R.L., Branchek, T.A. and Gluchowski, C. (1994a) *Eur. J. Pharmacol. (Mol. Pharmacol. Section)* **269**, 219-224.
19. Nielson, E.B., Suzdak, P.D., Andersen, K.E., Knutsen, L.J.S., Sonnewald, U. and Braestrup, C. (1991) *Eur. J. Pharmacol.* **196**, 257.
20. Kovacs, I., Maksay, G., and Simonyi, M. (1989) *Arzneim. Forsch/Drug Res.* **39**, 295-297.
21. Dhar, T.G.M., Borden, L.A., Tyagarajan, S., Smith, K.E., Branchek, T.A., Weinshank, R.L. and Gluchowski, C. (1994) *J. Med. Chem.* **37**, 2334-2342.

GABA: Receptors, Transporters and Metabolism
ed. by C. Tanaka & N.G. Bowery

DEVELOPMENTAL EXPRESSION OF THE GABA TRANSPORTER GAT4

Frantisek Jursky and Nathan Nelson

Roche Institute of Molecular Biology
Roche Research Center
Nutley, New Jersey 07110

Introduction

It has been known for many years that neurons and glia can accumulate neurotransmitters by a sodium-dependent transport processes. Neurotransmitters cotransport with sodium utilizing the energy stored in transmembrane electrochemical gradients generated by primary ion pumps. Studies on neurotransmitter uptake indicated the existence of multiple uptake systems, each relatively selective for a specific neurotransmitter. A family of Na^+/Cl^- neurotransmitter transporters was identified and cDNAs encoding several members of this family were cloned and sequenced (1-8). They can be grouped into four subfamilies of GABA transporters, monoamine transporters, amino acid transporters and "orphan" (NTT4) transporters (9,10). The transporters of the first three subfamilies show a common structure of twelve transmembrane helices with a single large loop in the external face of the membrane with putative glycosylation sites. An orphan subfamily (NTT4), somewhat remote from the other subfamilies of Na^+/Cl^--dependent transporters, was identified from mammalian sources (11,12). This subfamily of putative neurotransmitter transporters has no known function but is highly expressed in different parts of the brain. Their structure deviates from the others by having two potential glycosylation loops outside the membrane.

GABA is the predominant inhibitory neurotransmitter and it is widely distributed in the mammalian brain. We cloned cDNAs encoding four pharmacologically distinct GABA transporters from mouse brain (13). Homologous cDNAs were cloned from rat brain and other mammalian sources (14,15). We termed the transports GAT1 to GAT4 and we will use this nomenclature (mouse GAT3=rat GAT2 and mouse GAT4=rat GAT3, see ref. 13-15). The presence of GABA uptake activities of the four transporters was determined by measuring activity after injecting mRNA into *Xenopus* oocytes (13). The Kms for GABA uptake by the expressed GAT1 to GAT4 were 6, 79, 18 and 0.8 μM, respectively. GAT2 is also an effective betaine transporter with a Km of about 200 μM. GAT3 and GAT4 also transport β-alanine with Kms of 28 μM and 99 μM, respectively. They also transport taurine with Kms of 99 μM and 1.4 mM. While GAT1 and GAT4 gene expression were brain specific, GAT2 and GAT3 mRNAs were detected in other tissues such as liver and kidney, in which GAT3 mRNA was especially abundant (13,16). While GAT1 and GAT4 were more sensitive to inhibition by nipecotic acid, GAT2 and GAT3 were inhibited most effectively by betaine and β-alanine, respectively. Using specific antibodies we aim at localizing the specific transporters and to unravel their function. In this communication we studied the localization of GAT4.

Characterization of the antibodies against GAT4

The antibodies were affinity purified using specific fusion protein bound to Affigel. The specificity of the immunoreactivity was checked in several ways including omitting the primary antibody, or normal goat serum from the protocol and comparing the staining patterns obtained with both antibodies raised against two different GAT4 protein sequences (Fig. 1). Preincubation of the antibodies with the specific fusion proteins resulted in complete loss of the immunoreactivity. Preincubation with the maltose-binding protein or with unrelated fusion protein had no effect on staining. Western blot analysis of brain tissues revealed a single band at a position of about 70 kDa (Fig. 1.). The distribution of immunoreactivity was unique to GAT4 and did not coincide with the pattern obtained by antibodies against GAT1, the glycine transporters GLYT1 and GLYT2 or the orphan transporter NTT4 (17).

Expression of GAT4 in the adult CNS

Figure 1 depicts the immunoreactivity in sagital sections of rat brain using antibodies against the loop and C-terminus of GAT4. Both antibodies gave identical patterns. In the spinal cord, the area of the gray matter was strongly stained especially in laminae 2 and 3. In the cerebellum, strong immunoreactivity was present in deep cerebelar nuclei and a weak staining was detected in the granule cell layer. All parts of the brainstem showed from high to moderate presence of the GAT4 immunoreactivity. In midbrain, strongly stained regions were the central gray medial geniculate nuclei, substantia nigra and superficial gray matter of superior coliculus. No staining in the interpenducular nucleus of the midbrain was seen. In the pons and medulla, strong staining was visible in pontine nuclei, vestibular, facial nuclei, lateral reticular formation, hypoglossal, gracile nuclei and the whole rostrocaudal aspect of the spinal trigeminal nucleus. Staining was more intense in the areas of the medial longitudinal fasciculus and paramedian reticular nucleus. Other parts showed even staining of indistinguishable neurophils. In thalamus and hypothalamus immunoreactivity was strong and covered the whole rostro-caudal aspect of the thalamus and hypothalamus. Within the thalamus, moderately stained areas were the zona incerta, lateral habenula and posterior, posterioventral and posteriolateral thalamic nuclei. The entire hypothalamic area showed a high level of immunoreactivity with a generally stronger staining in comparison to one observed in the thalamus. In basal ganglia, a strong GAT4 immunoreactivity was present in the globulus palidus, entopenducular and subthalamic nucleus,

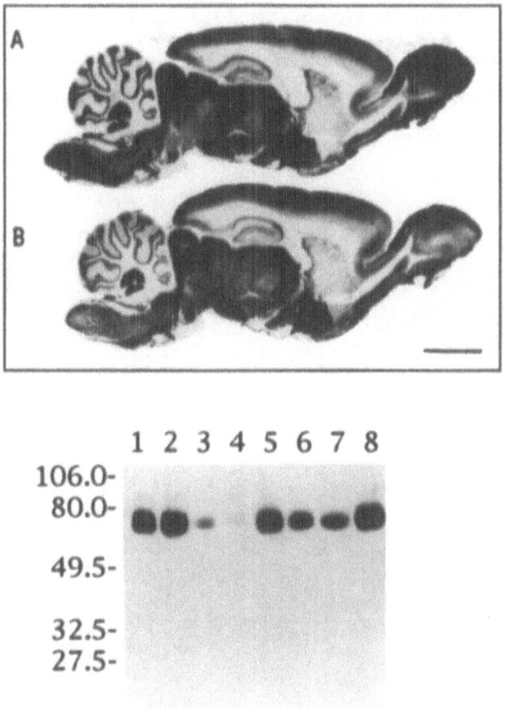

Figure 1. (Top) Comparison of staining patterns in the edult brain using the anti GAT4-C terminus (A) and anti GAT4-loop (B) antibodies. (Bottom) Western blot analysis of GAT4 -C-terminus immunoreactivity in dissected parts of the brain. 1-spinal cord, 2-medula+caudal pons, 3-cerebellum, 4-cerebellar lobe, 5-rostral pons+caudal diencephalon, 6-rostral diencephalon+basal forebrain, 7-frontal cortex, 8-olfactory bulb.

ventral palidum, medial and lateral preoptic region, substantia innominata, claustrum much less in fundus striati caudate putamen and nucleus accubens. In the septum, immunoreactivity was dense in the septofimbrial nucleus, medial septum from where immunoreactivity continued to the vertical and horizontal limb of the diagonal band. Immunoreactivity in the cerebral cortex was generally weak and organized in a bilaminary pattern. Dense neurophil staining around the principal neurons was frequently observed. Immunoreactivity was present in parietal, hindlimbic, retrosplenial, ethorinal cortex and visibly less present in piriform, cingulate and frontal cortex. Staining in the hippocampus was present more on the apical than on the basal side close to the stratum pyramidale in areas CA1-CA3, and much less in dentate gyrus. The olfactory bulb was a region with very strong presence of GAT4. The strongest GAT4 immunoreactivity was present in the glomerular region, protruding to the olfactory nerve layer. A strong signal was visible in the external and internal plexiform layers. Immunoreactivity was less intense in the mitral cell layer, internal granule layer, the epedyma and the region between the olfactory nucleus, and lateral olfactory tract.

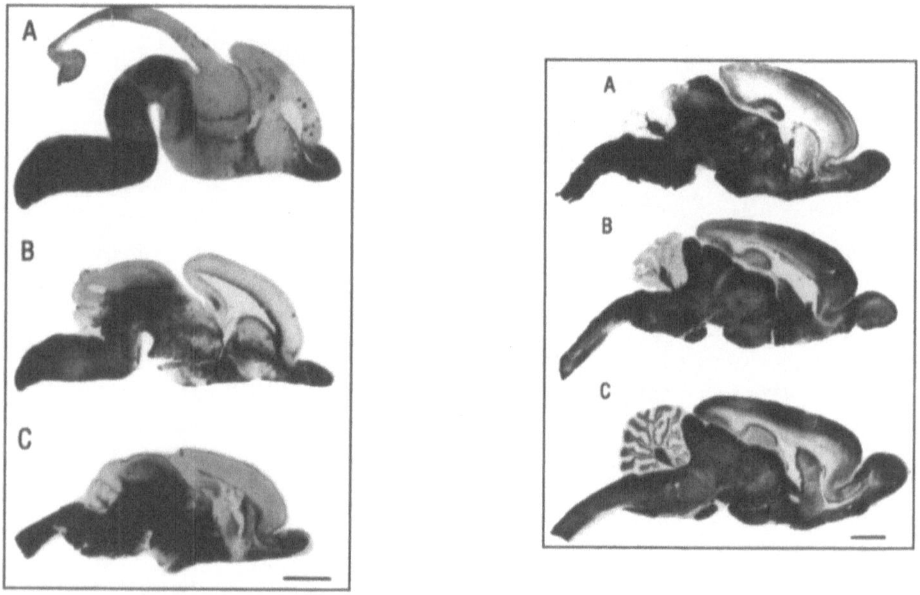

Figure 2. Distribution of GAT4 immunoreactivity in (left) E18 (A), E20 (B) and P0 (C) rat brain, and (right) in P5 (A), P15 (B) and P22 (C) rat brain.

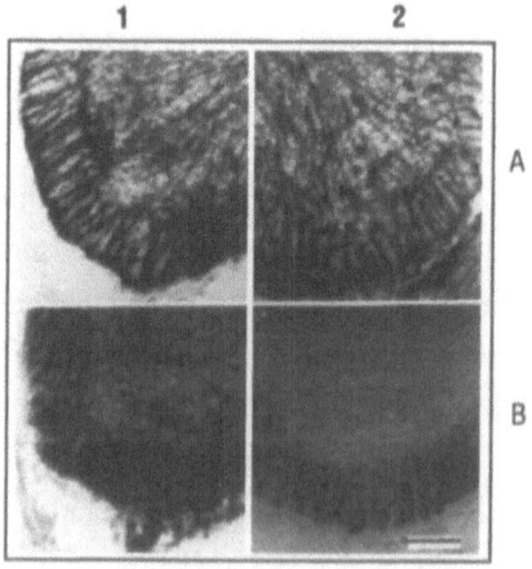

Figure 3. Comparison of the staining patterns of the anti GAT4-C terminus antibody (1A) with vimetin (2A), MAP1 (1B) and synaptophysis (2B) immunoreactivity in coronal sections thorough E16 rat spinal cord.

Developmental expression of the GAT4

The GABA transporter GAT4 exhibits the highest affinity for GABA than all of the other transporters (13). Yet, its specific distribution in adult brain gave no indication on possible specific function for this transporter. Therefore, we initiated a study of immunolocalization of GAT4 during the development of the rat brain. GAT4 immunoreactivity can be first detected in the ventral side of the spinal cord of E15 embryos and progresses continually with age to the dorsal horn. In the E16 embryo immunoreactivity is present at up to half of the spinal cord, and in E18 the whole area of the spinal cord is covered. During the prenatal stage, the immunoreactivity coincides with characteristic vimetin-positive radial glia fibers but not with the staining by synaptophysin or MAP1 (Fig. 3). There is a rapid change in immunoreactivity patterns after the animal is born. While in E20 rather strong radial patterns were observed, in PO only few of the radial glia fibers were immunoreactive with GAT4 antibody (Figs. 2, 4). GAT4 immunoreactivity is absent from the spinal cord floor plate, raphe and ventral commisure. In the olfactory bulb, GAT4 immunoreactivity appears in E16, and here too it coincides with vimetin immunoreactivity rather than synaptophysin.

Starting at E16, GAT4 appears in the rostral migratory system from where the staining continues laterally alongside the border between the basal ganglia and cortical wall (Fig. 5). This structure forms a continuation throughout the rostrocaudal aspect from the olfactory bulb to the hippocampus. The bulk of the staining in midbrain and diencephalon starts to appear in E18 (Fig. 2). Two lines of the immunoreactivity merge rostrally from the medulla and branching in the front of the thalamus. One line proceeds through the thalamus, the second continues to the ventricle and than it goes ventrally into the grooving palidum. Except for the cerebellum, hippocampus, cortex and colliculi, patterns very similar to adult are already visible in stages P0-P5 (Fig. 2). Staining in the hippocampus appears as two layers and remains at the same shape during the period E18-P5. In later stages, staining moves close to the pyramidal cell bodies (Fig. 2), and follows the laminae of the commisural input. Staining in the cerebral cortex and cerebellum becomes more dense in postnatal stages and roughly follows synaptogenesis. Interestingly at E20, GAT4 immunoreactivity is present in astrocyte-shaped cells in the white matter, especially in the anterior commisure. In this area, immunoreactivity is present in P15 animals but disappear after postnatal day 22.

Glial origin of the GAT4

Of all the known GABA transporters, GAT4 possesses the highest affinity for GABA (13). However, it transports β-alanine and its pharmacology does not follow strictly those of neuronal GABA transporters. Based on the ibotenic acid lesion in combination with in situ hybridization, GAT4 has been localized to neurons (14). Our result in embryonic brain and ultrastructural

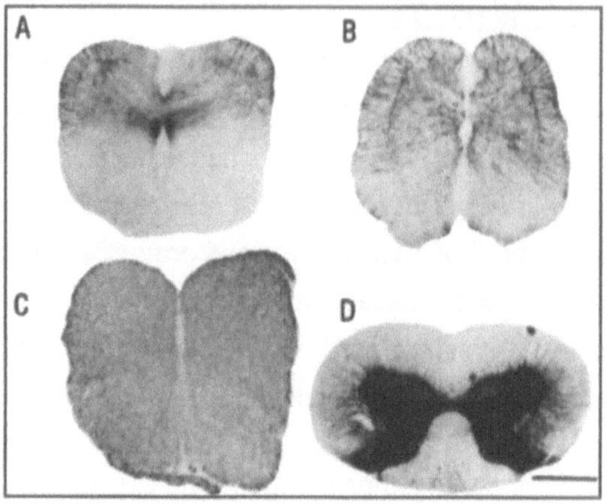

Figure 4. Developmental expression of GAT4 immunoreactivity in E16 (a), E18 (b) PO (c) and adult (d) spinal cord.

immunoanalysis reported by others (18) suggest glial localization. In fetal spinal cord, GAT4 immunoreactivity does not follow synaptophysin expression (Fig. 3) as do other putative neuronal neurotransmitter transporters such as GLYT2 or NTT4 (Jursky and Nelson, unpublished observations). Immunoreactivity coincides with vimetin-positive early radial glial cells (Fig. 3). Immunoreactivity first appears in single and fasciculated fibers or clusters of cells which are vimetin positive (Figs. 3, 4, 5). Similar clustering and fasciculation is described for radial glia markers (19). As the radial glial patterns disappear, bulk of the staining reach dense adult patterns in the period between P0 to P5 (Fig. 2). Double labeling showed that GAT4 immunoreactivity localized with Glial fibrilary acid protein immunoreactivity in some cells (Fig. 5). Immunorectivity in adult brain is, however, not present in the soma and exhibits dense patterns.

GAT4 immunoreactivity appears in the transition period of bipolar to monopolar radial glial cells

Radial glial cells are considered a transient type of glia which later differentiate into macroglial forms (20-22). Using a specific immunocytochemical marker (RC2), different morphological forms expressed in different periods of development have been identified (19). Bipolar forms which express very early (E9) are gradually replaced by monopolar radial forms after the development stage of E15 to 17. Monopolar cells emerge in the spinal cord and cerebellum, as early as E15, and in the hippocampus begining at E17 (19). GAT4 immunoreactivity is present in the spinal cord starting at E15 and significantly increases at E16 (this stage corresponds to E15 in ref. 19). A day later it appears in the olfactory bulb, between the striatal rudiment and lateral cortical wall, and at E18 it starts being expressed in the hippocampus. In the E16 spinal cord, strong GAT4 immunoreactivity is detected in palisadic longitudinal arrays corresponding to the monopolar radial glial form in the emerging ventrolateral funiculus (Fig. 4). GAT4 immunoreactivity is absent from the floor plate, supporting the recent findings that the floor plate contains distinct groups of cells that are phenotypically different from radial glia. Monopolar radial glia cells become evident in the spinal cord, after the major neurogenetic period, at the beginning of synaptogenesis. The correlation of GAT4 appearance with synaptogenesis indicates that this transporter may be involved in differentiation of neurons or glial cells and its expression may be connected with synapse formation. Strong immunoreactivity in E16 spinal cord is in contrast with the virtual absence of the radial pattern in the cerebellum or certain parts of the cerebral cortex. This observation led us to conclude that GAT4 is not expressed in all radial glial cells. Because patterns of expression in the embryonic brain correlates with the corresponding patterns in the adult brain, it suggests that GAT4 may be involved in the maturation of the brain regions in which it is present in the adult stage.

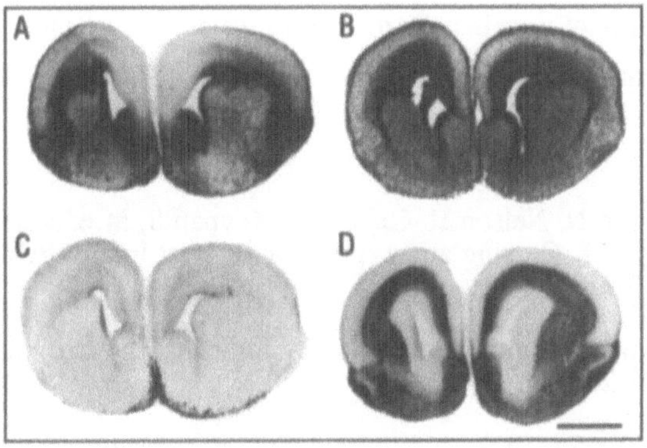

Figure 5. Comparison of the staining patterns of the anti GAT4-C terminus (A) with vimetin (B), glial fibrilary acid protein (C) and synaptophysin (D) immunoreactivity in coronal sections through E20 rat forebrain.

Simultaneously with the appearance in the radial fibers, GAT4 appears in numerous fascicles and cell clusters in the basal telencephalon, thalamus midbrain and white matter (Fig. 5). Cells in white matter are especially visible in the anterior commisure. They do not express MAP2 and disappear in about P22 animals. Colocalization of GAT4 and vimetin immunoreactivity suggests that these are immature glial elements perhaps also involved in guidance and neurite outgrowth. What might be the function of GAT4 in adult brain? Unusual distribution of GAT4 immunoreactivity in the adult brain suggests that GAT4 not only functions in the termination of the transmission. GAT4 in the CNS follows, in many places, the distribution of well known GABAergic neurons but generally GAT4 does not match the distribution of any single GABAr subunits or distribution known from radioactive agonist binding. Most striking is the accumulation of GAT4 in the spinal cord, brainstem, olfactory bulb and its low presence in the cerebellum hippocampus and cerebral cortex. Because the major inhibitory neurotransmitter in the caudal brain is glycine, it suggested that GABA and glycine may act in concert in these places. Several neurotransmitters have been recently found simultaneously in one cell, and neuronal coexistence of the GABA and glycine has been described (23). Indeed the patterns of GAT4 distribution is similar to that of the glycine transporters GLYT1ab (17,24).

Glial cells are generally considered as supporting cells for neurons providing trophic factors necessary for neuron survival (25,26). Neurotransmitters act as trophic factors during development (27), and several of them have been recently colocalized in single cells (28). Neurons exhibit much higher survival rate when cultured on homotypic rather than heterotypic astrocytes (29,30), and seem to require region specific glia-derived diffusible factors for neurite outgrowth (26). Therefore, specific localization of GAT4 together with the other substances may contribute to the capability of certain cells to support development, survival and proper function of specific groups of neurons.

References

1. Guastella J, Nelson N, Nelson H, Czyzyk L, Keynan S, Miedel MC, Davidson N, Lester HA, and Kanner, BI. Cloning and expression of a rat brain GABA transporter. Science 1990; 249:1303-1306.

2. Nelson H, Mandiyan S, and Nelson N. Cloning of the human brain GABA transporter. FEBS Lett 1990; 269:181-184.

3. Pacholczyk T, Blakely RD, and Amara SG. Expression cloning of a cocaine- and antidepressant-sensitive human noradrenaline transporter. Nature 1991; 350:350-354.

4. Shimada S, Kitayama S, Lin C-L, Patel A, Nanthakumar E, Gregor P, Kuhar M, and Uhl G. Cloning and expression of a cocaine-sensitive dopamine transporter complementary DNA. Science 1991 ; 254:576-578.

5. Kilty JE, Lorang D, and Amara SG. Cloning and expression of a cocaine-sensitive rat dopamine transporter. Science 1991 ; 254:578-579.

6. Hoffman BJ, Mezey E, and Brownstein MJ. Cloning of a serotonin transporter affected by antidepressants. Science 1991; 254:579-580.

7. Usdin TB, Mezey E, Chen C, Brownstein MJ, and Hoffman BJ. Cloning of the cocaine-sensitive bovine dopamine transporter. Proc. Natl. Acad. Sci. USA 1991; 88:11168-11171.

8. Blakely RD, Berson HE, Fremeau Jr. RT, Caron MG, Peek MM, Prince HK, and Bradley CC. Cloning and expression of a functional serotonin transporter from rat brain. Nature 1991; 354:66-69.

9. Nelson N. and Lill H. Porters and neurotransmitter transporters. J. Exp. Biol. 1994; 196:213-228.

10. Uhl G and Johnson PS. Neurotransmitter transporters: three important gene families for neuronal function. J. Exp. Biol. 1994; 196:229-236.

11. Uhl GR, Kitayama S, Gregor P, Nanthakumar E, Persico A, and Shimada S. Neurotransmitter transporter family cDNAs in a rat midbrain library: 'orphan transporters' suggest sizable structural variations. Mol. Brain Res. 1992; 16:353-359.

12. Liu Q-R, Mandiyan S, López-Corcuera B, Nelson H, and Nelson N. A rat brain cDNA encoding the neurotransmitter transporter with an unusual structure. FEBS Lett. 1993; 315:114-118.

13. Liu Q-R, López-Corcuera B, Mandiyan S, Nelson H, and Nelson N. Molecular characterization of four pharmacologically distinct γ-aminobutyric acid transporters in mouse brain. J. Biol. Chem. 1993; 268:2106-2112.

14. Clark JA, Deutch AY, Gallipoli PZ, and Amara SG. Functional expression and CNS distribution of a β-alanine-sensitive neuronal GABA transporter. Neuron 1992; 9:337-348.

15. Borden LA, Smith KE, Hartig PR, Branchek TA, and Weinshank RL. Molecular heterogeneity of the γ-aminobutyric acid (GABA) transport system. J. Biol. Chem. 1992; 267:21098-21104.

16. López-Corcuera, B, Liu Q-R, Mandiyan S, Nelson H, and Nelson N. Expression of a mouse brain cDNA encoding novel γ-amino-butyric acid transporter. J. Biol. Chem. 1992; 267:17491-17493.

17. Jursky F, Tamura S, Tamura A, Mandiyan S, Nelson H, and Nelson N. Structure, functionand brain localization of neurotransmitter transporters. J. Exp. Biol. 1994;196:283-295.

18. Saito N, Ikegaki N, Sakai N, Honda S. Distinct localization of GABA transporter subtypes and their functional modulation by PKC. Soc. Neurosci. Abstr. 1994 Vol 20; Part 1, p. 919

19. Misson J-P, Edwards MA, Yamamoto M, Caviness Jr. VS. Identification of radial glial cells within the developing murine central nervous system: studies based upon a new immunohistochemical marker. Dev. Brain Res. 1988; 44:95-108.

20. Choi BH, Kim RC, Lapham LW. Do radial glia give rise to both astroglial and oligodendroglial cells? Dev. Brain. Res. 1983; 8:119-130.

21. Hirano M, Goldman JE. Gliogenesis in rat spinal cord: evidence for origin of astrocytes and oligodendrocytes from radial precursors. J. Neurosci. Res. 1988; 21:155-167.

22. Cameron RS, Rakic P. Glial cell lineage in the cerebral cortex: A review and Synthesis. Glia. 1991; 4:124-137.

23. Triller A, Cluzeaud F, Korn H. g- aminobutyric acid-containing terminals can be apposed to glycine receptors at central synapses. J. Cell. Biol. 1987; 104:947-956.

24. Jursky F, Nelson N. Localization of glycine neurotransmitter transporter (GLYT2) reveals correlation with the distribution of glycine receptor. J. Neurochem 1995; 64:1026-1033.

25. Martin DL. Synthesis and release of neuroactive substances by glial cells. Glia 1992; 5:81-94.

26. Le Roux PD, Reh TA. Regional differences in glial-derived factors that promote dendritic outgrowth from mouse cortical neurons in vitro. J. Neurosci. 1994; 14:4639-4655.

27. Emerit MB, Riad M, Hamon M. Trophic effects of neurotransmitters during brain maturation. Biol. Neonate 1992; 62:193-201.

28. De Filipe J. Neocortical neuronal diversity: chemical heterogenity revealed by colocalization studies of classic neurotransmitters. Cereb Cortex 1993; 93:273-289.

29. Rousselet L, Fetler B, Chamak B, Prochiantz A. Rat mesencephalic neurons in culture exibit different morphological traits in the presence of media conditioned on mesencephalic or striatal astroglia. Dev Biol 1988; 129:495-504.

30. Autillo-Touati A, Rouget M, Araud D, Prochiantz A, Seite R. Astrocyte-regulated GABA-ergic striatal neurons development: an in vitro ultrastructural study. J Hirnforsch 1993; 34:291-297.

GABA: Receptors, Transporters and Metabolism
ed. by C. Tanaka & N.G. Bowery

FUNCTIONAL REGULATION OF GABA TRANSPORTERS BY PKC AND THEIR DISTINCT LOCALIZATION

Naoaki Saito and Chikako Tanaka*

Lab. Mol. Pharmacol., Biosignal Research Center, Kobe University, Kobe 657, Japan
*Hyogo Institute for Aging Brain and Cognitive Disorders, Himeji 670, Japan

Summary: We examined the distinct localization of these transporters in the synapse and also the possibility of the modulation of transporter activity by protein kinase C. Three GABA transporters showed distinct cellular and subcellular localization in the brain : GAT1 is distributed through the brain and localized in the nerve terminals, while GAT3 is localized in the glial processes in the restricted brain areas such as thalamus, pons and medulla. GAT2 is mainly concentrated in the non-neuronal tissues within the brain. In the GAT1 expressing cells, GAT1 activity was inhibited by the activation of PKC by phorbol ester. The GAT1 activity appeared to be modulated by the counterbalance of phosphorylation/dephosphorylation. The site-directed mutagenesis revealed that threonine residue in the intracellualr domain of GAT1 is resposible for the moduation

Introduction

γ-Aminobutyric acid (GABA) is a major inhibitory neurotransmitter and the GABAergic transmission is terminated by the rapid Na^+ and Cl^--dependent GABA uptake through GABA transporters (1). The GABA uptake systems have been divided into neuronal and glial uptake systems on the basis of pharmacological properties (1). Recent molecular cloning studies have revealed the existence of multiple subtypes of GABA transporters with similar structures but distinct pharmacological properties and distribution of their mRNAs (2, 3, 4, 5, 6). The GABA transporters (GAT1, GAT2 and GAT3; nomenclature according to Borden et al. (4)) are thought to be highly hydrophobic proteins with 12 potential transmembrane segments and alignment of the predicted amino acid sequences was shown more than 50 % homology between the GABA transporters. These subtypes of GABA transporter contains consensus phosphorylation sites for protein kinases including protein kinase C (PKC) (2, 3, 4, 5, 6) suggesting that the GABA transport process could be modulated by the phosphorylation such as

PKC and that the modulation of termination in the GABAergic synapse could be related with the control of synaptic plasticity. In the present study, we examined the light and electron microscopic localization of the three subtypes of GABA transporter and their modulation by phosphorylation and dephosphorylation.

Materials and methods

Antiserum production. The polyclonal antibodies were obtained and characterized as described by Ikegaki et al. (7). In brief, the carboxyl terminal region was selected as the specific sequence for each rat GABA transporter. The sequences were EQPQAGSSASKEAYI (residues 585-599) for GAT1, PMTSLLRLTELESNC (residues 588-602) for GAT2 and GDGTISAITEKETHF (residues 613-627) for GAT3 (3,4). The oligopeptides were coupled to keyhole limpet hemocyanin with *m*-maleimidobenzoic acid *N*-hydroxysuccinimide ester (8) and used as antigens. Antisera for each GABA transporter were raised in female Japanese White rabbits (2.5-3.5 kg). Initially, rabbits were given 300 mg of each antigen emulsified in Freund's complete adjuvant by multiple intracutaneous and intradermal injections. At 2 weeks interval, a booster injection of 100 - 150 mg antigen in Freund's incomplete adjuvant was given to each rabbit. Rabbits were bled 5 days after the fifth boost.

Purification of the antibodies. Each oligopeptide was conjugated to bovine serum albumin (BSA), then coupled to CNBr-activated Sepharose CL-4B (Pharmacia, NJ, USA) according to the manufacturer's standard procedure. Each antiserum was applied on the corresponding oligopeptide-BSA column and after washing with 0.01M phosphate buffered saline (PBS), the specific antibodies were eluted from the column with 0.2 M glycine buffer (pH. 2.5). The specificity of the antibodies was examined by immunoblot with or without the respective antigens as described previously (7).

Preparation of tissue sections. The following steps were carried at 4 °C unless otherwise stated. Frontal sections of the rat cerebellum were prepared and stained as described (9). Briefly, the brains or other peripheral tissues were fixed by perfusion with 4% paraformaldehyde and 0.2% picric acid in 0.1 M phosphate buffer (PB, pH. 7.4). After postfixation with the same fixative for 48 hr, the brains were washed in 30% sucrose in 0.1 M PB for 2 days and cut on a cryostat at 20-μm-thick for light microscopy. For electron microscopy, the brains washed in 30% sucrose in 0.1 M PB were frozen in liquid nitrogen and thawed, then cut into 40-μm-thick frontal sections on a vibratome.

Immunocytochemical procedures for the GABA transporters. The 0.01M PBS used here contained 0.03% Triton X-100 (PBS-T) for light microscopic immunocytochemistry but not for electron microscopy. The sections of the tissues were preincubated with 0.3% H_2O_2 and 5% normal goat serum in 0.01M PBS-T to block the endogenous peroxidase activity and nonspecific binding of the antibodies. The sections were incubated with the purified antibody against each GABA transporter in PBS-T containing 5% normal goat serum for 18 hr. After washing with PBS-T, the sections were incubated for an additional 4 hr with goat anti-rabbit IgG (MBL, Nagoya, Japan), then were incubated for 1.5 hr with rabbit peroxidase-antiperoxidase complex (ICN, Lisle, IL). After 3 rinses, the sections were developed with 0.02% diaminobenzidine (Sigma) and 0.2% nickel ammonium sulfate in 50 mM Tris-HCl (pH. 7.4) with 0.005% H_2O_2. The sections were observed and photographed under a Zeiss light microscope. For the immunofluorescent staining, the sections 20μm thick were incubated with purified antibody against each GABA transporter for 18 hrs then with FITC-labeled goat anti-rabbit IgG (MBL, Nagoya, Japan) for 2 hrs.

For the electron microscopy, the immunostained sections were washed in 0.01M PBS, postfixed for 1 hr in 2% osmium tetroxide in 0.01 M PBS, dehydrated in a graded series of ethanol, and then flat embedded on siliconized slides in Epon. After polymerization at 60°C for 48 hr, the selected areas were cut off and attached to Epon supports for further sectioning on a Reichert-Jung Ultracut E ultramicrotome. Ultrathin sections were cut and mounted on 400-mesh uncoated grids (MAXTAFORM), stained with 1% uranyl acetate in 50% ethanol, and observed and photographed with a HITACHI-7100 electron microscope.

Assay for GABA uptake in CHO cells expressing GAT1: The cDNA clone encoding GABA transporter was isolated by screening of a rat cDNA library and the plasmids containing 2.1kb XbaI/ApaI fragment of GAT1 were constructed by subcloning the inserts into pRC/CMV plasmids (Invitrogen). For the GABA transport assay, CHO cells transfected with the GAT1 in pRC/CMV by electroporation were harvested and washed with HEPES buffered saline (150 mM NaCl, 10 mM HEPES, 1 mM CaCl2, 10 mM glucose, 5 mM KCl, 1mM MgCl2 [pH 7.4]). The cells were pretreated with phorbol 12-myristate 13-acetate (TPA, 100 nM), 4α-phorbol 12, 13-didecanoate (100 nM), staurosporine (100 nM), okadaic acid (1 μM) or amiloride (100 μM) for 15 min, then incubated with HEPES buffered saline containing [3H] GABA (1 - 40 μM) (Du Pont/NEN), 100 mM aminooxyacetic acid and corresponding drugs for 15 min at 37°C. Assays were terminated with ice cold Na-free HEPES buffered saline in which Na^+ was replaced by equimolar Li^+ and the cells were washed three times with the same buffer. The amount of accumulated radioactivity was determined by liquid scintillation counting.

Site-directed mutagenesis: Oligonucleotide-directed site specific mutagenesis was performed using a kit from Amersham (UK), under conditions recommended by the supplier. For site-directed mutagenesis, the following oligonucleotides were used: 5'-TGC TTC AGA GCG CCC TTT AGG-3' (S562A GAT); 5'-CCC TTC CAT GCG TCC CGG TCA-3' (T46A GAT). The complete DNA inserts of all the mutants were sequenced to ensure that the correct mutations were introduced without unwanted side mutations.

Results

Tissue distribution of GAT1, GAT2 and GAT3

Immunoblot analysis revealed that both GAT1 and GAT3 were found only in the brain, while GAT2 was distributed in various tissues. Dense bands of GAT2 was detected in the brain and kidney and a moderately dense bands were found in the heart, pancreas and adrenal gland. Faint band were detected in the lung, liver and spleen.

Cellular distributions of GAT1, GAT2 and GAT3 in the brain

In frontal sections at the level of lateral habenula, the distinct localization of the three GABA transporters was evident. Intense immunoreactivity for GAT1 was seen in the hippocampus, cerebral cortex and lateral amygdaloid nuclei, and moderate immunoreactivity was seen in other regions, including thalamus, hypothalamus and caudate-putamen . White matter was devoid of staining for GAT1. In contrast, no obvious immunoreactivity for GAT2 was seen in a similar section of rat brain. Only faint and homogeneous immunoreaction was seen through the brain but the fringe of the cerebrum showed a dense immunoreaction . Under higher magnification, GAT2 was localized in the arachnoid membrane. GAT3 was found in the thalamus, lateral habenular nucleus, hypothalamus and medial amygdaloid nucleus, but GAT3 immunoreactivity was faint in the cerebral cortex, hippocampus, caudate putamen and white matter.

Distribution of GAT1, GAT2 and GAT3 in the eye

Immunoblot analysis showed much larger amounts of GAT1 and GAT3 in the posterior half of the eye than in the anterior half; whereas, GAT2 was distributed almost evenly in both halves. In the retina, intense GAT1 immunoreactivity was present in the inner plexiform layer (IPL), and weak immunoreactivity in the inner nuclear layer (INL) . No immunoreaction was found in the nerve fiber layer (NFL), ganglion cell layer (GCL), outer nuclear layer (ONL), photoreceptor layer (PRL) or retinal pigment epithelium layer (RPE). Intense GAT2 immunoreactivity was seen in the RPE, and moderately dense immunoreactivity in the NFL. Faint GAT2 immunoreactivity was present in the IPL, INL, OPL and ONL. GAT3 immunoreactivity was present in all the layers from ONL to NFL, but not in the PRL and RPE.

In non-retinal eye tissue, GAT2 immunoreactivity was present in the ciliary body epithelium. GAT2 immunoreactivity was seen in the perikarya but not in the nucleus of the ciliary body epithelium. GAT1 and GAT3 were not present in non-retinal eye tissue.

Light microscopic localization of GAT1 and GAT3 in the cerebellum

GAT1 was distributed throughout the cerebellar cortex with the most intense immunoreactivity in the molecular layer . GAT1 immunoreactivity was seen only in the neuropils but not in the neuronal soma. In the Purkinje cell layer, the dense immunoreaction was seen around the Purkinje cells. Moderate immunoreactivity was found in the granular layer and appeared to form small rings and network-like structure. In contrast, no intense immunoreaction of GAT3 was found in the molecular layer nor in the Purkinje cell layer, and small ring-like and network-like structure was stained by the antibody against GAT3. The white matter failed to show the immunoreactivity for GAT1 or GAT3. Under higher magnification, GAT1 immunoreactive fibers and terminal-like puncta were rich in the molecular layer. In the Purkinje cell layer, the intense GAT1 immunoreactivity was seen in the basket cell axons around the immunonegative somata of Purkinje cells and also confined to the pinceau area.

In the deep cerebellar nuclei, intense GAT3 immunoreactivity but not GAT1 immunoreactivity was found. Neuronal soma and white matter was not apparently stained. Under higher magnification, the GAT3 immunoreactivity was observed around the unstained large neurons and in the neuropils among the fiber bundles.

Electron microscopic localization of GAT1 and GAT3

In the molecular layer of the cerebellar cortex, GAT1 immunoreactivity was found in the presynaptic terminals and non-myelinated axon fibers. The GAT1 immunoreactivity was seen in the presynaptic terminals which make symmetrical synapses with the immunonegative dendrites (Fig. 1A). In the presynaptic terminal, GAT1 immunoreactivity was found around the synaptic vesicles and along the presynaptic membrane and mitochondrial membrane. GAT1 immunoreactive presynaptic terminals and axon fibers were also observed in the Purkinje cell layer but postsynaptic dendrites, neuronal soma and glial processes were unstained. GAT3 immunoreactivity was not found in the molecular layer.

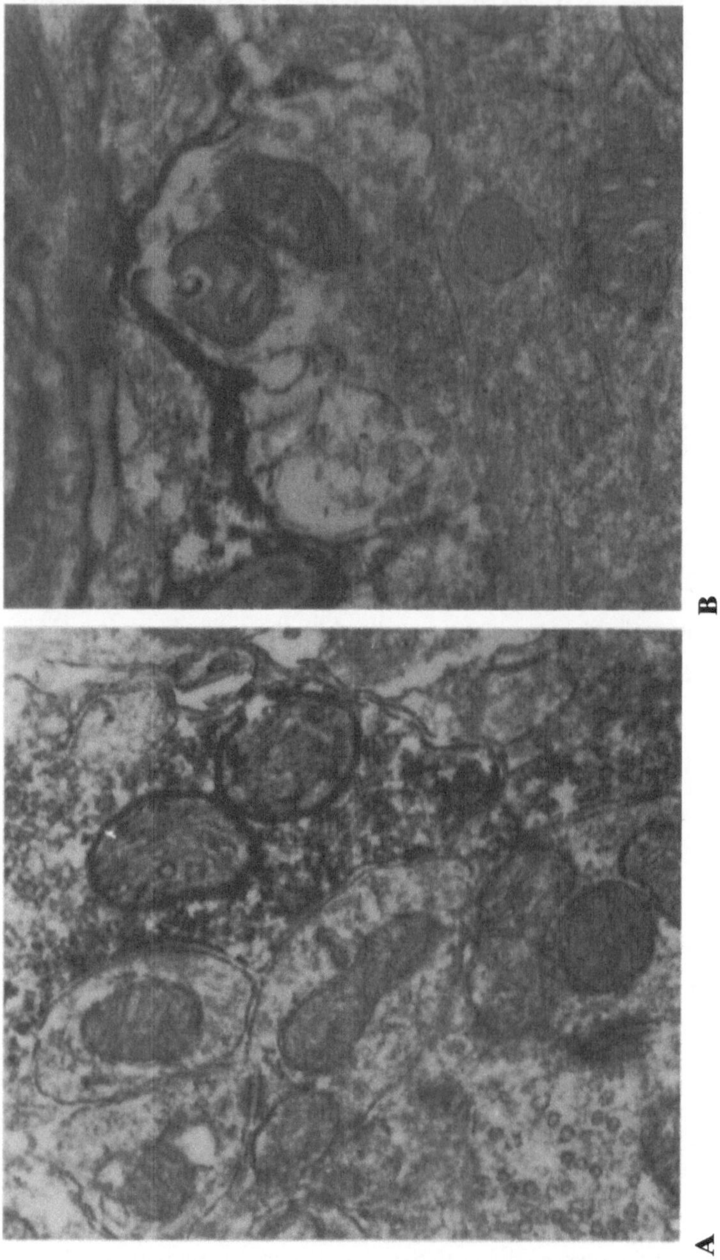

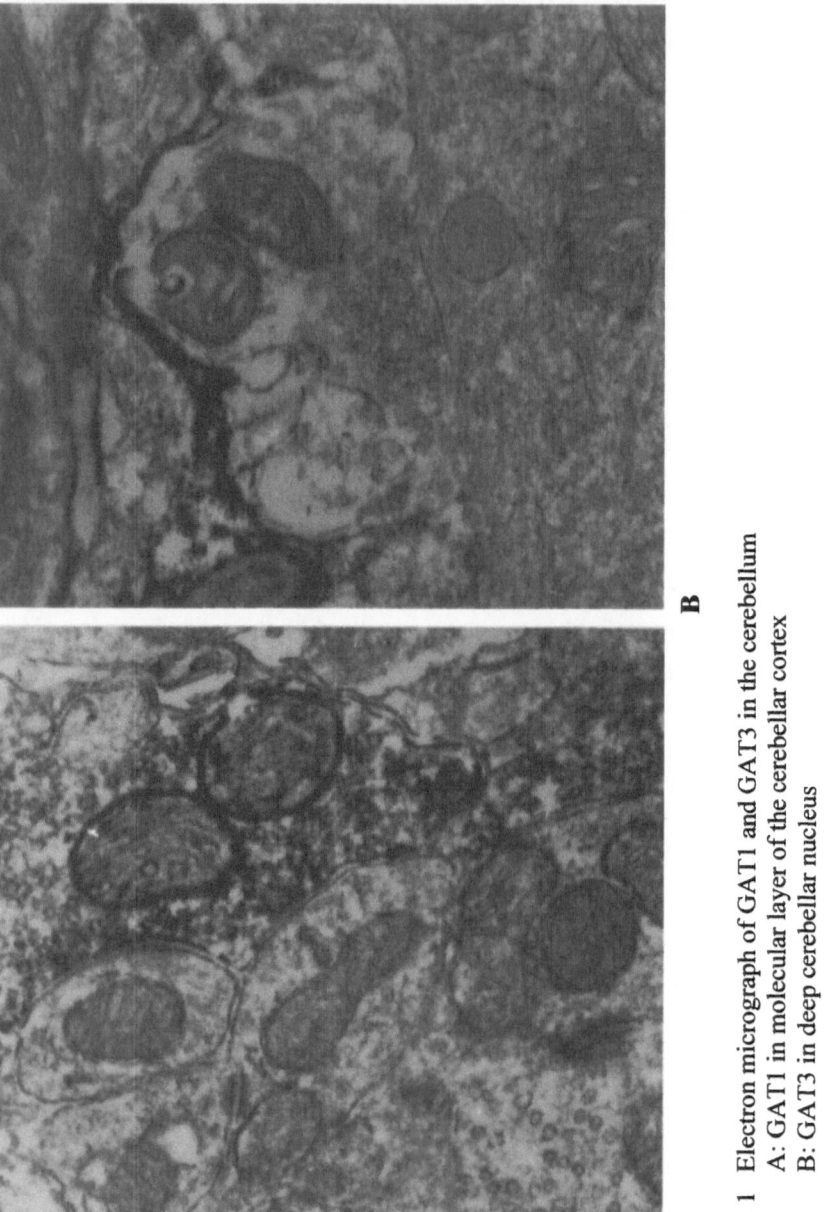

Fig. 1 Electron micrograph of GAT1 and GAT3 in the cerebellum
A: GAT1 in molecular layer of the cerebellar cortex
B: GAT3 in deep cerebellar nucleus

Both GAT1 and GAT3 immunoreactivity was seen as the small rings and network like structures in the granular layer of the cerebellar cortex. Electron microscopic observation in the granular layer of the cerebellar cortex revealed that GAT1 immunoreactivity was found in the nerve terminals surrounding the mossy fiber glomerulus. Swirls of immunonegative granule cell dendrites and immunopositive Golgi cell axons were seen to enwrap the central afferent mossy fiber axon terminal. Under higher magnification, the GAT1 immunoreactivity was seen in the Golgi cell axon terminal that made symmetrical contact with immunonegative granular cell dendrite. GAT3 immunoreactivity was localized in the glial processes that surrounded the immunonegative glomerulus . Neuronal cell bodies and axon terminals were not labelled with the antibodies against GAT3.

In the deep cerebellar nuclei, intense immunoreaction of GAT3 was found in the glial processes around the immunonegative synapses (Fig. 1B). Immunoreactive glial processes surrounded various synapses such as an axosomatic synapse and an axodendritic synapse in which the axon was intruded into a dendrite . GAT3 immunoreactivity was found not only on the glial membrane but also in the cytoplasm of the glial processes. Neuronal cell bodies and nerve terminals were not labelled with the antibodies against GAT3.

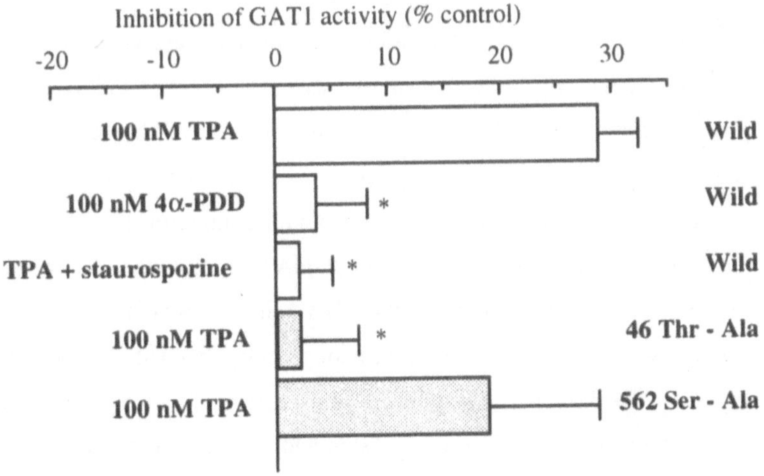

Fig. 2 Modulation of GAT1 activity by phorbol ester

Modulation of GABA transporter activity by phosphorylation/dephosphorylation

Application of 100 nM phorbol ester (TPA), a PKC activator, but not 4α-PDD, an inactive phorbol ester, decreased GABA uptake by about 30% and the inhibitory effect of TPA was prevented by staurosporine, a potent PKC inhibitor (Fig. 2). The effect of okadaic acid (OA), an inhibitor for protein phosphatases significantly inhibited GAT1 activity. The Eadie-Hofstee plot indicates that activation of PKC inhibited GAT1 activity by reducing the affinity for external GABA, without affecting the Vmax.

Site-directed mutagenesis experiments revealed that the mutation on Thr46 completely prevented the inhibitory effect of TPA on GABA transport (Fig. 2), while the mutation on Ser562 did not affect on the inhibitory effect of TPA, although the physiological properties of GAT1 such as the Km and Vmax values of GABA uptake were not significantly affected by the mutation.

Discussion

The present immunocytochemical results revealed the distinct expression of the three GABA transporters. Both GAT1 and GAT3 are found only in the neuronal tissues, whereas GAT2 is widely distributed in many tissues. Because GAT2 was found in ependymal cells and the arachnoid membrane that controls the metabolism of the cerebrospinal fluid and in the ciliary body epithelium that functions in the production of aqueous humor (10), GAT2 may have an important role in nutritional control of the nervous system.

GAT1 and GAT3 showed distinct distribution within the brain and the brain regions that were rich in GAT1 generally showed low expression of GAT3 and *vice versa*. The complimentary distribution of GAT1 and GAT3 was also observed in the neocortex and the hippocampus (7). In addition to the different regional distribution of these two transporters, GAT1 and GAT3 were expressed in the distinct neural compartment. The finding of the neuronal expression of GAT1 and glial expression of GAT3 appeared to be in good agreement with the classical pharmacological property of GABA transporters (11). The glial expression of GAT3, however, was inconsistent with the previous results obtained by in situ hybridization histochemistry (5). The discrepancy may be due to the less resolution by in situ hybridization, or due to the difference of the brain regions examined. However, as far as we examined, GAT3 was localized immunocytochemically in glial processes in various brain regions but not in neuronal cells.

The present results suggested that three types of the termination system of GABAergic transmission exist in the cerebellum; the first is the termination system only by neuronal uptake through GAT1, second is that only by glial uptake through GAT3 and third is that by both neuronal and glial through GAT1 and GAT3 respectively. In the deep cerebellar nuclei, GABA released from Purkinje cell terminals seemed to accumulate into glial cells by the uptake through GAT3 but not accumulate into nerve terminals through GAT1. This finding is in agreement with the previous report that mRNA of GAT1 was not found in the Purkinje cells by in situ hybridization (12). As the expression of GAT1 mRNA was very high in the molecular layer (12), it is suggested that GABAergic inhibition on the dendrites of Purkinje cell by basket or stellate cells is terminated by the uptake of GABA into nerve terminals by GAT1. The GABAergic inhibition on the mossy fiber glomerulus by Golgi cells appeared to be terminated by both GAT1 and GAT3. The different GABA uptake system may be related with the characteristics of each GABAergic neuron. Purkinje cells project in longer distance than basket and stellate cells and it is possible that Purkinje cells do not require the reuptake of GABA into nerve terminals to use GABA as a transmitter again.

The effect of phosphorylation/dephosphorylation on GAT1 activity was also demonstrated in the study. The GAT1 activity is inhibited by the activation of PKC by reducing the affinity of GAT1 for external GABA, without affecting the Vmax. It is considered that the GAT1 activity is controlled by the counterbalance of its phosphorylation and dephosphorylation. Furthermore, the mutation study strongly suggests that the phosphorylation of Thr46 by PKC is responsible for the PKC effect on GAT1, and that the dephosphorylation of Thr46 counterbalances the phosphorylation-induced inhibition of the GAT1 activity.

In conclusion, the termination process of GABAergic synapse by GABA transporter is different in each GABA synapse and the activity of each GABA transporter could be modulated by the activation of protein kinases or protein phosphatases. The development of the specific inhibitor for each GABA transporter or protein kinases may make it possible to regulate each GABAergic neurotransmission selectively.

Acknowledgments

This work was supported by a grant from the Ministry of Education, Science, Sports and Culture in Japan, Yamanouchi Foundation for Research on Metabolic Disorders, and Kato Memorial Bioscience Foundation and Sunbor Grant Foundation.

References

1. Iversen LL and Kelly JS . Uptake and metabolism of γ-aminobutyric acid by neurones and glial cells. Biochem. Pharmac. 1975;24:933-938.

2. Liu QR, Lopez CB, Mandiyan S, Nelson H and Nelson N . Molecular characterization of four pharmacologically distinct α-aminobutyric acid transporters in mouse brain. J. Biol. Chem. 1993;268:2106-2112.

3. Guastella J, Nelson N, Nelson H, Czyzyk L, Keynan S, Miedel MC, Davidson N, Lester HA and Kanner BI . Cloning and expression of a rat brain GABA transporter. Science 1990;249:1303-1306.

4. Borden LA, Smith KE, Hartung PR, Branchek TA and Weinshank RL . Molecular heterogeneity of the γ-aminobutyric acid (GABA) transport system. J. Biol. Chem. 1992;267:21105-21111.

5. Clark JA, Deutch AY, Gallipoli PZ and Amara SG . Functional expression and CNS distribution of a beta-alanine-sensitive neuronal GABA transporter. Neuron 1992;9:337-348.

6. Lopez-Corcuera B, Liu Q-R, Mandiyan S, Nelson H and Nelson N . Expression of a mouse brain cDNA encoding novel γ-aminobutyric acid transporter. J. Biol. Chem. 1992;267:17491-17493.

7. Ikegaki N, Saito N, Hasima M and Tanaka C . Production of specific antibodies against GABA transporter subtypes (GAT1, GAT2, GAT3) and their application to immunocytochemistry. Mol. Brain Res. 1994;47-54.

8. Kishimoto K, Saito N, Ogita K and Kikkawa U . Preparation and use of protein kinase C subspecies-specific anti-peptide antibody. Methods in Enzymology. Hunter and Sefton ed. 1991;Academic Press. New York.

9. Saito N, Kose A, Ito A, Hosoda K, Mori M, Hirata M, Ogita K, Kikkawa U, Ono Y, Igarashi K, Nishizuka Y and Tanaka C . Immunocytochemical localization of βII subspecies of protein kinase C in rat brain. Proc. Natl. Acad. Sci. USA 1989;86:3409-3413.

10. Mito T and Delamere NA . Alteration of active Na-K transport on protein kinase C activation in cultured ciliary epithelium. Invest. Ophthalmol. Vis. Sci. 1993;34:539-546.

11. Iversen LL and Johnston GAR . GABA uptake in rat central nervous system: comparison of uptake in slices and homogenates and the effects of some inhibitors. J. Neurochem. 1971;18:1939-1950.

12. Rattray M and Priestley J . Differential expression of GABA transporter-1 messenger RNA in subpopulations of GABA neurones. Neurosci. Lett. 1993;156:163-166.

GABA: Receptors, Transporters and Metabolism
ed. by C. Tanaka & N.G. Bowery

GABAB RECEPTOR CONTROL OF TRANSMITTER RELEASE IN THE SPINAL CORD

H. Teoh, M. Malcangio and N.G. Bowery

Department of Pharmacology, The School of Pharmacy, University of London
29/39 Brunswick Square, London, WC1N 1AX, U.K.

Summary : Gamma-aminobutyric acid (GABA) activates $GABA_B$ receptors to reduce the release of endogenous amino acids and neuropeptides from the rat isolated hemisected dorsal horn slice, evoked by electrical stimulation of primary dorsal afferents. CGP36742, CGP55845A and CGP52432 reversed the inhibitory effect that (–)-baclofen had on the stimulated outflow of GABA, glutamate and substance P-like immunoreactivity. CGP56999A was without effect on the baclofen-induced inhibition of glutamate release but was an effective antagonist on GABA and substance P release processes. Thus, these data only provide weak evidence for heterogeneity of $GABA_B$ receptors on nerve terminals within the spinal cord dorsal horn.

Introduction

Gamma-aminobutyric acid (GABA) is believed to mediate its inhibitory effects by acting on $GABA_A$ and $GABA_B$ receptors (1,2). The spinal cord locus of action of baclofen in its use as an antispasticity agent highlights the significance of $GABA_B$ receptors in this region of the central nervous system (3). The receptors responsible for mediating the action of baclofen appear to be located on presynaptic terminals where they modulate the release of neurotransmitters (4,5,6,7). Activation of $GABA_B$ receptors can reduce the release of endogenous amino acids (AAs) and neuropeptides from rat spinal cord and this modulatory action on transmitter release appears to be important not only in modifying afferent motorneurone output but also in decreasing nociceptive sensory transmission (6,7). Previous work has demonstrated that $GABA_B$ receptor stimulation reduces the release of substance P, calcitonin gene-related peptide, glutamate, aspartate and GABA, from the isolated dorsal horn, evoked by electrical stimulation of primary afferent fibres (6,7,8,9).

Raiteri and colleagues have recently suggested that within the higher centres, there exists at least 4 subtypes of $GABA_B$ receptors which are located on distinct presynaptic terminals (10). Since these authors have indicated that GABA, glutamate and peptide-containing terminals were each different we have examined a variety of $GABA_B$ antagonists to determine whether the same distinctions are apparent on terminals in the rat spinal cord. The antagonists used were CGP36742 (3-aminopropyl-n butyl phosphonic acid), CGP52432 ([3-[[(3,4-dichlorophenyl)methyl]amino]propyl]diethoxymethyl phosphonic acid, CGP55845A (3-[1-(S)-(3,4-dichlorophenyl) ethyl] amino-2(S)hydroxypropyl-P-benzyl-phosphonic acid) and CGP56999A (-{[1-(R)-(3-carboxyphenyl)ethyl]amino}-2-(S)-hydroxy-propyl]cyclohexyl-methyl-phosphonic acid).

Materials and Methods

Hemisected lumbosacral dorsal horn slices (≈200-300μm thick, ≈2cm long) with intact dorsal roots were obtained from adult male Wistar rats (≈250g) as previously described (7,8,9,11). Each slice was continuously superfused at $1mlmin^{-1}$ with oxygenated (95% O_2 + 5% CO_2) Krebs'-bicarbonate solution for an hour at room temperature. This solution contained (in mM) : NaCl, 118; KCl, 4; KH_2PO_4 , 1.2; $NaHCO_3$, 25; $MgSO_4$, 1.2; $CaCl_2$, 2.5 and glucose, 11. In some experiments, this bathing solution was replaced with a modified Krebs' solution following the equilibration period in order to detect the release of substance P-like immunoreactivity (SP-LI; see reference 7).

At the end of the equilibration period, consecutive three-minute (AA) or eight-minute (SP-LI) samples were collected. In each study, three fractions were obtained before a 20V, 0.5msec stimulus at 1Hz was applied during collection of the fourth sample. In the AA experiments, the stimulation was repeated after an interval of nine minutes. All samples were maintained at –80 °C until they were analyzed biochemically. AA contents were determined by high performance liquid chromatography coupled with fluorescence detection (11) while SP-LI was measured by radioimmunoassay (7).

The mean concentrations of GABA, glutamate and SP-LI measured in the first three samples were taken to be the spontaneous outflow. The amount of AAs in the following fractions were then expressed as a percentage of this basal level. The first stimulated response in each case was identified as S1 and the second as S2; their relationship thereafter expressed as an S2/S1 ratio. SP-LI results were calculated as fmol released by electrical stimulation after subtracting the basal outflow. The data shown are expressed as mean±s.e.mean. Statistical significance was determined by comparing the results in each experiment with the appropriate controls using the Mann-Whitney 'U' test.

Results

Electrical stimulation of dorsal afferents in this preparation significantly increased the release of GABA, glutamate and SP-LI (7,11). The mean control S2/S1 values for GABA and glutamate were 0.64 and 0.72 respectively. CGP36742, CGP52432 and CGP55845A (1–30μM) significantly increased the evoked release of GABA and glutamate as well as reversed the inhibitory effect of (–)-baclofen on their outflow (Fig. 1 and 2). Whilst CGP56999A (0.01–10μM) had no significant effect on the stimulated release of glutamate in the presence or absence of (–)-baclofen (Fig. 3), it did however reverse the inhibitory action of (–)-baclofen on the evoked release of GABA (Fig. 4).

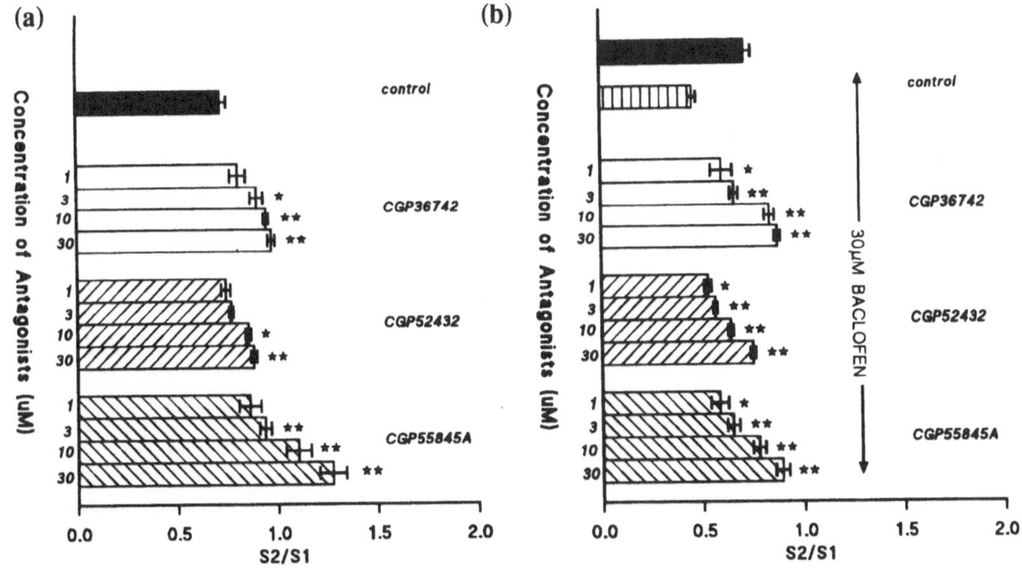

Figure 1. (a) Effects of CGP36742 (□), CGP52432 (//) and CGP55845A (\\) on the S2/S1 ratio for glutamate release. Drug effects were compared with the control responses. (b) Antagonism of (–)-baclofen effect on glutamate release by these antagonists (n=6–7). The effects of each antagonist were compared with the (–)-baclofen response (n=7). ★ $p<0.05$; ★★ $p<0.01$.

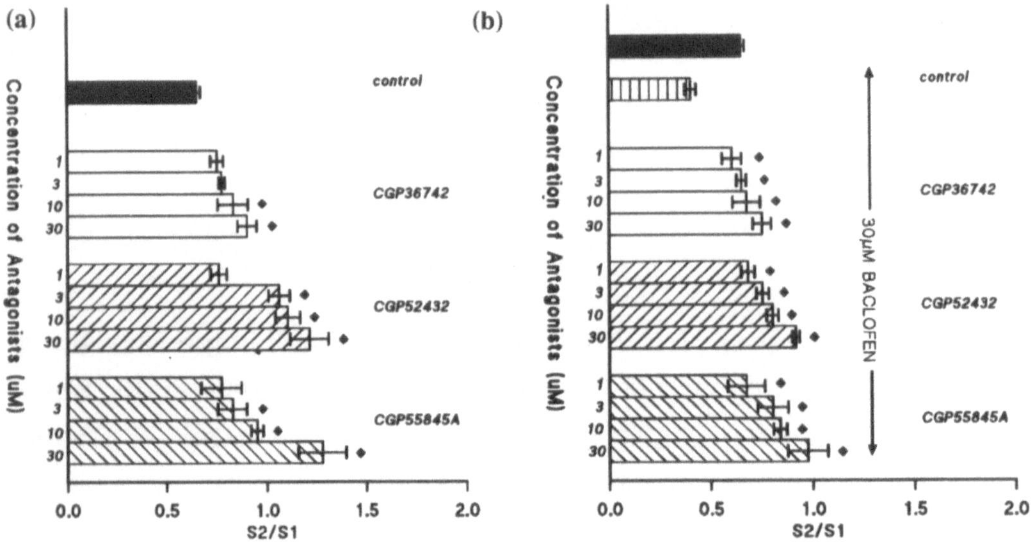

Figure 2. (a) Effects of CGP36742 (□), CGP52432 (//) and CGP55845A (\\) on the S2/S1 ratio for GABA release. Drug effects were compared with the control responses. (b) Antagonism by these antagonists on the inhibitory (–)-baclofen effect on GABA release (n=3). The effects of each antagonist were compared with the (–)-baclofen responses (n=3). ◆ p=0.05.

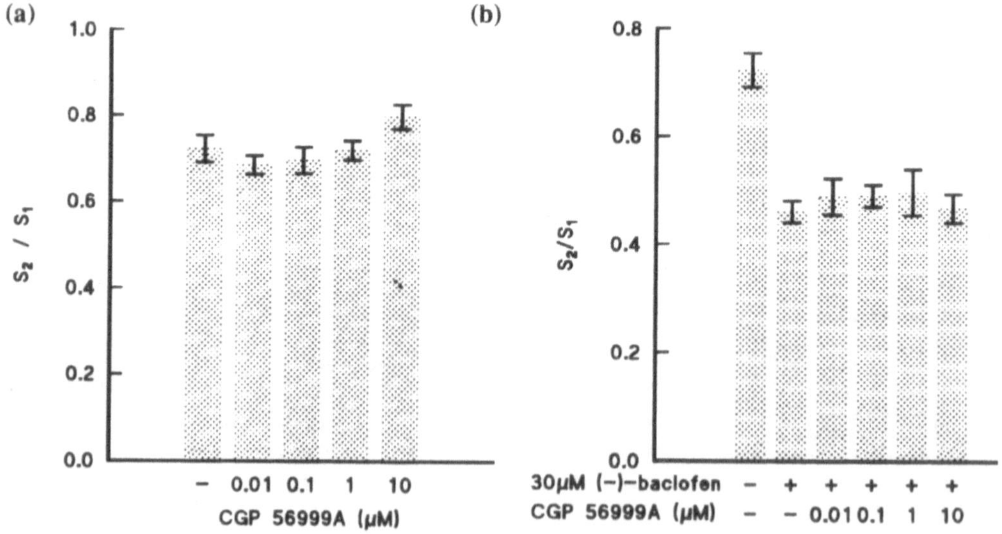

Figure 3. (a) Effect of CGP56999A on the S2/S1 ratio for glutamate release. Drug effects were compared with the control responses. (b) Antagonism by CGP56999A on the inhibitory (–)-baclofen effect on glutamate release (n=6). The effects of each antagonist were compared with the (–)-baclofen responses (n=7).

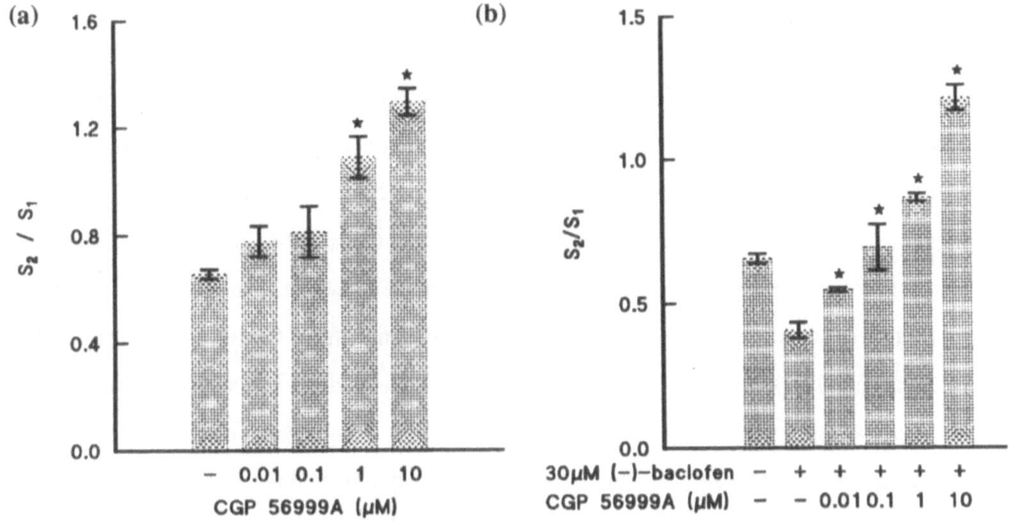

Figure 4. (a) Effect of CGP56999A on the S2/S1 ratio for GABA release. Drug effects were compared with the control responses. (b) Antagonism by CGP56999A on the inhibitory (–)-baclofen effect on GABA release (n=4). The effects of the antagonist were compared with the (–)-baclofen responses (n=3). ★ $p<0.05$.

CGP35348 and CGP36742 have been shown to concentration-dependently reverse the effect of (–)-baclofen on the electrically-evoked release of SP-LI (7). In the present studies CGP56999A (100μM) also antagonised the (–)-baclofen-induced inhibitory effect on stimulated SP-LI outflow (Fig. 5).

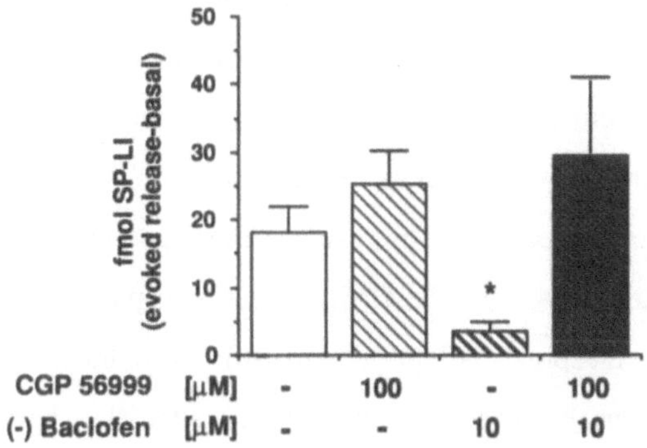

Figure 5. Effect of CGP56999A on the electrically-evoked release of SP-LI and the inhibitory effect of (–)-baclofen (n=3–4). ★ $p<0.05$ when compared with the control results.

Discussion

Previous work with rats neonatally treated with capsaicin suggest that a substantial amount of the glutamate and SP-LI overflow measured in this preparation originates from capsaicin-sensitive fibres while the source of GABA is likely to be the intrinsic dorsal horn neurones (12). It is therefore tempting to propose that a significant proportion of the glutamate and SP-LI released upon electrical stimulation arises from the nociceptive-sensitive primary afferents which terminate in the superficial dorsal horn (13) where they receive axoaxonal synaptic signals from GABA interneurones (14). Indeed, behavioural experiments have implicated glutamate and substance P as putative nociceptive agents (15,16) while GABA and its analogues appear to exert antinociceptive actions (17). Hence, the latter responses could be attributed in part to the inhibitory effects that the GABAergic network has on the outflow from glutamate- and substance P-containing cells in the spinal cord dorsal horn (6,7).

Data from superfusion release studies on synaptosomes have not only suggested that the presynaptic $GABA_B$ receptors in the rat central nervous system are heterogeneous, but also that $GABA_B$ autoreceptors in the spinal cord and the higher centres differ pharmacologically (10). The various receptor groups were not, however, distinguishable in a cortical slice preparation when the same $GABA_B$ antagonists were used (18). In the present study, CGP36742, CGP52432 and CGP55845A did not seem to express any selectivity for $GABA_B$ sites controlling the release of GABA or glutamate in the rat dorsal horn. Previous work on this preparation has demonstrated that CGP36742 also reverses the inhibitory effect that (–)-baclofen has on SP-LI release (7). Our CGP52432 results are therefore in conflict with those of Gemignani et al. (10). In our hands, CGP56999A antagonised the (–)-baclofen-induced inhibition of GABA and SP-LI release whilst not affecting that of glutamate. Hence, these observations lend support to the possible differentiations between the $GABA_B$ autoreceptor and the $GABA_B$ heteroreceptors that regulate glutamate and neuropeptide release in the rat spinal cord dorsal horn (10). In contrast, Waldmeier and colleagues (18) were unable to distinguish between the effects that CGP56999A had on the GABAergic and glutamatergic systems. The reasons behind the discrepancies between our results and those reported previously are not known but could possibly reside in the different preparations and stimulations used.

Conclusion

In conclusion, our results further demonstrate that high intensity electrical stimulation of primary afferent fibres enhances the outflow of GABA, glutamate and SP-LI from the rat spinal cord dorsal horn. The release of glutamate and substance P within this area appears to be modulated by GABA$_B$ receptors and this could be one of the underlying mechanisms for the antinociceptive effects exerted by baclofen (17). The data derived from CGP56999A also provide marginal support for the suggestion that GABA$_B$ receptors regulating the release of GABA, glutamate and substance P in the region of the superficial dorsal horn of the rat may be different from each other.

Acknowledgements

We gratefully thank CIBA-Geigy, Basel, Switzerland for the generous gifts of GABA$_B$ ligands.

References

1. Hill, DR and Bowery, NG. ^{3}H-baclofen and ^{3}H-GABA bind to bicuculline-insensitive GABA$_B$ sites in rat brain. Nature 1991; 290 : 149–152.

2. Bowery, NG, Hill, DR and Hudson, AL. Characteristics of GABA$_B$ receptor binding sites in rat whole brain synaptic membranes. Br J Pharmacol 1983; 78 : 191–206.

3. Bowery, NG. GABA$_B$ receptor pharmacology. Ann Rev Pharmacol Toxicol 1993; 33 : 109–147.

4. Bowery, NG, Hill, DR, Hudson, AL, Doble, A, Middlemiss, DN, Shaw, J and Turnbull, M. (–)Baclofen decreases neurotransmitter release in the mammalian CNS by a novel GABA receptor. Nature (1980); 283 : 92–94.

5. Conzelmann, U, Meyer, DK & Sperk, G. Stimulation of receptors of γ-aminobutyric acid modulates the release of cholecystokinin-like immunoreactivity from slices of rat neostriatum. Br J Pharmacol 1986; 89 : 845–852.

6. Kangrga, I, Jiang, MC and Randić, M. Actions of (–)-baclofen on rat dorsal horn neurons. Brain Res 1991; 562 : 265–275.

7. Malcangio, M and Bowery, NG. γ-aminobutyric acid$_B$ but not γ-aminobutyric acid$_A$ receptor activation, inhibits electrically evoked substance P-like immunoreactivity release from rat spinal cord in-vitro. J Pharmacol Exptl Ther 1993; 266 : 1490–1496.

8. Malcangio, M and Bowery, NG. Effect of the tachykinin NK_1 receptor antagonists, RP 67580 and SR 140333, on electrically-evoked substance P release from rat spinal cord. Br J Pharmacol 1994; 113 : 635–641.

9. Malcangio, M, Libri, V, Teoh, H, Constanti, AC and Bowery, NG. Chronic (–)baclofen or CGP 36742 alters $GABA_B$ sensitivity in rat brain and spinal cord. Neuroreport 1995; 6(2) : 399–403.

10. Gemignani, A, Paudice, P, Bonanno, G and Raiteri, M. Pharmacological discrimination between γ-aminobutyric acid type B receptors regulating cholecystokinin and somatostatin release from rat neocortex synaptosomes. Mol Pharmacol 1994; 46 : 558–562.

11. Teoh, H, Fowler, LJ and Bowery, NG. Effect of lamotrigine on the electrically-evoked release of endogenous amino acids from slices of dorsal horn of the rat spinal cord. (in press; Neuropharmacol).

12. Teoh, H, Malcangio, M, Fowler, LJ and Bowery, NG. Evidence for glutamic acid, aspartic acid and substance P but not GABA from primary afferent fibres in rat spinal cord. (submitted).

13. Cervero, F and Iggo, A. The substantia gelatinosa of the spinal cord. A critical review. Brain 1980; 103 : 717–772.

14. Barber, RP, Vaughan, JE, Saito, K, M[c]Laughlin, BJ and Roberts, E. GABAergic terminals are presynaptic to primary afferent terminal in the substantia gelatinosa of the rat spinal cord. Brain Res 1978; 141 : 35–55.

15. Yashpal, K and Henry, JL Substance P analogue blocks SP-induced facilitation of a spinal nociceptive reflex. Brain Res Bull 1984; 13 : 597–600.

16. Aanonsen, LM, Lei, S-Z and Wilcox, GL. Excitatory amino acid receptors and nociceptive neurotransmission in rat spinal cord. Pain 1990; 41 : 309–321.

17. Sawynok, J. GABAergic mechanisms of analgesia : An update. Pharmacol Biochem Behav 1987; 26 :463–474.

18. Waldmeier, PC, Wicki, P, Feldtrauer, J-J, Mickel, SJ, Bittiger, H & Baumann, PA. GABA and glutamate release affected by $GABA_B$ receptor antagonists with similar potency : no evidence for pharmacologically different presynaptic receptors. Brit J Pharmacol 1994; 113 : 1515–1521.

GABA: Receptors, Transporters and Metabolism
ed. by C. Tanaka & N.G. Bowery

ROLE OF $GABA_A$ AND $GABA_B$ RECEPTORS IN Ca^{++} HOMEOSTASIS AND TRANSMITTER RELEASE IN CEREBELLAR GRANULE NEURONS

Arne Schousboe[1)], Lisbeth Elster[1)], Inge Damgaard[1)], Henrik Pallos[1)], Povl Krogsgaard-Larsen[2)] and Julianna Kardos[3)]

PharmaBiotec Res. Center, [1)]Dept. of Biol. Sci. and [2)]Dept. of Medicinal Chem., Royal Danish School of Pharmacy, DK-2100 Copenhagen, Denmark and [3)]Central Res. Inst. for Chem., Hungarian Acad. Sci., H-1525 Budapest, Hungary.

Summary: Cerebellar granule neurons cultured from dissociated cerebellum of 7-day-old mice or rats express $GABA_A$ receptors with different affinities for GABA as well as $GABA_B$ receptors. These receptors are coupled to regulation of transmitter release as well as intracellular Ca^{++} homeostasis. Transmitter release elicited by moderate depolarization can be inhibited either by $GABA_A$ or $GABA_B$ receptor agonists whereas transmitter release evoked by a more pronounced depolarization can only be inhibited by $GABA_B$ receptor agonists. However, if the neurons express low affinity $GABA_A$ receptors these receptors can mediate an inhibitory effect even under these conditions. While a coupling between $GABA_A$ receptors and Ca^{++} homeostasis has not been consistently observed, $GABA_B$ receptors are clearly coupled to regulation of Ca^{++} influx. The inhibitory action of (R)-Baclofen on K^+ induced increase in intracellular Ca^{++} ($[Ca^{++}]_i$) as well as transmitter release was dependent on coupling to G-proteins although this G-protein coupling appears to be rather complex. Finally, an interaction between $GABA_A$ and $GABA_B$ receptors seems to exist with regard to regulation of transmitter release.

Introduction

Regulatory mechanisms for glutamatergic activity can be conveniently studied in cultured cerebellar granule neurons in which evoked glutamate release is inhibited by GABA [1,2,3]. Moreover, since addition of kainic acid to the culture medium selectively destroys contaminating GABAergic neurons [4,5], the cultures after such treatment represent a monotypic neuronal preparation of glutamatergic cells. Such granule neurons, deprived of inhibitory GABAergic connections, may constitute a model in which the effect of externally applied hyperpolarizing agents can be characterized [2,3,6]. These neurons are also suitable for studies of changes in the intracellular Ca^{++} homeostasis induced by depolarizing agents. Thus, it has been shown that depolarization with either a high concentration of K^+ or with glutamate leads to changes in intracellular Ca^{++} ($[Ca^{++}]_i$) where a distinction can be made between the cell body and the processes [7].

The present review shall deal with regulatory mechanisms for expression of GABA receptors on these neurons and the mechanisms by which these receptors may regulate Ca^{++} homeostasis and neurotransmitter release.

Expression of GABA receptors

Numerous studies (see Ref. [8]) have shown that $GABA_A$ receptor expression can be influenced by exposure of neurons to $GABA_A$ receptor agonists. In cerebellar granule neurons exposure to GABA or the $GABA_A$ receptor agonist THIP (4,5,6,7-tetrahydroisoxazolo-[5,4-c]pyridin-3-ol) induces expression of certain subunits of the $GABA_A$ receptors leading to appearance of low affinity GABA receptors in addition to high affinity GABA receptors [1,2,6,9-11]. This induction of GABA receptors which is dependent on protein synthesis [2] is associated with expression of the respective receptor subunit mRNAs [12]. It has recently been observed that also $GABA_B$ receptors as measured by binding of [^{3}H]Baclofen can be induced by treatment of granule cells in culture with THIP [13]. In this case the K_D was

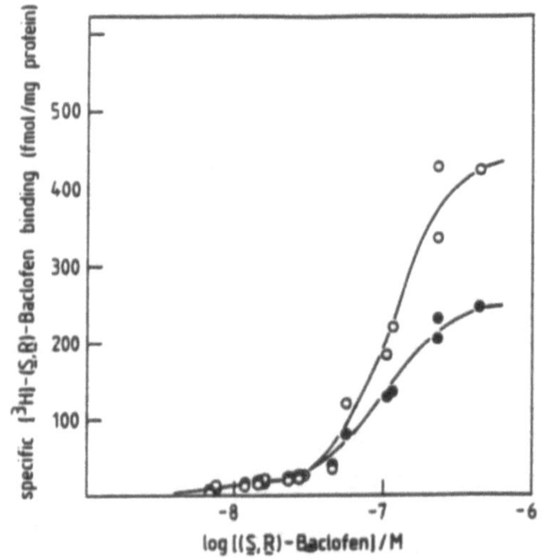

Figure 1. The effect of THIP-treatment (150 μM THIP) on the number of [^{3}H](S,R)-Baclofen binding sites in cerebellar granule cells in culture. Control (•) and THIP-treated (○) cells were cultured for 7 days from dissociated cerebella from 7-day-old mice and binding was subsequently determined as described by Kardos *et al.* [13]. From Kardos *et al.* [13].

not affected but the B_{max} was increased by 75% (Fig. 1). The presence of $GABA_B$ receptors on these neurons is in agreement with previous studies [14-18] but it has not been reported previously that the expression of this receptor can be enhanced by the $GABA_A$ receptor specific agonist THIP [19]. It seems, however, that exposure of these neurons to THIP may lead to an enhanced cellular differentiation with regard to cell plasma membrane proteins as also other proteins such as adhesion molecules, synaptic marker proteins and voltage gated Ca^{++} channels are induced by THIP [20,21].

Ca^{++} homeostasis

Exposure of cerebellar granule neurons to depolarizing concentrations of potassium or excitatory amino acids leads to an increase in intracellular Ca^{++} [7,13,22]. This increase in $[Ca^{++}]_i$ can be inhibited by Baclofen as shown in Table I. It can be seen

Table 1. Effect of (R)-Baclofen on K^+ induced increase in $[Ca^{++}]_i$ in granule cells cultured in the absence or presence of THIP.

Culture condition	Increase in $[Ca^{++}]_i$ (%) Control	Baclofen	Inhibition (%)
- THIP	211± 4	170±12*	20±3
+ THIP	230±11	183± 8*	20±4

Cerebellar granule neurons were cultured for 7 days from dissociated cerebella from 7-day-old mice [13] in the presence or absence of THIP. Measurements of $[Ca^{++}]_i$ were performed after addition of 40 mM K^+ and (R)-Baclofen (100 μM) as indicated using the fluorescent Ca^{++} chelator fluo-3 [23]. Results are means ± SEM obtained using 24 cultures in each of the different preparations. The resting intracellular Ca^{++} concentration ($[Ca^{++}]_{i,R}$) was 125 ± 13 and 93 ± 10 nM in control and THIP-treated cultures, respectively. Means of control and treated groups were compared using Student's t-test for non-paired samples. The asterisks indicate a statistically significant ($P < 0.025$) inhibitory action of Baclofen.

that in spite of the fact that cells cultured in the presence of THIP have an increased number of $GABA_B$ receptors (cf. above), the ability of Baclofen to inhibit the K^+ induced increase in $[Ca^{++}]_i$ was found to be independent of the culture condition. It should, however, be pointed out that treatment of the cells with pertussis toxin which inactivates the G-protein coupling of $GABA_B$ receptors [18] led to an abolishment of the inhibitory action of Baclofen in cells grown in plain culture media but had no effect in THIP treated cultures [13]. This indicates that the G-protein coupling of the $GABA_B$ receptors may be different in granule neurons cultured in plain culture media and in cells grown in the presence of THIP. In other words, the $GABA_B$ receptors induced by THIP may have unique properties (cf. below).

Transmitter release

Depolarization coupled transmitter release which can be conveniently studied by the use of [^{3}H]D-aspartate to label the neurotransmitter glutamate pool in the granule neurons [7] can be

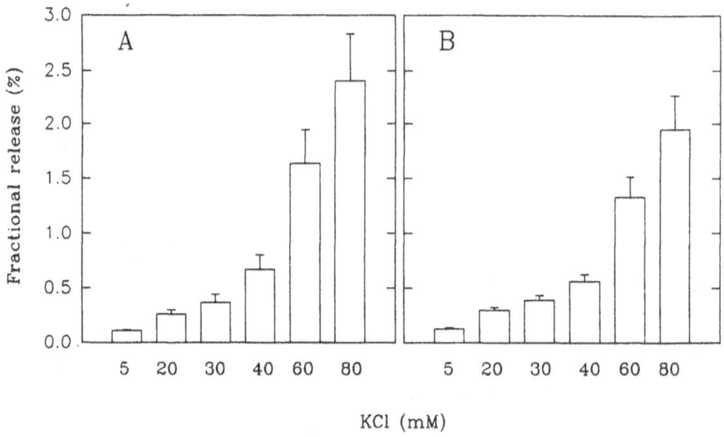

Figure 2. Potassium stimulated transmitter release in cerebellar granule cells cultured for 7-8 days in the absence (A) or presence (B) of THIP (150 μM). Cells were prepared from dissociated cerebella from 7-day-old mice and cultured as described by Kardos *et al.* [13]. After preloading (30 min) with [^{3}H]D-aspartate (2 μCi/culture) release from the transmitter glutamate pool was followed by continuous superfusion of the cultures [13] stimulating for 30 s with increasing concentrations of KCl maintaining the osmolarity by isosmotic reduction of the concentration of NaCl in the superfusion medium. Results are averages of SEM values indicated by vertical bars (I. Damgaard, H. Pallos & A. Schousboe, unpublished).

modulated by $GABA_A$ and $GABA_B$ receptors in a manner which is dependent on the depolarizing condition. Fig. 2A,B shows the relationship between the release of [^{3}H]D-aspartate and the depolarization induced by K^+ in control and THIP treated cultures. It is seen that below a threshold K^+ concentration of 20 mM no transmitter release can be detected while K^+ concentrations higher than 20 mM lead to an increasing amount of transmitter release. This pattern appears to be the same in the two sets of cultures in spite of the fact that expression of voltage gated Ca^{++} channels is influenced by the exposure of the cells to THIP [21]. The transmitter release elicited by 30 mM K^+ has been shown to be inhibited by activation of high affinity $GABA_A$ receptors coupled

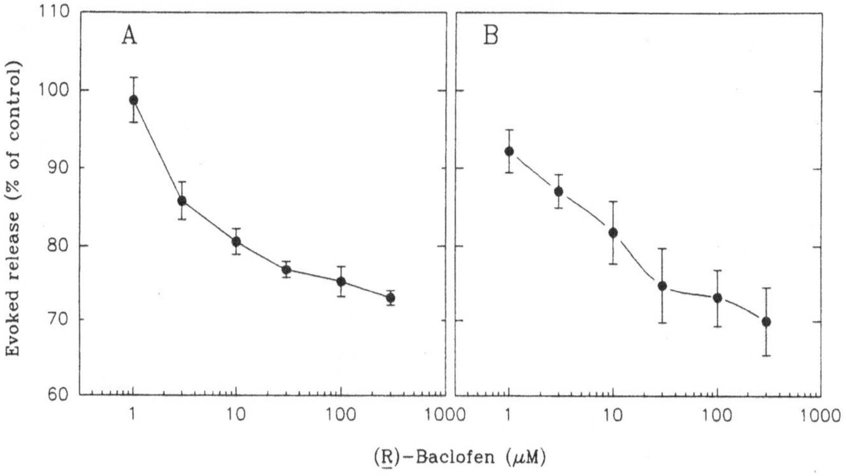

Figure 3. Dose dependence of (R)-Baclofen-mediated inhibition of KCl (30 mM) stimulated [^{3}H]D-aspartate release from mouse cerebellar granule cells cultured for 7 days in culture media in the absence (A) or presence (B) of THIP (150 μM). Cells were loaded with [^{3}H]D-aspartate (2 μCi/culture) 30 min prior to superfusion experiments which were performed as detailed by Kardos *et al.* [13]. Each dose of (R)-Baclofen was tested separately on 6 to 15 cultures (mean ± SEM) from 2 to 6 different preparations of cells. Inhibition is expressed as percentage of the average of two pairs of control stimulations applied before and after the stimulations performed in the presence of (R)-Baclofen (H. Pallos, I. Damgaard & A. Schousboe, unpublished).

to benzodiazepine receptors in a picrotoxinin dependent fashion [3]. However, the larger transmitter release elicited by 50 mM K^+ (cf. Fig. 2) can not be inhibited by these receptors [1,3] but can only be inhibited by activation of the low affinity $GABA_A$ receptors in a picrotoxinin independent fashion [3]. This latter finding has led to the proposal that low affinity $GABA_A$ receptors may be functionally coupled to Ca^{++} channels [3,21] but attempts to demonstrate such a functional coupling have not been conclusive [24].

The K+ evoked transmitter release can also be inhibited by (R)-Baclofen [13,14,17,18]. As shown in Fig. 3A,B the dose-response curves for the inhibitory action of (R)-Baclofen on transmitter release induced by 30 mM K+ were almost identical in granule neurons cultured in the absence or presence of THIP. This may reflect the finding that the affinity of the $GABA_B$ receptors for Baclofen was

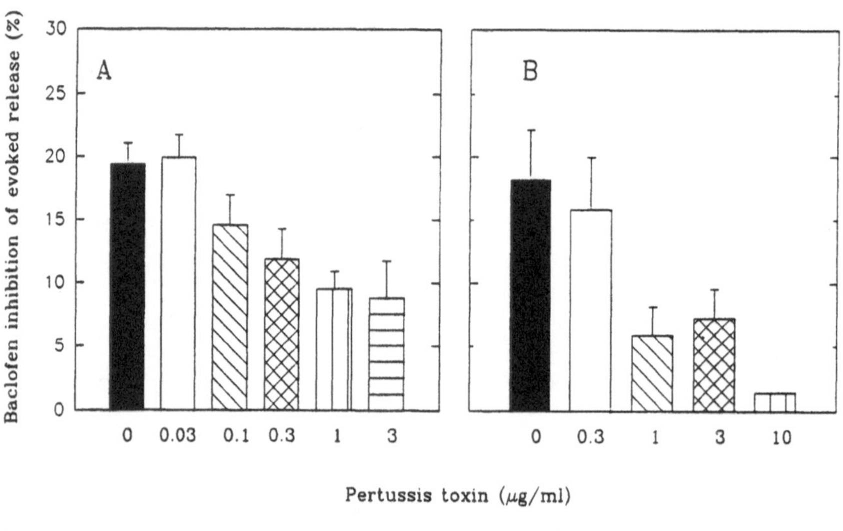

Figure 4. Dose-dependency of pertussis toxin inactivation of (R)-Baclofen (10 μM) mediated inhibition of KCl (30 mM) stimulated [^{3}H]D-aspartate release from cerebellar granule cells cultured for 7 days in the absence (A) or presence (B) of 150 μM THIP. PTX was added to the cultures 14-20 h before beginning the release experiment. PTX affected neither the basal nor the depolarization-induced responses. Results are averages of experiments with SEM values shown as vertical bars. Inhibition is expressed as percentage of control stimulations (cf. Fig. 3) (H. Pallos, I. Damgaard & A. Schousboe, unpublished).

independent of the presence of THIP in the culture media (see Fig. 1). The IC_{50} value for (R)-Baclofen was similar to that reported by Huston *et al.* [17]. It was previously reported that the sensitivity of this action of Baclofen to pertussis toxin was different in cultures grown in the presence or absence of THIP [13]. This has been further investigated and the results are shown in Fig. 4A,B. It can be seen that under both culture conditions the inhibitory action of Baclofen on K^+-induced transmitter release was affected by exposure of the neurons to pertussis toxin. However, the sensitivity to pertussis toxin was different in the two sets of cultures in agreement with the previous observation [13]. This finding is in keeping with the observation that the sensitivity to pertussis toxin of the Baclofen mediated inhibition of K^+-induced aberrations in Ca^{++} homeostasis was different in THIP treated and non-treated cultures.

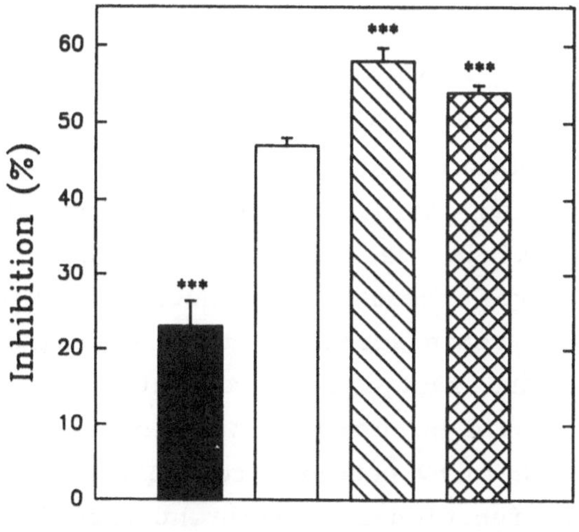

Figure 5. Effect of GABA receptor agonists on KCl (40 mM) stimulated [^{3}H]D-aspartate release from cerebellar granule cells cultured for 7 days in culture media without addition of THIP. Filled column: 50 μM (R)-Baclofen, open column: 100 μM isoguvacine, hatched column: 50 μM (R)-Baclofen plus 100 μM isoguvacine and cross-hatched column: 100 μM GABA. Results are averages of 26 experiments with SEM values shown as vertical bars. Inhibition is expressed as percentage of control-stimulations (cf. Fig. 2). Asterisks indicate a statistically significant difference from inhibition mediated by isoguvacine ($P < 0.001$; Student's t-test).
From Kardos *et al.* [13].

Finally, it should be pointed out that an interaction between $GABA_A$ and $GABA_B$ receptors might exist in the granule neurons. Fig. 5 shows that the inhibitory action of 100 μM GABA on evoked transmitter release could be mimicked by (R)-Baclofen (50 μM) applied together with isoguvacine (100 μM). However, the inhibitory action of the two selective agonists together was smaller than the sum of the effects of the two agonists applied individually (Fig. 5). This lack of additivity might suggest a functional coupling between $GABA_A$ and $GABA_B$ receptors leading to a disinhibitory action on $GABA_A$ receptors by $GABA_B$ receptors. Such a disinhibition was also suggested by the observation that the ability of GABA to inhibit transmitter release was largely unaffected by PTX in spite of the fact that the $GABA_B$ receptors were partly inactivated by pertussis toxin [13]. Moreover, a disinhibitory action of $GABA_B$ receptors on the $GABA_A$ receptor may also be partly responsible for the fading of the GABA-induced $^{36}Cl^-$ uptake in rat cerebellar granule neurons [25]. Disinhibitory effects of $GABA_B$ receptors are well documented in several regions of the brain ([26] and references cited therein) suggesting that a disinhibitory interaction of $GABA_A$ and $GABA_B$ receptors could be responsible for many reportedly 'excitatory' GABA actions ([27-31] and references cited herein). It thus appears that granule neurons may possess $GABA_B$ receptors which by inhibiting the $GABA_A$ receptor mediated inhibition would help driving the granule cell input to Purkinje cells, and consequently enhance the inhibitory output from these latter neurons. This would be an emergency mechanism reminiscent of that entrained by external conditions that demand fast, tight regulation of skeletal motor outflows [32]. It should be noted that the disinhibitory action of $GABA_B$ receptors achieved by coupling to $GABA_A$ receptors present on the same neurons is functionally different from the disinhibitory actions of GABA mediated by GABAergic interneurons (cf. Ref. [33]).

Acknowledgements

The expert secretarial assistance by Ms. Hanne Danø is gratefully acknowledged. The work was supported by grants from the Danish State Biotechnology Program (1991-95), the Lundbeck Foundation, the

Hungarian Research Council (OTKA 1762) and the CEC BIOTEC (BIO2 CT-930224) Program. Ciba-Geigy, Basel, Switzerland is cordially thanked for the supply of (R)-Baclofen.

References

1. Meier E, Drejer J, Schousboe A. GABA induces functionally active low-affinity GABA receptors on cultured cerebellar granule cells. J Neurochem 1984;43:1737-1744.

2. Belhage B, Hansen GH, Meier E, Schousboe A. Effects of inhibitors of protein synthesis and intracellular transport on the GABA-agonist induced functional differentiation of cultured cerebellar granule cells. J Neurochem 1990b;55:1107-1113.

3. Belhage B, Damgaard I, Saederup E, Squires RF, Schousboe A. High- and low-affinity GABA-receptors in cultured cerebellar granule cells regulate transmitter release by different mechanisms. Neurochem Int 1991;19:475-482.

4. Drejer J, Schousboe A. Selection of a pure cerebellar granule cell culture by kainate treatment. Neurochem Res 1989;14:751-754.

5. Schousboe A, Pasantes-Morales H. Potassium-stimulated release of ^{3}H-taurine from cultured GABAergic and glutamatergic neurons. J Neurochem 1989:53:1309-1315.

6. Belhage B, Hansen GH, Schousboe A. GABA agonist induced changes in ultrastructure and GABA receptor expression in cerebellar granule cells is linked to hyperpolarization of the neurons. Int J Devl Neurosci 1990a;8:473-479.

7. Belhage B, Rehder V, Hansen GH, Kater SB, Schousboe A. ^{3}H-D-Aspartate release from cerebellar granule neurons is differentially regulated by glutamate- and K^+-stimulation. J Neurosci Res 1992;33:436-444.

8. Schousboe A, Redburn DA. Modulatory actions of gamma aminobutyric acid (GABA) on GABA type A receptor subunit expression and function. J Neurosci Res 1995;41:1-7.

9. Belhage B, Meier E, Schousboe A. GABA-agonists induce the formation of low affinity GABA-receptors on cultured cerebellar granule cells via preexisting high affinity GABA receptors. Neurochem Res 1986;11:599-606.

10. Hansen GH, Belhage B, Schousboe A. Effect of a GABA agonist on the expression and distribution of $GABA_A$ receptors in the plasma membrane of cultured cerebellar granule cells: an immunocytochemical study. Neurosci Lett 1991;124:162-165.

11. Elster L, Hansen GH, Belhage B, Fritschy JM, Möhler H, Schousboe A. Differential distribution of $GABA_A$ receptor subunits in soma and processes of cerebellar granule cells: Effects of maturation and a GABA agonist. Int J Devl Neurosci 1995; in press.

12. Kim HY, Sapp DDW, Olsen RW, Tobin AJ. GABA alters $GABA_A$ receptor mRNAs and increases ligand binding. J Neurochem 1994;62:2334-2337.

13. Kardos J, Elster L, Damgaard I, Krogsgaard-Larsen P, Schousboe A. Role of $GABA_B$ receptors in intracellular Ca^{2+} homeostasis and possible interaction between $GABA_A$ and $GABA_B$ receptors in regulation of transmitter release in cerebellar granule neurons. J Neurosci Res 1994;39:646-655.

14. Travagli RA, Ulivi M, Wojcik WJ. γ-Aminobutyric acid-B receptors inhibit glutamate release from cerebellar granule cells: consequences of inhibiting cyclic AMP formation and calcium influx. J Pharm Exp Ther 1991;258:903-909.

15. Kardos J, Kovacs I. Binding interaction of gamma aminobutyric acid A and B receptors in cell culture. Neuroreport 1991;2:541-543.

16. DeErasquin G, Grooker G, Costa E, Woscik WJ. Stimulation of high affinity γ-aminobutyric $acid_B$ receptors potentiates the depolarization induced increase of intraneuronal ionised calcium content in cerebellar granule neurons. Molec Pharmac 1992;42:407-414.

17. Huston E, Scott RH, Dolphin AC. A comparison of the effect of calcium channel ligands and $GABA_B$ agonists and antagonists in transmitter release and somatic calcium channel currents in cultured neurons. Neuroscience 1990;38:721-729.

18. Huston E, Gullen G, Sweeney MI, Pearson H, Fazeli MS, Dolphin AC. Pertussis toxin treatment increases glutamate release and dihydropyridine binding sites in cultured rat cerebellar granule neurons. Neuroscience 1993;52:787-798.

19. Krogsgaard-Larsen P, Johnston GAR, Lodge D, Curtis DR. A new class of GABA agonists. Nature 1977;268:53-55.

20. Meier E, Hertz L, Schousboe A. Neurotransmitters as developmental signals. Neurochem Int 1991;19:1-15.

21. Hansen GH, Belhage B, Schousboe A. First direct electron microscopic visualization of a light spatial coupling between $GABA_A$-receptors and voltage sensitive calcium channels. Neurosci Lett 1992; 137:14-18.

22. Bouchelouche P, Belhage B, Frandsen Aa, Drejer J, Schousboe A. Glutamate receptor activation in primary cultures of cerebellar granule cells increases cytosolic free Ca^{2+} by mobilisation of cellular calcium and activation of Ca^{2+} influx. Exp Brain Res 1989; 76:281-291.

23. Wahl P, Schousboe A, Honoré T, Drejer J. Glutamate induced increase in intracellular Ca^{2+} in cerebral cortex neurons is transient in immature cells but permanent in mature cells. J Neurochem 1989;53:1316-1319.

24. Elster L, Saederup E, Schousboe A, Squires RF. ω-Conotoxin binding sites and regulation of transmitter release in cerebellar granule neurons. J Neurosci Res 1994;39:424-429.

25. Kardos J. $^{36}Cl^-$ flux measurements on $GABA_A$receptor-activated chloride exchange. Multiple mechanisms of the chloride channel inactivation. Biochem Pharmac 1989;38:2587-2591.

26. Mott DD, Darrell VL. Facilitation of the induction of long-term potentiation by $GABA_B$ receptors. Science 1991;252:1718-1720.

27. Mitchell R. A novel GABA receptor modulates stimulus induced glutamate release from cortico-striatal terminals. Eur J Pharmacol 1980;67:119-122.

28. Levi G, Gallo V. Glutamate as a putative transmitter in the cerebellum: stimulation by GABA of glutamic acid release from specific pools. J Neurochem 1981;37:22-31.

29. Nielsen EØ, Aarslew-Jensen M, Diemer NH, Krogsgaard-Larsen P, Schousboe A. Baclofen-induced, calcium-dependent stimulation of *in vivo* release of D-[^{3}H]aspartate from rat hippocampus monitored by intracerebral microdialysis. Neurochem Res 1989:14:321-326.

30. Cherubini E, Gaiarsa JL, Ben-Ari Y. GABA: an excitatory transmitter in early postnatal life. TINS 1991;14:515-519.

31. DeLorey TM, Olsen RW. γ-Aminobutyric $acid_A$ receptor structure and function. J Biol Chem 1992;267:16747-16750.

32. Fair CM (1992): Cortical Memory Functions. Boston: Birkhäuser, 1992.

33. Roberts E. Living systems are tonically inhibited, autonomous optimizers, and disinhibition coupled to variability generation is their major organizing principle: Inhibitory command-control at levels of membrane, genome, metabolism, brain, and society. Neurochem Res 1991;16:409-421.

GABA: Receptors, Transporters and Metabolism
ed. by C. Tanaka & N.G. Bowery

INTERACTIONS BETWEEN ENDOGENOUS L-DOPA AND $GABA_A$ SYSTEMS FOR CARDIOVASCULAR CONTROL IN THE RAT NUCLEUS *TRACTUS SOLITARII*

Yoshimi Misu, Jin-Liang Yue, Yoko Okumura, Takeaki Miyamae and Yoshio Goshima

Department of Pharmacology, Yokohama City University School of Medicine,
Yokohama 236, Japan

Summary: Depressor responses to L-DOPA 30 ng microinjected into depressor sites were inhibited by GABA 3-30 ng and by nipecotic acid 100 ng whereas potentiated by bicuculline 10 ng. By microdialysis, basal L-DOPA release was reduced by muscimol 10 to 30 μM perfused via a probe by 50-60 % whereas increased by bicuculline 30 μM by 40%. Pressor responses to 300 ng GABA were reduced by 50% by a competitive L-DOPA antagonist, L-DOPA methyl ester 1 μg. Meanwhile, L-DOPA 1-100 nM perfused released GABA by 35-60%. This release was not mimicked by D-DOPA and dopamine 10 nM, not affected by inhibition of central DOPA decarboxylase, and was markedly reduced by Ca^{2+} deprivation or tetrodotoxin 1 μM.

Introduction

We have proposed that L-3,4-dihydroxyphenylalanine (L-DOPA) is a neurotransmitter in the central nervous system (CNS) (1-5). Endogenous L-DOPA is released in a transmitter-like manner from in vitro and in vivo striata (1, 2) and from areas related to blood pressure (BP) regulation in the lower brainstem of rats (4, 5). In addition, exogenously applied L-DOPA produces by itself in vitro presynaptic (1, 2) and in vivo postsynaptic (3-5) responses. All of these responses are stereoselective in common with many types of receptors and are antagonized by L-DOPA methyl ester, a competitive antagonist for L-DOPA (6). L-DOPA and L-DOPA methyl ester produce no displacement of selective binding of catecholaminergic α_1, α_2, β, D_1 and D_2 ligands in membrane preparations from rat brain (6). A recognition site for L-DOPA may exist. In the striatum, L-DOPA is a precursor for dopamine and seems to be an endogenous potentiater for postsynaptic DA D_2-

receptors related to locomotor activities and for presynaptic β-adrenoceptors to facilitate evoked dopamine release (2). These three characteristics may be involved in effectiveness in parkinsonism. In the lower brainstem, L-DOPA produces by itself dose-dependent postsynaptic cardiodepressor responses in the nucleus tractus solitarii (NTS) (3-5) and the caudal ventrolateral medulla and cardiopressor responses in the rostral ventrolateral medulla (2). There exist neurons that may contain L-DOPA as an end product in various regions of CNS including these three areas (2, 7).

The dorsomedial region of NTS contains the first synapse of the baroreceptor reflex arc. Thus, it plays a critical role in cardiovascular regulation. Neurotransmitters of the primary baroreceptor afferents terminating in NTS have not been conclusively identified, although there are various candidates such as L-glutamate (2). We have proposed that L-DOPA is a neurotransmitter of the primary baroreceptor afferents (2-5), since in addition to postsynaptic responses to exogenous L-DOPA itself (3), endogenous basal L-DOPA release during microdialysis of NTS is in part tetrodotoxin (TTX)-sensitive and Ca^{2+}-dependent, the release by aortic nerve stimulation is TTX-sensitive, the release by high K^+ is Ca^{2+}-dependent, the release by phenylephrine-induced baroreceptor stimulation is abolished by bilateral sino-aortic denervation, and unilateral aortic denervation decreases tyrosine hydroxylase(TH)- and L-DOPA-, but not dopamine- and dopamine-β-hydroxylase-immunoreactivities in the ipsilateral NTS. Furthermore, endogenously released L-DOPA is involved in baroreceptor reflex and tonically functions to activate depressor neurons in NTS. The tonic L-DOPA system is altered in NTS and RVLM of spontaneously hypertensive rats compared to age-matched Wistar-Kyoto rats (5).

In addition, various neurotransmitter substances have been shown to modulate baroreceptor reflex responses at NTS. GABA is a main inhibitory neurotransmitter in CNS and a $GABA_A$ mechanism tonically functions to inhibit the baroreflex pathway in NTS. Microinjected GABA, muscimol, a selective $GABA_A$ agonist, and nipecotic acid, a GABA uptake inhibitor, produced hypertension and tachycardia (8, 9). Depressor and bradycardic responses to electrical aortic nerve stimulation were attenuated by GABA and muscimol, whereas those were potentiated by bicuculline, a selective $GABA_A$ antagonist (8). Thus, if our hypothesis of L-DOPA as a neurotransmitter of the primary baroreceptor afferents is truly the case, the $GABA_A$ system has to inhibit presynaptic and postsynaptic components of L-DOPA in NTS.

GABA tonically functions to inhibit postsynaptic cadiodepressor responses to L-DOPA microinjected in NTS

GABA-induced hypertension and L-DOPA-induced hypotension

In urethane-anesthetized rats, mean BP and heart rate (HR) were 75 ± 1 mmHg and 354 ± 7 beats/min (n = 58). GABA 3-300 ng microinjected into depressor sites of NTS produced dose-dependent increases in mean BP and HR (n = 4-7). The peaks were seen within a few s and recovered to a basal level 2 min after microinjection. These are consistent with previous findings (8). On the other hand, L-DOPA 10, 30 and 60 ng microinjected into depressor sites of NTS caused dose-dependent decreases in mean BP and HR (n = 7-18). The cardiodepressor responses are consistent with a previous finding and are postsynaptic in nature, since these are also seen in rats with intraventricular 6-hydroxydopamine (3). L-DOPA at a moderate dose of 30 ng produced repetitive and constant hypotension (22 ± 2 mmHg) and bradycardia (10 ± 2 beats/min) (n = 18) and these responses were not modified by prior microinjection of saline vehicle.

Effects of $GABA_A$ receptor-related drugs on depressor and bradycardic responses to L-DOPA

Depressor and bradycardic responses to 30 ng L-DOPA were inhibited dose-dependently by the prior microinjections of 3, 10 and 30 ng GABA, when pressor responses returned to the basal level.

In addition, nipecotic acid 100 ng itself produced peak hypertension (10 ± 2 mmHg) and tachycardia (8 ± 2 beats/min) within 1 min, which returned to the basal level within 2 min after injection. This is consistent with a previous finding (9). Depressor and bradycardic responses (24 ± 3 mmHg and 10 ± 3 beats/min) to L-DOPA 30 ng were inhibited by nipecotic acid 100 ng (12 ± 3 mmHg and 3 ± 1 beats/min, $p < 0.05$, n = 5, respectively), when the pressor responses returned to the basal level.

Furthermore, bicuculline 10 ng itself produced decreases in mean BP (21 ± 4 mmHg) and HR (18 ± 4 beats/min), which reached a peak a few s and returned to the basal level 3-4 min after microinjection. This finding is consistent with a previous finding (8). Depressor and bradycardic responses (21 ± 2 mmHg and 6 ± 2 beats/min) to L-DOPA 30 ng were potentiated by prior microinjection of bicuculline 10 ng (29 ± 4 mmHg and 13 ± 3 beats/min, $p < 0.05$, n = 6, respectively), when depressor responses returned to the basal level.

These findings indicate that endogenously released GABA in depressor sites of NTS tonically functions to produce hypertension and tachycardia via activation of $GABA_A$ receptors. This idea is consistent with previous findings (8, 9). Further, the present findings suggest that endogenously released GABA in NTS functions tonically to inhibit postsynaptic depressor responses to exogenously microinjected L-DOPA.

The highest GABA dose used at 300 ng produced increases in mean BP (18 ± 3 mmHg) and HR (17 ± 3 beats/min) (n = 4). When the tonic L-DOPA function is abolished by L-DOPA methyl ester 1 μg, a competitive L-DOPA antagonist (6), unilaterally microinjected 1 min previously, pressor and tachycardiac responses to GABA 300 ng was reduced by a half (8 ± 1 mmHg and 9 ± 3 beats/min, $p < 0.05$, compared to GABA alone, n = 4, respectively). Unilateral microinjection of this antagonist at 1 μg completely antagonizes postsynaptic depressor responses to L-DOPA 100 ng (3), antagonizes depressor and bradycardic responses to electrical aortic nerve stimulation and reduces the reflex bradycardia elicited by phenylephrine-induced baroreceptor stimulation (4). Furthermore, endogenously released L-DOPA tonically functions to activate depressor sites in NTS, since microinjections of L-DOPA methyl ester in the bilateral NTS (2 μg x 2) produce hypertension and tachycardia, which is abolished by prior i.p. injection of α-methyl-*p*-tyrosine (100 mg/kg), a TH inhibitor, with a marked decrease in basal L-DOPA release (4), whereas GABA tonically inhibits depressor and bradycardic responses to aortic nerve stimulation (8). Taking into consideration these findings (3, 4, 8), this antagonist-induced inhibition of pressor responses to GABA shows a probability that some part of GABA-induced hypertension may occur under the intact condition of tonic function of endogenously released L-DOPA in NTS.

GABA tonically functions to inhibit L-DOPA release during microdialysis of NTS

Basal L-DOPA release

Using microdialysis of NTS in anesthetized rats and a high performance liquid chromatography (HPLC) with an electrochemical detector, the basal release of endogenous L-DOPA was constantly detectable in 40 min dialysate samples and reached a stable level 160 min after the start of perfusion with Ringer solution. The mean absolute values of three successive samples before drug perfusion was 12.7 ± 2.6 fmol (n = 24). The basal L-DOPA release is in part TTX-sensitive and Ca^{2+}-dependent (4). L-DOPA is released at least partially via neuronal activity in NTS.

Effects of $GABA_A$ receptor-related drugs on L-DOPA release

Muscimol 10 and 30 μM perfused via probes gradually decreased basal L-DOPA release and the decrease reached a peak 2 h after the perfusion. At the peak, muscimol 3 μM produced no effect (96 ± 9%, n = 4), compared to control (93 ± 16%, n = 4), when the mean absolute value was taken as 100%. The peak decrease was 48 ± 10% at muscimol 10 μM ($P < 0.05$, compared to control, n = 4) and 42 ± 15% at 30 μM ($P < 0.05$, n = 3).

In addition, bicuculline 30 μM gradually increased basal L-DOPA release and the increase reached a peak 2 h after perfusion. The peak increase was 137 ± 12% ($P < 0.05$, n= 4).

These findings suggest that endogenously released GABA tonically functions to inhibit L-DOPA release via activation of $GABA_A$ receptors at least partially from some neurons in NTS, since the basal release of L-DOPA is in part TTX-sensitive and Ca^{2+}-dependent (4, 5).

The present findings support the idea that a $GABA_A$ system tonically functions to inhibit presynaptic and postsynaptic components of the L-DOPA system within NTS. This idea further support our working hypothesis that L-DOPA is a neurotransmitter of baroreceptor afferents terminating in NTS (1, 2). GABA-induced hypertension at a level of NTS seems to be in part due to inhibition of these presynaptic and postsynaptic components of the L-DOPA system.

L-DOPA itself increases basal GABA release during microdialysis of NTS

Basal GABA release

Using microdialysis and HPLC with a fluorescence spectrophotometer, endogenous GABA was constantly detectable in dialysate samples for 20 min successively collected. The basal release of GABA reached a stable level 140 min after the start of perfusion with Ringer solution. The absolute value of GABA release before drug perfusion was 266 ± 25 fmol (n = 6) in the absence and 248 ± 36 fmol (n =24) in the presence of NSD-1015 200 μM, a central DOPA decarboxylase inhibitor, respectively. No difference was seen between the two groups.

Effects of L-DOPA-related drugs on GABA release

In the absence of NSD-1015, perfusion of L-DOPA 10 nM for 20 min increased GABA release at a peak by 37 ± 4 % from a basal level (n = 3), when the absolute value before L-DOPA perfusion was taken as 100 %. This release reached the peak 40 min after the start of L-DOPA perfusion and returned to the basal level 20 min after reperfusion with normal Ringer solution. In the presence of NSD-1015 200 μM, L-DOPA 1, 10 and 100 nM concentration-dependently released GABA by 34 ± 14% (n = 4), 43 ± 10% (n = 7) and 60 ± 18% (n = 5). Increase in GABA release by L-DOPA 10 nM was not affected by NSD-1015, suggesting that increase in GABA release is due to an action of L-DOPA itself but not due to conversion to dopamine.

This idea was further supported by the finding that dopamine 10 nM produced no GABA release (-1 ± 2%, n = 5, $P < 0.05$, compared to L-DOPA 10 nM).

In addition, the action of L-DOPA is stereoselective in nature in common with many types of receptors, since D-DOPA 10 nM produced no GABA release (-3 ± 5%, n = 3). It should be explored whether or not L-DOPA acts on the recognition site to release GABA. However, L-DOPA methyl ester can not be used as a tool, since this ester is readily hydrolyzed to L-DOPA in this type of experiments (2). This antagonist is competitively effective against L-DOPA when the ester is continuously supplied by superfusion in brain slices (2, 6) and is also effective within a few min when the ester is microinjected into NTS (4), the rostral and caudal ventrolateral medulla (2).

Effects of Ca^{2+} deprivation and TTX perfusion on L-DOPA-evoked release of GABA

In the presence of NSD-1015, L-DOPA(10 nM)-induced increase in GABA release was almost completely inhibited by Ca^{2+} omission plus 12.5 mM Mg^{2+} addition (6 ± 9%, n = 5, $p < 0.05$, compared to L-DOPA 10 nM alone) or 1 μM TTX perfusion (6 ± 7%, n = 5, $P < 0.05$). L-DOPA may release GABA from GABAergic cell bodies and/or axons within NTS. L-DOPA-induced GABA release seems to be reflection of a mutual compensatory mechanism between "L-DOPAergic" and GABAergic systems within NTS of the baroreceptor reflex pathway. Tonic transmitter-like GABA release, however, was not evident, since basal GABA release in a sample 140 min after the start of perfusion was not modified by Ca^{2+} deprivation (335 ± 66 fmol , n = 5) and by TTX perfusion (263 ± 36 fmol, n = 5) simultaneously performed at the start of perfusion.

These findings further provide novel evidence for a neurotransmitter and/or neuromodulator role of L-DOPA itself.

Conclusion

GABA seems to act tonically on $GABA_A$ receptors to inhibit presynaptic L-DOPA release and postsynaptic depressor responses to L-DOPA in rat NTS. This inhibition seems to be in part involved in GABA-induced hypertension in NTS. L-DOPA by itself releases transmitter-like endogenous GABA within NTS. These findings further support our hypothesis that L-DOPA is a neurotransmitter of the primary baroreceptor afferents terminating in rat NTS.

References

1. Misu Y, Goshima Y. Is L-dopa an endogenous neurotransmitter? Trends Pharmacol Sci 1993;14: 119-123.

2. Misu Y, Ueda H, Goshima Y. Transmitter-like actions of L-DOPA. Adv Pharmacol 1995; 32: 427-459.

3. Kubo T, Yue J-L, Goshima Y, Nakamura S, Misu Y. Evidence for L-DOPA systems responsible for cardiovascular control in the nucleus tractus solitarii of the rat. Neurosci Lett 1992; 140: 153-156.

4. Yue J-L, Okamura H, Goshima Y, Nakamura S, Geffard M, Misu Y. Baroreceptor-aortic nerve-mediated release of endogenous L-3,4-dihydroxyphenylalanine and its tonic depressor function in the nucleus tractus solitarii of rats. Neuroscience 1994; 62: 145-161.

5. Yue J-L, Okumura Y, Miyamae T, Ueda H, Misu Y. Altered tonic L-3,4 dihydroxyphenylalanine systems in the nucleus tractus solitarii and the rostral ventrolateral medulla of spontaneously hypertensive rats. Neuroscience 1995; 67: 95-106.

6. Goshima Y, Nakamura S, Misu Y. L-Dihydroxyphenylalanine methyl ester is a potent competitive antagonist of the L-dihydroxyphenylalanine-induced facilitation of the evoked release of endogenous norepinephrine from rat hypothalamic slices. J Pharmacol Exp Ther 1991; 258: 466-471.

7. Tison F, Mons N, Rouet-Kamara S, Geffard M, Henry P. Endogenous L-DOPA in the rat dorsal vagal complex: an immunocytochemical study by light and electron microscopy. Brain Res 1989; 497: 260-270.

8. Kubo T, Kihara M. Evidence for γ-aminobutyric acid receptor-mediated modulation of the aortic baroreceptor reflex in the nucleus tractus solitarii of the rat. Neurosci Lett 1988; 89: 156-160.

9. Catelli JM, Giakas WJ, Sved AF. GABAergic mechanisms in nucleus tractus solitarius alter blood pressure and vasopressin release. Brain Res 1987; 403: 279-289.

GABA: Receptors, Transporters and Metabolism
ed. by C. Tanaka & N.G. Bowery

POTENTIATION BY PROTEIN KINASE C ACTIVATION OF GABA RELEASE FROM *XENOPUS* OOCYTES INJECTED WITH RAT BRAIN mRNA

Y. Kataoka[1], S. Kan[2], S. Mameya[2], H. Shibaguchi[2], K. Yamashita[2], M. Niwa[3] and K. Taniyama[2,4]

[1]Department of Hospital Pharmacy, Faculty of Medicine Kyushu University, Fukuoka 812 and Department of Pharmacology [3]I and [2]II, Nagasaki University School of Medicine, Nagasaki 852, Japan, [4]Correspondence

Summary: The *Xenopus laevis* oocytes injected with rat brain mRNA preloaded with [^{3}H]GABA showed spontaneous release of [^{3}H]GABA, approximately 0.5 % of GABA content in the oocytes per 1 min. Stimulation with Ca^{2+} ionophore (A23187) produced increase in the release of [^{3}H]GABA from the mRNA-injected oocytes. 12-*O*-tetradecanoylphorbol 13-acetate (TPA) (10 nM - 300 nM), but not 4α-phorbol-12,13-didecanoate (300 nM) potentiated the A23187-evoked release of [^{3}H]GABA from the mRNA-injected oocytes, in a concentration dependent manner, as in the case of rat cerebellum, thereby indicating that the functional GABAergic nerve terminals were expressed in the oocytes by injecting mRNA extracted from the rat brain. The A 23187-evoked release of [^{3}H]GABA was reduced by treatment with anti-synaptophysin. Thus, synaptophysin may play a role in the exocytotic release of GABA.

The release of various kinds of neurotransmitters from nerve tertminals is apparently triggered by Ca^{2+} entering into presysnaptic terminals, and synaptic vesicles are fused to plasma membrane sites, the active zone, although acetylcholine (ACh) is released from cholinergic nerve terminals in vesicular and non-vesicular manners (1). The release of ACh has been examined in the electric lobe of *Torpedo marmorata* (2) and the giant synapse of the squid *Loligo pealei* (3, 4). Since nerve terminals in mammalian tissues are too minute in size to study mechanisms of neurotransmitter release, cells such as COS and chromaffin cells, *Drosophila*, and yeast serve as substitutes for nerve terminals (5-8). *Xenopus laevis* oocytes injected with mRNA or cDNA are a pertinent tissue for studying the function of receptors, ion channels and transporters (9, 10). The machinery of nerve terminals of cholinergic (11, 12) and glutamatergic (13) neurons has been expressed in *Xenopus* oocytes by injecting mRNA from the electric lobes of *Torpedo* and rat cerebellum, respectively. This review is concerned with properties of GABA release apparatus expressed in the *Xenopus* oocytes by injecting rat brain mRNA and possible involvement of synaptophysin, a fusion protein in GABA release from the oocyte.

Release of [^{3}H]GABA from *Xenopus laevis* oocytes injected with rat brain mRNA

Oocytes preloaded with 0.1 μM [^{3}H]GABA were incubated in modified Barth's solution containing 10 μM aminooxyacetic acid and 1 mM L-2,4-diaminobutyric acid, and then [^{3}H]GABA release from each oocyte was measured by the batch method. Fractional rate (amount of released [^{3}H]GABA per [^{3}H]GABA content in oocyte) of spontaneous [^{3}H]GABA release from oocytes injected with rat brain mRNA was approximately 0.005/min. This value was similar to that of brain (14). When the mRNA-injected oocytes were stimulated with combination of 10 μM A23187 and 2 mM $CaCl_2$ at 40 min intervals, the first release of [^{3}H]GABA was approximately 1.5 times of the spontaneous release seen immediately before stimulation, and the second stimulated release of [^{3}H]GABA was approximately 1.6 times that of the spontaneous release. Water-injected oocytes did not respond to stimulation with combination of 10 μM A23187 and 2 mM $CaCl_2$. Stimulation with A231897 alone did not alter the release of [^{3}H]GABA, thereby indicating that the release of [^{3}H]GABA from the mRNA-injected oocytes was external Ca^{2+}-dependent, as in the case of native synapses. Thus, the apparatus of GABA uptake, possibly transporter, and release were expressed in the mRNA-injected oocytes.

Potentiation of stimulation-evoked release of GABA by protein kinase C activation

Spontaneous endogenous GABA release from the slices of deep cerebellar nucleus was 2.22 ± 0.11 pmol/min/mg tissue (N=6) 60 min after superfusion with Ca^{2+}-free medium. Stimulation with combination of 0.1 μM A23187 and 2 mM $CaCl_2$, but not the application of A231897 alone, increased the release of endogenous GABA from the slices by approximately 1.5 times of spontaneous release. Pretreatment with 12-*O*-tetradecanoylphorbol 13-acetate (TPA) at the concentrations of 10 nM to 300 nM for 30 min potentiated the release of endogenous GABA stimulated with combination of A23187 (0.1 μM) and $CaCl_2$ (2 mM), in a concentration dependent manner, while pretreatment with 4α-phorbol-12,13-didecanoate (4α-PDD) (300 nM) did not significantly alter the stimulated release (Fig. 1). Thus, the released GABA from the deep cerebellar nucleus may originate from vesicles within the GABAergic nerve terminals. Same effects of TPA and 4α-PDD have been observed on the release of GABA evoked by high K^+, but not by ouabain (15). In the rat deep cerebellar nucleus, there are neuronal cell bodies surrounded exclusively by protein kinase C immunoreactive GABAergic Purkinje cell axon terminals, which possess densely immunoreactive synaptic vesicles (16).

The release of [^{3}H]GABA stimulated with A23187 (10 μM) and $CaCl_2$ (2 mM) from mRNA-injected oocytes was also potentiated by pretreatment with TPA, but not by 4α-PDD (Fig. 1), as observed in the slices of rat deep cerebellar nucleus. These results suggest that the vesicular apparatus was reconstituted in the oocytes by injecting rat brain mRNA.

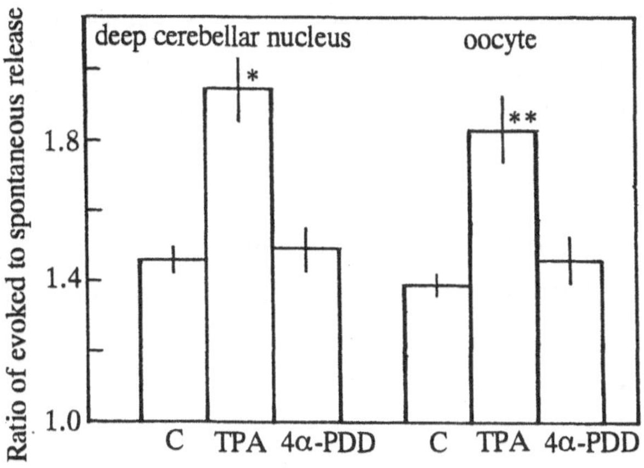

Figure 1. Potentiation of stimulation-evoked release of GABA by protein kinase C activation. TPA (300 nM) and 4α-PDD (300 nM) were applied 30 min before and during stimulation with combination of A23187 (0.1 μM) and $CaCl_2$ (2 mM) in the slices of deep cerebellar nucleus (N=7) and with combination of A23187 (10 μM) and $CaCl_2$ (2 mM) in the mRNA-injected oocytes (N=6). * and ** Significantly different from respective control (C) value ($P<0.05$).

Possible involvement of synaptophysin in vesicular release of GABA

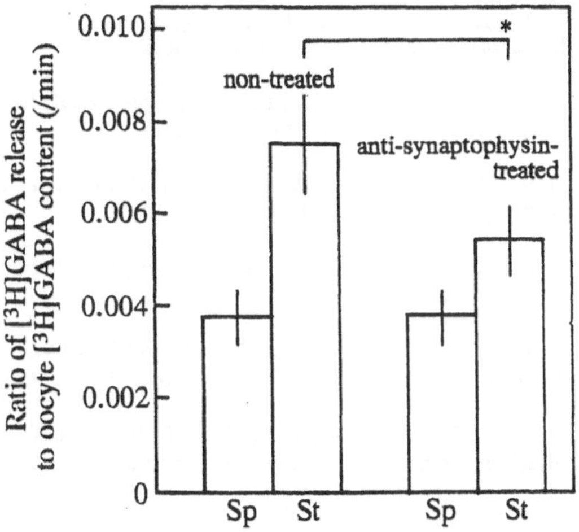

Figure 2. Release of [^{3}H]GABA stimulated with combination of A23187 (10 μM) and Ca^{2+} (2 mM) from mRNA-injected oocytes treated with antibody to synaptophysin. Sp represents spontaneous release immediately before the stimulation (St). Each column represents the mean ± SEM value from 12 oocytes (non-treated) and 9 oocytes (anti-synaptophysin-treated). * Significantly different from value in the non-treated oocytes ($P<0.05$).

There are specific proteins associated with exocytosis, including synapsins, synaptotagmin, synaptophysin, synaptobrevin (VAMP), SNAP-25 and rabphillin, within the nerve terminals (17-19). Involvement of synaptophysin in the vesicular release of GABA was examined in the mRNA-injected oocytes. The antibody to synaptophysin was injected together with rat brain mRNA, and incubated at 19°C in modified Barth's solution, for 2-4 days. The release of [^{3}H]GABA stimulated with combination of A23187 (10 μM) and Ca^{2+} (2 mM) from the oocytes treated with antibody to synaptophysin reduced to approximately 50 % of release from the non-treated oocytes (Fig. 2). Thus, synaptophysin may involve in the vesicular release of GABA. Similar finding has been shown in the glutamate release from oocytes injected with rat cerebellar mRNA (13). *Xenopus* oocytes is useful to investigate molecular events of the neurotransmitter secretion, such as the process between vesicle docking and fusion, proteins participating in the exocytotic release of neurotransmitters.

Acknowledgment

This work was supported by grants from the Ministry of Education, Science and Culture, Japan.

References

1. Cooper JR, Meyer EM. Possible mechanisms involved in the release and modulation of release of neuroactive agents. Neurochem Int 1984; 6: 419-433.

2. Dunant Y, Israël M. The release of acetylcholine. Scientific America 1985; 252: 40-48.

3. Bommert K, Charlton MP, DeBello WM, Chin GJ, Betz H, Augustine GJ. Inhibition of neurotransmitter release by C2-domain peptide implicates synaptotagmin in exocytosis. Nature 1993; 363: 163-165.

4. Hunt JM, Bommert K, Charlton MP, Kistner A, Habermann E, Augustine GJ, Betz H. A post-docking role for synaptobrevin in synaptic vesicle fusion. Neuron 1994; 12: 1269- 1279.

5. Bennett MK, García-Arrarás JE, Elferink LA, Peterson K, Fleming AM, Hazuka CD, Scheller RH. The syntaxin family of vesicular transport receptors. Cell 1993; 74: 863-873.

6. DiAntonio A, Schwarz TL. The effect on synaptic physiology of synaptotagmin mutations in *Drosophila*.. Neuron 1994; 12: 909-920.

7. Protopopov V, Govindan B, Novic P, Gerst J E. Homologs of the synaptobrevin/VAMP family of synaptic vesicle proteins function on the late secretory pathway in *S. cerevisiae*. Cell 1993; 74: 855-861.

8. Shoji-Kasai Y, Yoshida A, Sato K, Hoshino T, Ogura A, Kondo S, Fujimoto Y, Kuwahara R, Kato R, Takahashi M. Neurotransmitter release from synaptotagmin-deficient clonal variants of PC12 cells. Science 1992; 256: 1820-1823.

9. Snutch TP. The use of *Xenopus* ocytes to probe synaptic communication. Trends Neurosci 1988; 11: 250-256.

10. Zhang JF., Randall AD, Ellinor PT, Horne WA, Sather WA, Tanabe T, Schwarz TL, Tsien RW. Distinctive pharmacology and kinetics of cloned neuronal Ca^{2+} channels and their possible counterparts in mammalian CNS neurons. Neuropharmacology 1993; 32: 1075-1088.

11. Cavalli A, Eder-Colli L, Dunant Y, Loctin F, Morel N. Release of acetylcholine by *Xenopus* oocytes injected with mRNAs from cholinergic neurons. EMBO J 1991; 10: 1671-1675.

12. Cavalli A, Dunant Y, Leroy C, Meunier F-M, Morel N, Israël M. Antisense probes against mediatophore block transmitter release in oocytes primed with neuronal mRNAs. Eur J Neurosci 1993; 5: 1539-1544.

13. Alder J, Lu B, Valtorta F, Greengard P, Poo M-M. Calcium-dependent transmitter secretion reconstituted in *Xenopus* oocytes: requirement for synaptophysin. Science 1992; 257: 657-661.

14. Valdes F, Orrego F. Electrically induced, calcium-dependent release of endogenous GABA from rat brain cortex slices. Brain Res 1978; 141: 357-363.

15. Shuntoh H, Taniyama K, Tanaka C. Involvement of protein kinase C in the Ca^{2+}-dependent vesicular release of GABA from central and enteric neurons of the guinea pig. Brain Res 1989; 483: 384-388.

16. Kose A, Saito N, Ito H, Kikkawa U, Nishizuka Y, Tanaka C. Electron microscopic localization of type I protein kinase C in rat Purkinje cells. J Neurosci 1988; 8: 4262-4268.

17. Greengard P, Valtorta F, Czernik AJ, Benfenati F. Synaptic vesicle phosphoproteins and regulation of synaptic function. Science 1993; 259: 780-785.

18. Söllner T, Rothman JE. Neurotransmission: harnessing fusion machinery at the synapse. Trends Neurosci 1994; 17: 344-348.

19. Trifaró J-M, Vitale ML. Cytoskeletion dynamics during neurotransmitter release. Trends Neurosci 1993; 16: 466-472.

GABA: Receptors, Transporters and Metabolism
ed. by C. Tanaka & N.G. Bowery

THE MOLECULAR BIOLOGY OF $GABA_A$ RECEPTORS AND ITS APPLICATIONS

Eric A. Barnard

Molecular Neurobiology Unit, Royal Free Hospital School of Medicine, London NW3 2PF, UK.

The repertoire of subunit types

In the initial cDNA cloning of $GABA_A$ receptor subunits (1,2) it was apparent that different families of subunits contribute to these receptors: the α1, α2 and α3 subunits discovered were much more similar to each other than to a β subunit and hence these were designated as in two families, α and β (2). The range has been extended in studies, mainly by Peter Seeburg and co-workers and in some cases by others (reviewed in refs. 3,4) to a present total of 15 related mammalian subunits (plus an additional avian one), each from a different gene. These polypeptides are all of ~ 50,000 daltons in size, and each carries 4 presumed transmembrane hydrophobic segments (TM1-4). Figure 1 illustrates the 5 different families into which these fall structurally and their relationships. While three β subunits have been found in the mammals so far studied, a "β4" subunit cDNA was cloned from the chicken (5), this being about equidistant in sequence to the mammalian β1, β2 and β3. However, while β2 and β3 homologues can be found from chicken brain cDNA (6,7), neither a chicken β1 nor a mammalian β4 have been obtainable by cross-hybridisation and it is provisionally assumed that β1 in the bird has undergone more variation to become "β 4".

In mammals, three γ subunits have also been identified. In the chicken γ1 (8) and γ2 (9) subunit cDNAs have been cloned, these being ~ 90% identical in protein sequence to rat γ1 and γ2 respectively. However, a further chicken γ subunit found (10) has been assigned as γ4 by its lack of close correspondence to the mammalian γ3 subunit (69% amino-acid identity, in contrast to ≥ 90% for other chicken-mammal pairs). It might yet be that this is the chicken γ3 subunit and is unusually divergent, but the alternative that a homologue of the chicken γ4 exists in mammals has not been excluded. The regional distribution of γ4 mRNA in the chicken brain does not

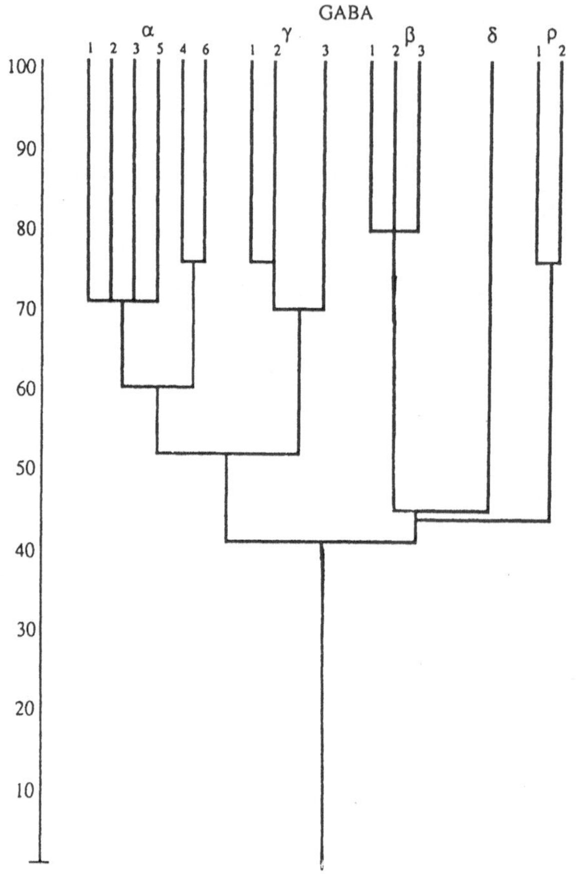

Figure 1. A dendrogram showing the relationships of the isoforms of the α, β, etc. subunit types of the $GABA_A$ receptor. The vertical scale represents the percentage of amino-acid identity of the subunits bracketed by the horizontal lines. Alignments are compared by a computer program which uses all sequence homology features. All sequences are from the rat except ρ (human).

correspond to that of γ3 in the mammal (10) and a comparison of γ3 and γ4 Pharmacological properties which should be diagnostic, is not yet available.

This heterogeneity is increased by alternative exon splicing of the pre-mRNA, which generates from one gene two differently-distributed forms of the γ2 subunit (9,11-13) and likewise for the β2 subunit (7), in each case with longer and shorter products designated "L" and "S" and differing by a short peptide in the long intracellular loop between TM3 and TM4. Therefore, in assessing possible combinations of subunit types to form a $GABA_A$ receptor, we must consider in a given species 6 α forms, 4 β forms (including β2L and β2S), 5 γ forms (including γ2L and γ2S) and 1 δ form. In addition there are two known ρ subunits (solely or primarily in the retina), but these (so far as is known) form homomeric receptors and do not participate in combinations with the aforementioned types (14). Hence, a pool of 16 subunit types (at least) in forming $GABA_A$ receptors by multiple selection is made from a combination, plus 2 ρ subunits used separately.

Establishment of the Subunit Number per Receptor Molecule

To investigate the construction of $GABA_A$ receptor subtypes from this repertoire of subunits, it is necessary first to establish the total number of subunits in each receptor molecule, then to know whether this number is constant for all the native compositions, and finally to know the stoichiometry of the subunit types within that number. Regarding the number of subunits per receptor, the suggestion has often been made that this will be the same (five subunits) as for another transmitter-gated ion channel where the composition has been unequivocally established: the $GABA_A$ receptor subunits share a low but definite (~ 25%) amino-acid sequence homology with the subunits of the nicotinic acetylcholine receptors, both being in the same superfamily of the transmitter-gated ion channels (1,15). That receptor occurs in *Torpedo* electric organ at such a high density in large post-synaptic membrane sheets that it is possible to prepare membranes containing a surface lattice of the receptors, from which a low-resolution 3-D structure of the molecule could be obtained by electron optical diffraction techniques by N. Unwin and colleagues (16). Those studies showed the nicotinic receptor to be pentameric, with the ion channel located in the centre of five homologous subunits (having the stoichiometry $\alpha_2 \beta\gamma\delta$). Likewise for the neuronal type of the nicotonic receptor, indirect methods (possible because only two subunit classes, α and β, occur therein) applied to its recombinant subunits when co-expressed have indicated the composition $\alpha_2 \beta_3$, again in a pentamer

(17,18). For the $GABA_A$ receptors, the situation is necessarily more complex, since the unique situation in the T*orpedo* post-synaptic membranes does not recur and since there are several classes of subunits involved in highly variable ways. It is, therefore, preferable to use the natural $GABA_A$ receptor population from the brain rather than a selected expressed recombinant composition which may or may not be representative of that population, and to make direct analyses thereon, since these will not be limited by an assumption of the subunit classes to be taken as co-assembling.

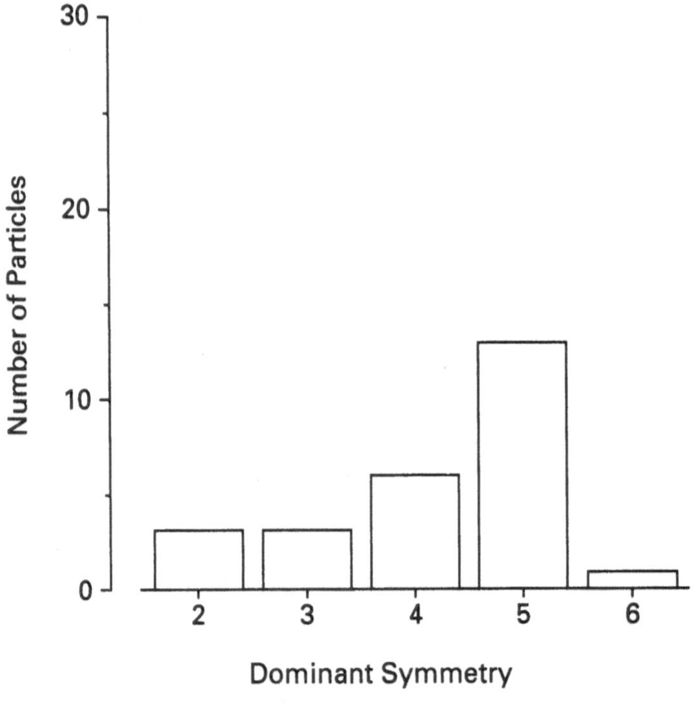

Figure 2. Analysis of the rotational symmetry of a population of pure $GABA_A$ receptor molecules. The electron microscopic images were subjected to harmonic analysis and to rotational averaging. [For details see Nayeem et al. (19)]. The dominant symmetry for each particle was used in plotting the histogram. Particles with only one-fold apparent symmetry, which is trivial, were rejected as being tilted or distorted. The small group seen here with two-fold symmetry was also shown to be similarly artefactal. The distribution seen around the peak at five-fold symmetry is consistent with 100% being pentameric, taking into account the random error of the method.

Accordingly, we have purified the $GABA_A$ receptors from pig brain cortex (with full retention of its allosteric binding properties) (19). The dispersed molecules of the receptors were analysed, using their electron microscope images, by harmonic analysis (Fig.2). This method yields a power spectrum for each particle with a peak at its dominant symmetry. Fig.2 illustrates that this symmetry is five-fold, over the population of particles analysed. Further, the negatively-stained image which is seen in all the receptor particles due to entry of uranyl acetate into the central pore of a rosette corresponds very well to the image observed similarly with negatively-stained *Torpedo* receptor particles (16). As in the latter, the five subunits in each GABA receptor (19) enclose a central pore in the membrane and an adjacent vestibule. The particles isolated from the brain will comprise a variety of $GABA_A$ receptor subtypes. We can only say that at least the majority are pentameric, since a deviating small minority would not be detected in the experimental noise. However, in view of the concurrence noted with other receptors in the same superfamily, it is presumed that the observation made holds for all of the $GABA_A$ receptors.

The Potential Subtype Numbers

A very wide range of ligands acts in a modulatory capacity at the benzodiazepine (BZ) site, involving many different heterocyclic types. (Hence, it seems more logical to broaden its designation, e.g. as a "BZ/ω" site, as has been proposed (20)). It has been found that these BZ/ω ligands especially the non-benzodiazepine structures, provide much better discrimination between multiple $GABA_A$ receptor subtypes than any other pharmacology at present available. We can, therefore, start to define subtypes on the basis of this site, although this approach should later be extended when suitable series of ligands for other sites on the receptor become available.

Most native $GABA_A$ receptors are potentiated by BZ/ω-site positive agonists but some have been detected which are not (21-23). Studies of recombinant co-expression have shown that the joint presence of α and γ subunits is required for the BZ/ω modulatory effects (24-27). For the full receptor activity, however, including native-receptor-like Hill coefficients, α,β and γ2 subunits are required (28-31). Moreover, the γ1 subunit can produce decreased or atypical BZ/ω sensitivities, with inverse agonists becoming partial positive agonists (26,31). Again, if α6 is combined with β and γ2 subunits, the normal BZ/ω site is inactive (although a different form of this site, strongly binding Ro15-4513 and some other inverse agonists, is active there) (23,32,33). The same holds true for α4 (34). The diazepam-insensitivity of some native $GABA_A$ receptors in the cerebellar granule cells is attributable, therefore, to α

6-containing receptors. In the thalamus (the major site of α4 expression (3) it can be attributed to α4-containing receptors and possibly likewise at some other sites where α4 mRNA is found at low levels. In fact, in rat cerebellar granule cells evidence from various immuno- cytochemical techniques at high resolution suggests the presence of more than one receptor type therein, including α6 β δ-containing and α6 β γ2-containing receptors (35,36); if γ subunits are completely replaced by δ, BZ potentiation is lost (when tested in co-expression with α1 and β subunits) (27). It has been reported (36,37) that some of the native δ-containing receptors, e.g. from the rat cerebellum, indeed do not bind the typical BZ/ω site-ligands with high affinity.

Another possible cause of BZ-insensitivity could be the existence in restricted brain locations of receptors containing only α and β subunits. The immuno-coprecipitation results reported to date on rat brain extracts do not exclude this (37).

The Subunit Sub-types In One Receptor

The majority of $GABA_A$ receptors contain, on the evidence reviewed above, α, β and γ subunits, while the total of the subunits per molecule is five (Fig.2). Hence the receptors in this set can obviously have one of 3 general compositions: $(\alpha)_2\,(\beta)_2\,\gamma$ *or* $(\alpha)_2\,\beta\,(\gamma)_2$ *or* $\alpha\,(\beta)_2\,(\gamma)_2$. (Here, parentheses are used to indicate that the subscript numeral then shown represents counting of the isoforms present in one molecule and not the isoform identity). Such additional cases as $(\alpha)_3\,\beta\,\gamma$ *or* $\alpha\,\beta\,(\gamma)_3$ would have been theoretically possible, but measurements of an electrophysiological property determined quantitatively by the number of tagged recombinant subunits of each type forming the channel (38), in the case of co-expression of the α3 β2 and γ2 subunits, have excluded the presence of 3 of any of those types in one receptor molecule. The next logical step, therefore, in enumerating the potential combinations of subunits is to ask whether two isoforms of α *or* of β *or* of γ can occur in one receptor molecule, e.g. to produce compositions of the type $(\alpha 1\ \alpha 2)\,(\beta)_2\,\gamma$. In the case of the α subunits, there is a variety of evidence for such co-occurence of isoforms, in a minority of $GABA_A$ receptors. This has come firstly from co-precipitation of a second α isoform when a brain-derived population of $GABA_A$ receptors is treated with an antibody specific for a first α isoform. Receptors containing at least the pairs α1α2, α1α3, α1 α5, α2α3 and α3α5 have been detected thus (each in a minority of the population containing the respective individual isoforms) (39-43). A second method is the use of isoform-specific antibodies, applied in light or electron microscopy. Thus, by confocal laser microscopy with double or triple immunofluorescent staining, Fritschy et al. (44) found that on the membranes of certain neurones α1 α3 γ2 were co-localised (as well as α1γ2 and α3γ2 separately on other neurones). Thirdly, the co-occurrence of isoform mRNAs in one cell nucleus can be sought by *in situ* hybridisation, but this method, due to the resolution limit, requires either the use of very large neuronal cell bodies or cases where a majority of the neurones seen carry both labels. It was found thus, in the large cell bodies of motor neurones in the rat spinal cord, that a minority contain α2 α3 β3 γ2 receptor mRNAs (45). For the γ subunits, the similar use of isoform-specific antibodies has, on brain extracts or purified receptor preparations, shown evidence for the frequent co-occurrence of γ2 with γ3 and of γ2L with γ2S (37,46). Further, the δ subunit can frequently co-precipitate with a γ2 subunit and then enhances the diazepam binding (42), and this agrees with γ2 and δ being co-localised by immuno-staining at certain locations (44,47). Also, *In vitro* co-expression has shown (48) that the combination α1β1γ2δ can assemble and has distinctive channel characteristics and enhanced (relative to α1β1γ2) diazepam potentiation. Hence the combination of γ2 with δ must be included in enumerating probable receptor compositions. In some other cases, however, δ did not occur with a γ subunit (36). γ1 did not occur (37) with another γ nor with δ.

The Probable Subunit Stoichiometry

Since two isoforms of the α subunit can sometimes occur in one receptor, as reviewed above, the receptors in general are considered as having two α places in the pentamer. Likewise, since two γ isoforms can co-occur, it is presumed that there are also two γ places in the pentamer . Since δ has been observed (see above) to replace either one γ or two γ subunits but never α or β subunits, it is further deduced that γ or δ subunits can occupy the γ places, in different receptors. That is, the most likely stoichiometry is:

$(\alpha)_2\ \beta\ (\gamma)_2$, with 1 or 2 γ subunits being in some cases replaceable by δ.

Two isoforms of β have not yet been found in any receptor. Benke et al. (49) compared by immunoprecipitation brain fractions containing β1,β2 or β3 receptors and found that no combination of these in any one receptor population could be detected. Since γ1 (which is rare) seems always to occur without another γ or δ (37), it cannot be excluded that also $(\alpha)_2\ (\beta)_2\ \gamma 1$ occurs instead of $(\alpha)_2\ \beta\ (\gamma 1)_2$. For the γ 2-containing receptors, however, Backus et al. (38) have shown, using marker mutants and their electrophysiological effects in the recombinant α3β2γ2 receptor, that the $(\alpha)_2\beta(\gamma)_2$ composition best fitted the properties found. The $(\alpha)_2\ \beta\ (\gamma)_2$ stoichiometry is further supported by the evidence that has been noted above for the major role of both α and γ subunits in the interaction at the BZ/ω site. This has suggested that this site is at the interface between an α and a γ subunit (25). This again implies equal numbers of α and γ subunits in the complex (at least for receptors containing γ2, i.e. the great majority).

The Theoretical Maximum Number of Receptor Subtypes

We start with the assumed stoichiometry of $(\alpha)_2\ \beta\ (\gamma/\delta)_2$, deduced above to be probable for at least the great majority of the $GABA_A$ receptors . In expression of the recombinants in cultured cells or the oocyte, it has been found that any ternary combination (e.g. $\alpha_i + \beta_j + \gamma_k$) so far tested can yield a functional receptor in the membrane. Hence the limit to the number of ternary subtypes *in vivo* appears not to be set by barriers to the co-assembly of some, but by the program for gene expression of different isoforms in a given cell. However, when an α+β+γ set is co-expressed, the ternary combination assembles (as far as the subunits are available) to the exclusion of binary combinations (50). There are (see above) in a given species 6 α isoforms and 4 β isoforms (including the alternate-exon variants of β). There are no obligate combinations or exclusions of αβ pairings known from the co-distribution data, at the present resolution limits. However, some exclusions are known (see above) at the γ/δ position. The evidence available so far permits 10 of the possible permutations at that position (drawn from the 5 γ or δ isoforms contributing there). These are : $(\gamma 1)_2$, $(\gamma 2L)_2$, $(\gamma 2S)_2$, γ2L·γ2S, γ2L·δ, γ2S·δ, $(\gamma 3)_2$, γ2L·γ3, γ2S·γ3, $(\delta)_2$. It cannot yet be determined if all of these 10 occur in native receptors. The theoretical number of ternary combinations containing α,β,γ and/or δ subunits is therefore simply derived as follows:

	α_i α_j		β		γ/δ	
Choices:	6 6		4		10	
Combinations:	21	x	4	x	10	= 840

(α_i and α_j denote any of the possible non-identical or identical α isoforms, but without regard for the order; i.e. $\alpha_1\alpha_2$ is equivalent to $\alpha_2\alpha_1$ here).

The enumeration is obtained on the basis that, for a given identity of the subunit set which will form one receptor, there will only be one arrangement and stoichiometry in the native molecule *in situ*. This is found to be so with all other heteromeric proteins containing tightly-bound subunits; for example, there is only one order of the subunits $(\alpha)_2\ \beta\ \gamma\ \delta$ present in the population of *Torpedo* acetylcholine receptors (16) and further, using those subunits we do not find that the same receptor type, in a variety of skeletal muscles, can contain another stoichiometry. That , and the circular order of subunits around the rosette, is fixed by the interactions between the interfaces of different subunits. Therefore we do not count all possible permutations (n=36) of 2 α isoforms present, but only those for a fixed order (n=21). Likewise, when a given set of α, β and γ isoforms is to be used to form a receptor, we do not count alternate stoichiometries with one or with two β subunits in the molecule. However, as noted above, we cannot exclude that γ1 assembles differently from γ2 or γ3 and can accept 2 β subunits, but by the same arguments this would give the addition of one form with $(\beta)_2\ \gamma1$, and the elimination of another with $\beta\ (\gamma1)_2$, so the total will be unaltered. If two different isoforms of β could occur with γ1, then the theoretical maximum would be somewhat increased. However, γ1 mRNA has a relatively limited distribution in the brain (51) and probably exists in very few different subtypes.

The ρ subunits have not been observed to co-assemble with α, β, or γ and have been regarded as forming homo-oligomers (14). There is no evidence as yet that ρ1 and ρ2 co-occur. We therefore must provisionally add only $(\rho_1)_5$ and $(\rho_2)_5$ to the total, to give **842** as the maximum.

In addition, we cannot completely exclude that some of the BZ/ω-insensitive receptors, if in regions where the δ mRNA (relatively rare) is absent, contain only α and β subunit types. This has been suggested as a possibility for some cells in the thalamus where only α1, α4 and β2 mRNAs were detectable (51). If so, that could be due either to the absence of opportunity there to incorporate a γ or δ subunit, or to a unique combination of particular α and β isoforms. That type of possibility could increase the theoretical maximum beyond 842, but the number of types of such binary combinations (if they exist) must be very small, given the ubiquity (51) of the γ1 + γ2+ γ3 + δ mRNAs when taken together.

While the theoretical maximum of receptor subtypes is therefore ~ 850 (but not thousands), the theoretical minimum is 18 (the number of subunit isoforms so far known), since no obligate exclusive combinations have been observed in the localisation studies. That is, there is no isoform that only occurs in one combination with another such. Between the extremes of 18 and 850, we can only conclude at present that a number far above the minimum of 18 actually occurs, given the localisation and co-precipitation studies referenced above. Similarly, the number is likely to be well below 850. Much further analysis by the methods noted above will be needed, especially by high-resolution spatial co-localisation techniques, to specify the actual number and identities of the $GABA_A$ receptor subtypes which occur *in vivo*.

The Significance of the Receptor Multiplicity

The diversity of the $GABA_A$ receptors is interpreted as the means of providing a large number of functionally distinct pathways of fast inhibition in the CNS and a basis for their plasticity and adaptativeness. Each change of one isoform can in principle introduce an altered receptor with different parameters of binding or of channel activity or of response to endogenous modulators. One principal transmitter for fast inhibitory responses is used ubiquitously, GABA, and diversity in the circuitry using these responses is created by the *combinatorial principle* seen above to be creating many different forms of the receptors for GABA. Segregation of these isoforms in neurone types, within one brain structure, which have individual programs for neuropeptide expression has already begun to emerge (52). There are also sometimes requirements for different trafficking of the newly-synthesized receptors within a given neurone, to control separately the subsets of GABA receptors (53) located on dendrites (proximal *or* distal) vs the soma, etc., for which different isoforms can be used.

In addition, the use of 16 genes (plus the use of several alternative exons) permits a variety of gene-expression programs to be used in constructing different inhibitory circuits. Thus subtypes occur (and spliced forms) in $GABA_A$ receptor developmental stages of some brain regions which are different to those in their adult stage (54-56). For example, α6 mRNA, seen in the adult only in cerebellar granule cells, occurs at embryonic days 15-17 in the rat instead in the thalamus and cortex (55). Further, in some brain regions or developmental stages GABA has been reported as trophic or excitatory (57,58) instead of inhibitory, which could require a change of subtype. From such cases and considering also rare specialised neuronal types in the adult brain, it is possible that subunit combinations occur which are not detected with the relatively crude techniques available for immuno-precipitation plus Western blotting or mRNA visualisation. Those techniques will therefore, tend to considerably under-state the true heterogeneity of the $GABA_A$ receptors.

In summary, the multiplicity of the $GABA_A$ receptor population reflects the complexity of the structure and of the development of the innumerable types of neuronal circuits in which they participate.

Acknowledgments. Parts of this article are taken from a recent article by the author: E.A. Barnard; Advances in Biochemical Psychopharmacology 1995;48:1-16

References

1. Schofield PR, Darlison MG, Fujita N, Burt DR, Stephenson FA, Rodriguez H, Rhee LM, Ramachandran J, Reale V, Glencorse TA, Seeburg PH, Barnard EA. Sequence and functional expression of the $GABA_A$ receptor shows a ligand-gated receptor superfamily. Nature 1987;328:221-227

2. Levitan ES, Schofield PR, Burt DR, Rhee LM, Wisden W, Kohler M, Fujita N, Rodriguez H, Stephenson FA, Darlison MG, Barnard EA , Seeburg PH. Structural and functional basis for $GABA_A$ receptor heterogeneity. Nature 1988;335:76-79

3. Wisden W, Seeburg PH. $GABA_A$ receptor channels: from subunits to functional entities. Curr Opinion Neurobiol 1992;2:263-269

4. Macdonald RL, Olsen W. $GABA_A$ receptor channels . Annu Rev Neurosci 1994;17:569-602

5. Bateson AN, Lasham A, Darlison MG. Gamma-aminobutyric acid-A receptor heterogeneity is increased by alternative splicing of a novel beta-subunit gene transcript. J Neurochem 1991;56:1437-1440

6. Bateson AN, Harvey RJ, Bloks CCM, Darlison MG. Sequence of the chicken GABA-A receptor beta-3 subunit cDNA. Nucleic Acids Res 1990;18:5557.

7. Harvey RJ, Chinchetru MA, Darlison MGD. Alternative splicing of a 51-nucleotide exon that encodes a putative protein kinase C phosphorylation site generates two forms of the chicken γ-aminobutyric $acid_A$ receptor β2 subunit. J Neurochem 1994;62:10-16

8. Glencorse TA, Darlison MG, Barnard EA, Bateson AN. Sequence and novel distribution of the chicken homologue of the mammalian γ-aminobutyric $acid_A$ receptor γ1 subunit. J Neurochem 1993;61:2294-2302.

9. Glencorse TA, Bateson AN, Darlison MG. Sequence of the chicken $GABA_A$ receptor gamma2-subunit cDNA. Nucleic Acids Res 1991;10:7157

10. Harvey RJ, Kim HC, Darlison MG. Molecular cloning reveals the existence of a fourth gamma subunit of the vertebrate brain $GABA_A$ receptor. FEBS Lett 1993;331:211-216.

11 Whiting P, McKernan RM, Iversen LL. Another mechanism for creating diversity in gamma-aminobutyrate type A receptors: RNA splicing directs expression of two forms of gamma 2 subunit, one of which contains a protein kinase C phosphorylation site. Proc Natl Acad Sci USA 1990;87:9966-9970.

12 Kofuji P, Wang JB, Moss SJ, Huganir RL, Burt DR. Generation of two forms of the gamma-aminobutyric acid-A receptor gamma-2-subunit in mice by alternative splicing. J Neurochem 1991;56:713-715.

13 Glencorse TA, Bateson AN, Darlison MG. Differential localization of two alternative spliced $GABA_A$ receptor γ_2-subunit mRNAs in the chick brain. Eur J Neurosci 1992;4:271-277.

14. Cutting GR, Curristin S, Zoghbi H, O'Hara B, Seldin MF, Uhl GR. Identification of a putative γ-aminobutyric acid (GABA) receptor rho2 cDNA and colocalization of the genes encoding rho2 (GABRR2) and rho1 (GABRR1) to human chromosome 6q14-q21 and mouse chromosome 4. Genomics 1992;12:801-806

15. Barnard EA. Receptor classes and the transmitter-gated ion channels. Trends Biochem Sci 1992;17:368-374

16. Toyoshima C, Unwin N. Ion channel of acetylcholine receptor reconstructed from images of postsynaptic membranes. Nature 1988;336:247-250

17. Anand R, Conroy WG, Shoepfer R, Whiting P, Lindstrom J. Neuronal nicotinic acetylcholine receptors expressed in Xenopus oocytes have a pentameric quaternary structure. J Biol Chem 1991;266:11192-11198

18. Cooper E, Couturier S, Ballivet M. Pentameric structure and subunit stoichiometry of a neuronal nicotinic acetylcholine receptor. Nature 1991;350:235-238

19. Nayeem N, Green TP, Martin IL, Barnard EA. Quaternary structure of the native $GABA_A$ receptor determined by electron microscope image analysis. J Neurochem 1994;62:815-818.

20. Langer SZ, Arbilla S. Imidazopyridines as a tool for the characterization of benzodiazepine receptors: a proposal for pharmacological classification as omega receptors. Pharmacol Biochem Behav 1988;29:763-766.

21 Osmanovic SS, Shefner SA. γ-Aminobutyric acid responses in rat locus coeruleus neurones in vitro: a current-clamp and voltage-clamp study. J Physiol 1990;421:151-70

22 Wong G, Gu Z-Q, de Costa B, Skolnick P. Labelling of diazepam-sensitive and-insensitive benzodiazepine receptors with tert-butyl-8-chloro-5,6-dihydro-5-methyl-6-oxo-4H-imidazo [1,4] benzodiazepine 3-carboxylate (ZG-63). Eur J Pharmacol 1993;247:57-63.

23 Turner DM, Sapp DW, Olsen RW. The benzodiazepine/alcohol antagonist Ro 15-4513: binding to a $GABA_A$ receptor subtype that is insensitive to diazepam. J Pharmacol Exp Ther 1991; 257:1236-1243.

24 Pritchett DB, Sontheimer H, Shivers BD, Ymer S, Kettermann H, Schofield PR, Seeburg PH. Importance of a novel $GABA_A$ receptor subunit for benzodiazepine pharmacology. Nature 1989;338:582-585.

25 Stephenson FA, Duggan MJ, Pollard S. The γ2 subunit is an integral component of the γ-aminobutyric acid A receptor but the α1 polypeptide is the principle site of the agonist benzodiazepine photoaffinity reaction. J Biol Chem 1990;265:21160-21165.

26 Puia G, Vicini S, Seeburg PH, Costa E. Influence of recombinant γ-aminobutyric acid$_A$ receptor subunit compositions on the action of allosteric modulators of γ-aminobutyric acid-gated Cl currents. Mol Pharmacol 1991;39:691-696.

27 Shivers BD, Killisch I, Sprengel R,Sontheimer H, Kohler M, Schofield PR, Seeburg PH. Two novel GABA$_A$ receptor subunits exist in distinct neuronal subpopulations. Neuron 1989;3:327-337.

28 Angelotti TP, Uhler MD, Macdonald RL. Assembly of GABA$_A$ receptor subunits: analysis of transient single-cell expression utilizing a fluorescent substrate/marker gene technique. J Neurosci 1993;13:1418-1428

29 Sigel E, Baur R, Trube G, Mohler H, Malherbe P. The effect of subunit composition of rat brain GABA$_A$ receptors on channel function. Neuron 1990;5:703-711

30 Draguhn A, Verdoorn TA, Ewert M, Seeburg PH, Sakmann B. Functional and molecular distinction between recombinant rat GABA$_A$receptor subtypes by Zn^{2+}. Neuron 1990;5:781-788

31. Wafford KA, Bain J Whiting PJ, Kemp JA. Functional comparison of the role of the γ subunit in recombinant human γ-aminobutyric acid$_A$/benzodiazepine receptors. Mol Pharmacol 1993;44:437-442.

32 Wieland HA, Lüddens H, Seeburg PH. A single histidine in GABA$_A$ receptors is essential for benzodiazepine agonist binding. J Biol Chem 1992;267:1426-1429.

33 Kleingoor C, Ewert M, von Blankenfeld G, Seeburg PH, Kettermann H. Inverse but not full benzodiazepine agonists modulate recombinant $\alpha_6\beta_2\gamma_2$ GABA$_A$ receptors in transfected human embryonic kidney cells. Neurosci Lett 1991;130:169-172.

34 Wisden W, Herb A, Wieland H, Keinanen K, Luddens H, Seeburg PH. Cloning, pharmacological characteristics and expression pattern of the rat GABA$_A$ receptor α4 subunit. FEBS Lett 1991;289:227-230.

35 Baude A, Sequier JM, McKernan RM, Oliver KR, Somogyi P. Differential subcellular distribution of the α6 subunit versus the α1 and $\beta_{2/3}$ subunits of the GABA$_A$/benzodiazepine receptor complex in granule cells of the cerebellar cortex. Neuroscience 1992;51:739-748.

36 Caruncho HJ, Costa E. Double-immunolabeling analysis of in $GABA_A$ receptor subunits in label fracture replicas of cultured cerebellar granule cells. Receptors and Channels 1994; 2:143-153.

37 Quirk K, Gillard NP, Ragan CI, Whiting PJ, McKernan RM. gamma-Aminobutyric acid type A receptors in the rat brain can contain both γ2 and γ3 subunits, but γ1 does not exist in combination with another γ subunit. Molec Pharmacol 1994;45:1061-1070

38 Backus KH, Arigoni M, Drescher U, Scheurer L, Malherbe P, Mohler H, Benson JA. Stoichiometry of a recombinant $GABA_A$-receptor deduced from mutation-induced rectification. NeuroReport 1993;5:285-288.

39 Duggan MJ, Pollard S, Stephenson FA. Immunoaffinity purification of $GABA_A$ receptor α-subunit iso-oligomers. Demonstration of receptor populations containing α1α2, α1α3 and α2α3 subunit pairs. J Biol Chem 1991:266;24778-24784.

40 Lüddens H, Killisch I, Seeburg PH. More than one alpha variant may exist in a $GABA_A$ benzodiazepine receptor complex. J Receptor Res 1991;11:535-551.

41 McKernan RM, Quirk K, Prince R, Cox PA, Gillard NP, Ragan CI, Whiting P. $GABA_A$ receptor subtypes immunopurified from rat brain with α subunit-specific antibodies have unique pharmacological properties. Neuron 1991;7:667-676.

42 Mertens S, Benke D, Mohler H. $GABA_A$ receptor populations with novel subunit combinations and drug binding profiles identified in brain by α5-and δ -subunit-specific immunopurification. J Biol Chem 1993;268:5965-5973.

43 Endo S, Olsen RW. Antibodies specific for α subunit subtypes of $GABA_A$ receptors reveal brain regional heterogeneity. J Neurochem 1993;60:1388-1398.

44 Fritschy JM, Benke D, Mertens S, Oertel WH, Bachi T, Mohler H. Five subtypes of type A γ-aminobutyric acid receptors identified in neurons by double and triple immunofluorescence staining with subunit-specific antibodies. Proc Nat Acad Sci USA. 1992;89:6726-6730.

45 Wisden W, Gundlach AL, Barnard EA, Seeburg PH, Hunt SP. Distribution of $GABA_A$ receptor subunit mRNAs in rat lumbar spinal cord. Mol Brain Res 1991;10:179-183.

46 Khan ZU, Gutierrez A, Deblas AL. Short and long form γ2 subunits of the $GABA_A$/benzodiazepine receptors. J Neurochem 1994;63:1466-1467.

47 Benke K, Mertens S, Trzeciak A, Gillessen D, Mohler H. Identification and immunohistochemical mapping of $GABA_A$ receptor subtypes containing the δ -subunit in rat brain. FEBS Lett 1991;283:145-149

48 Saxena NC, Macdonald RL. Assembly of $GABA_A$ receptor subunits: role of the δ-subunit. J Neurosci 1994;14:7077-7086

49 Benke D, Fritschy JM, Trzeciak A, Bannerworth W, Mohler H. Distribution , prevalence and drug-binding profile of $GABA_A$-receptor subtypes differing in β-subunit isoform. J Biol Chem 1994;269:27,100-27,107.

50 Angelotti TP, Macdonald RL. Assembly of GABAA receptor subunits: α1β1 and $\alpha_1\beta_1\gamma_{2S}$ subunits produce unique ion channels with dissimilar single-channel properties. J Neurosci 1993;13:1429-1440

51 Wisden W, Laurie DJ, Monyer H, Seeburg PH. The distribution of 13 $GABA_A$ receptor subunit mRNAs in the rat brain. I. Telencephalon, diencephalon, mesencephalon. J Neurosci 1992;12:1040-1062

52 Gao B, Fritschy JM. Selective allocation of $GABA_A$-receptors containing the α1-subunit to neurochemically distinct subpopulations of hippocampal interneurons. Eur J Neurosci 1994;6:837-853.

53 Buhl EH, Halasy K, Somogyi P. Diverse sources of hippocampal unitary inhibitory postsynaptic potentials and the number of synaptic release sites. Nature 1994;368:823-828.

54 Laurie D, Wisden W, Seeburg PH. The distribution of thirteen $GABA_A$ receptor subunit mRNA in the rat brain. III. Embryonic and postnatal development. J Neurosci 1992;12:4151-4172.

55 Poulter MO, Barker JL, O'Carroll AM, Lolait SJ, Mahan LC. Differential and transient expression of $GABA_A$ receptor α-subunit mRNAs in the developing rat CNS. J Neurosci 1992;12:2888-2900.

56 Bovolin P, Santi MR, Memo M, Costa E, Grayson DR. Distinct developmental expression of rat α_1, α_5, γ_{2S}, γ_{2L} γ-aminobutyric $acid_A$ receptor subunit mRNAs in vivo and in vitro. J Neurochem 1992;59:62-72.

57 Cherubini E, Giairsa J-L, Ben-Ari Y. GABA: an excitatory transmitter in early postnatal life. Trends Neurosci 1991;14:515-519.

58 Meier E, Hertz L, Schousboe A. Neurotransmitters as developmental signals. Neurochem Int 1991;19:1-15.

GABA: Receptors, Transporters and Metabolism
ed. by C. Tanaka & N.G. Bowery

MODELLING FUNCTIONAL DOMAINS OF THE $GABA_A$ RECEPTOR CHLORIDE CHANNEL

R.W. Olsen, G.B. Smith and S. Srinivasan

Department of Pharmacology, UCLA School of Medicine,
Los Angeles, CA 90095-1735, USA.

Summary: Secondary structure predictions of domains surrounding residues identified in ligand binding in the α1 subunit of $GABA_A$ receptors suggest a large structural motif that may be coupled to channel gating. We recently identified amino acids (F65 and H102 in bovine α1 subunit) as sites of photoaffinity labelling with GABA and benzodiazepine (BZ) ligands respectively. The two domains, TIDVFFR(61-67) for GABA and TPDTFFH(96-102) for BZ show sequence homology, suggesting that this portion of the BZ binding site may have arisen from a modified agonist site. Secondary structure predictions suggest that these two sequences in the extracellular domain are likely to exist as short strands of β sheet, which flank a sequence including two W residues which is highly conserved in all of the ligand-gated ion channel receptor superfamily. The whole βαβ-β domain, with ligand binding at the N-terminal end and allosteric ligand binding at the other end, is suggested to transduce binding energy into a conformational change that results in ion channel opening.

Introduction

$GABA_A$ receptors are major members of the ligand-gated ion channel receptor superfamily. The cDNAs for all of the subunits for the various members of the superfamily are evolutionarily related, sharing a common sequence homology and predicted topological structure for the polypeptides, and presumably similar structures for the proteins. These are thought to be heteropentamers with each subunit contributing to the wall of the central ion channel; the subunits are composed of a putative large N-terminal extracellular domain and four transmembrane domains with a large intracellular loop between the third and fourth membrane-spanning regions [1-4].

The $GABA_A$ receptor is actually a family of receptors, in which 6 α, 4 β, 4 γ, 1 δ, and 2 ρ subunits and splicing variants of many of them suggests are combined as 20 or so heteropentameric isoforms that differ in biological and pharmacological characteristics [5,6, chapter by Barnard, this volume; chapter by Mohler, this volume].

Potential drug development strategies based on pharmacological heterogeneity of $GABA_A$ receptor subtypes remain exciting. Information on the exact structure of the drug binding domains of the receptors would be useful in this regard. In the absence of X-ray crystallography information on these membrane-bound proteins, structural information has been limited to studies with affinity labeling/ microsequencing, site-directed mutagenesis, and domain-specific antibody localization. Available evidence suggests that information gained with any member of the LGIC superfamily is applicable to the others. This article describes the biochemical properties of the $GABA_A$ receptors, identification of some residues involved in ligand binding, and a speculative model for secondary and tertiary structure in the area of the protein participating in ligand-gating of the ion channel.

Purification and properties of the $GABA_A$ receptors from mammalian brain

$GABA_A$ receptors that bind benzodiazepines, i.e., the vast majority of $GABA_A$ receptors, can be purified to homogeneity by benzodiazepine affinity chromatography [7]. Fig. 1 shows SDS-PAGE of the protein purified from rat brain [8] showed four bands on protein staining: 52 kDa and 56 kDa (originally designated the α and β subunits, respectively); 47 kDa (breakdown-products of those of higher Mr, or perhaps the γ2 subunit); 36 kDa (a non-receptor polypeptide co-purified [9]. The 52 kDa band, like in crude brain membranes (Mohler chapter, this volume), is photoaffinity labeled by the benzodiazepine [^{3}H]flunitrazepam [7,8]; the 56 kDa band is photoaffinity labeled by the GABA ligand [^{3}H]muscimol [10].

Figure 1. SDS-PAGE of BZ affinity-purified bovine cortex GABA$_A$ receptor showing protein staining and photoafinity labelling with [^{3}H]muscimol and [^{3}H]flunitrazepam.

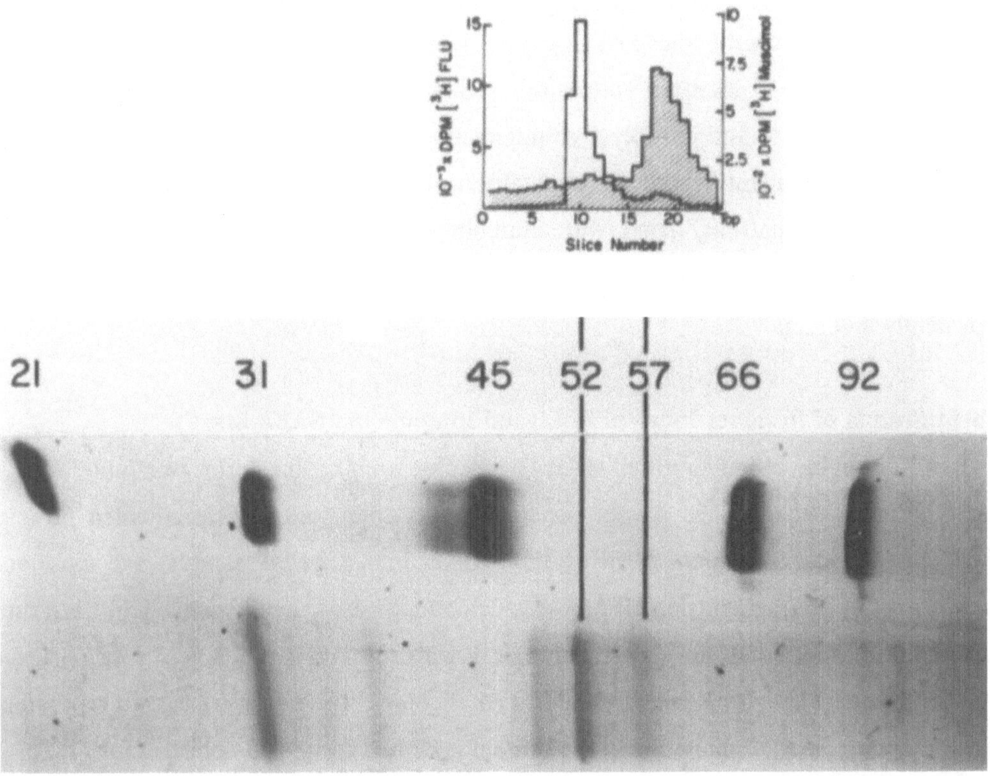

Partial sequencing of the purified protein after proteolytic digestion and fragment purification led to the first cloning of a GABA$_A$ receptor cDNA [1]. Expression of GABA-activated chloride channels from the two cDNAs isolated was consistent with the suggestion that the two subunits seen on SDS-PAGE (α at 52, β at 56) were necessary and sufficient to produce the receptor, and that the protein had an $\alpha_2\beta_2$ tetrameric structure [7]. In fact, the story is much more complex.

Just as a multitude of cDNA clones related to the first two were isolated [5; Barnard chapter, this volume], likewise myriad receptor polypeptide bands on SDS-PAGE were detected. Several discrete polypeptides were identified by [^{3}H]flunitrazepam and [^{3}H]muscimol photolabeling of crude brain homogenates and affinity-purified preparations. Likewise, antibodies to peptides based on the original two α and β cDNAs were found to recognize multiple bands on Western blots of crude brain homogenates and benzodiazepine affinity-purified receptors [11-13; Mohler chapter, this volume]. A highly purified rat preparation, appearing as two fuzzy bands on protein staining of SDS-PAGE, showed four bands stained with a generic anti-β subunit antibody [14], and four different bands stained with subunit-specific antibodies for α1, α2, α3, and α5 (previously called α4 [15]). In fact, there were actually at least ten polypeptide subunits present; furthermore, some of the α subunits, as well as δ and possibly γ co-migrated with the so-called 'β' band. This meant that the previous identification of photoaffinity labeled subunits was inconclusive; co-migration of photoaffinity label with antibody staining was inadequate, and protein sequencing was needed to identify the subunit(s), as well as the active site residues that were covalently modified.

Identification of Residues Involved in Ligand Binding: the GABA Site

Benzodiazepine affinity-purified $GABA_A$ receptor from bovine cortex was found to give no N-terminal sequence, so partial sequencing was attempted with fragmented protein. Following tryptic digestion of a [^{3}H]muscimol photolabeled sample, we observed approximately 50 fragments on HPLC, of which about 4-8 were radioactive [16]. We then subjected a large sample to preparative SDS-PAGE, observing labeling in four apparent bands, 58 and 57 kDa (heaviest), 54 and 51 kDa (lighter). The four were cut out separately, electroeluted, the SDS diluted, and the protein adsorbed to DEAE-Sephacel for digestion with chymotrypsin. The products were eluted, the salt diluted, and dried for HPLC. All the bands yielded multiple fragments as indicated by absorbance, with each showing two major radioactive peaks. One radioactive peak, present in all four bands, gave no recognizable sequence. A second major peak found only in 57 and 58 kDa bands at the same elution time was pooled to yield 200 pmol, and gave a good clean sequence for at least 5 residues, with about 60-90 pmol for the beginning fractions. The sequence -TIDVF- coincides with a sequence deduced from the cDNA for α subunits (α1, α2, α3, and α5: residues 61-65 for α1), following the chymotryptic substrate Y, in the N-terminal putative extracellular domain. The radioactivity was found in fraction 5 (F), corresponding to F65 of bovine α1 subunit. The homologous F residue is found in all α subunits and γ2 and δ, while the rest of the sequence diverges; the substrate Y is conserved in all subunits, but the β sequence would be -TLTMYF-. The affinity labeling thus suggests that cow F65 is probably positioned near a GABA binding site, although this conclusion is qualified by our lack of knowledge of the actual chemical reaction involved in the photoaffinity labeling with muscimol [16].

Interestingly, a single point mutation of the rat α1 subunit at F64 (correponds with bovine F65) produced a sharp decrease in agonist and antagonist affinities when coexpressed with β2 and γ2 subunits [17]. This is consistent with a role for this α subunit residue in the GABA binding pocket. It remains possible that β subunits could also be affinity labeled with muscimol (e.g., our second peak). Further, a single binding site might be shared by two subunits. In support of this, two sets of residues (157-160 and 202-205) in the β2 subunit were shown by mutagenesis to affect functional activation by GABA [18].

The Benzodiazepine Binding Site

Benzodiazepine drugs exert their central nervous system effects by binding to the $GABA_A$ receptor complex and allosterically enhancing the action of GABA. Although a γ subunit is needed for a $GABA_A$ receptor isoform to bind benzodiazepines and for these drugs to enhance GABA responses [5], the nature of the α subunit determines the benzodiazepine site specificity [19] and the α subunit (α1, α2, α3, α5) appears to be photoaffinity labeled by [^{3}H]flunitrazepam [11-13].

Benzodiazepine affinity-purified bovine $GABA_A$ receptor was photoaffinity labeled with [^{3}H]flunitrazepam and subjected to preparative SDS-PAGE, yielding a major (cortex) or single band (cerebellum). This was electroeluted and digested with *S. aureus* V8 protease and the digest separated on a second preparative gel, then transferred to PVDF. A radioactive fragment of about 30 kDa was sequenced and gave a mixed sequence. Reasoning that the sequence, derived from a highly purified single polypeptide from highly purified receptor protein, must be present in the α1 subunit, the data were examined for a match with the deduced amino acid sequence. Two possible matches were found in α1: I94-103 with the label in the H102 and P97, and I223-231 with the label in the Y231 and F226. In order to distinguish between these two possibilities, peptide mapping of deglycosylated and proteolyzed photolabeled subunits was performed. The digests yielded products that could not have been derived from I223-231 but could have from I94-103. Thus the most likely site of covalent attachment of the benzodiazepine label was H102 [20].

Our results are consistent with other observations. The site of [^{3}H]flunitrazepam photoincorporation was found earlier to lie between residues 59-148 of the α1 subunit (includes H102, but not Y231), based on CNBr cleavage and immunoblotting with sequence-specific antisera [21]. More pertinent, the same single residue H102 (rat H101) markedly changed benzodiazepine agonist affinity and efficacy as demonstrated by mutagenesis: the α4 and α6 subunits form recombinant receptors which are insensitive to benzodiazepines, and their mRNAs are found in brain areas enriched in diazepam-insensitive binding of the inverse agonist [^{3}H]Ro15-4513 and [^{3}H]muscimol [19, 22, 23]. The α4 and α6 subunits contain R in the position coresponding to bovine α1H102. Replacing H101 of the rat α1 subunit with R produced a subunit which when coexpressed with β2 and γ2 subunits gave GABA currents which were insensitive to benzodiazepine agonists. The H101R-containing mutant α1 produced oligomers which still bound Ro15-4513 with normal affinity and antagonist Ro15-1788 (flumazenil) with reduced affinity, but was totally diazepam-insensitive. The reverse mutation of α6 R to H 'restored' benzodiazepine binding [19]. In addition, a naturally-occurring variant of the α6 subunit with N at this position was isolated from a strain of rats showing behavioral high susceptibility to motor impairment by benzodiazepines and alcohol; the allelic variant of α6 produced intermediate affinity for agonists when expressed with β and γ [23].

Additional residues in α subunits, some distance away, between the disulfide loop and the first membrane-spanning region M1, G200 and V211, as well as T162, had effects on benzodiazepine selectivity [19]. Because the changes produced were ones of relative binding affinity for agonist binding subtypes, these regions may or may not actually participate in the binding pocket itself.

A Speculative Model of a Portion of The Agonist Binding Site in Ligand-gated Ion Channels [24]

As shown in Fig. 2, the two α subunit regions shown to be involved in the GABA and benzodiazepine binding sites share a common motif: TXDXFF with four residues identically shared between the two and also conserved in both regions for all $GABA_A$ α subunits. Further, the T residue shows conservation in other ligand-gated ion channel receptor subunits. The homology suggests that this portion of the benzodiazepine site may be a modified form of an equivalent domain in the GABA site. The H101 residue implicated in benzodiazepine binding aligns with the R66 residue in the GABA-binding domain, which is interesting in the sense that the α subunits that do not bind benzodiazepine agonists have an R at the 101 equivalent position as noted above.

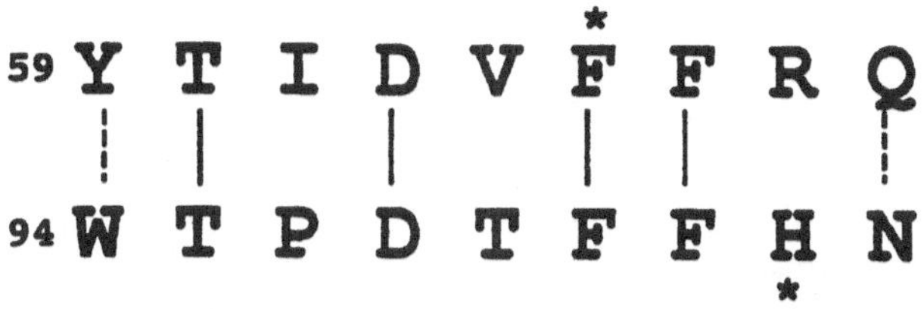

Figure 2. Alignment of GABA$_A$ receptor α1 subunit sequences showing sequence homology about the two proposed binding regions.

Secondary structural predictions [25,26] for these regions predict a β-strand conformation for the two TXDXFF motif segments, with alternation of polar and nonpolar residues. The residues in the alternating hydrophilic positions, which could be exposed at the protein surface, are also the residues which are conserved in both segments and which are involved in ligand binding. Both segments show a tendency to form β-strands in all ligand-gated ion channel subunit subtypes (not shown) and the conserved motif suggests that they may be oriented in parallel to one another.

The entire sequence of approximately 25 residues between these two regions also contains highly conserved amino acids. This intervening region is demarcated in the GABA$_A$ α subunit by W69 and W94 (highlighted in Fig. 3). These W residues are absolutely conserved in all superfamily subunits. In addition, residues corresponding to D71, R73, and L74 of the α1 subunit (also highlighted in Fig. 3) are also conserved in virtually all super-family members. This region is predicted to adopt a turn, followed by a short helix of about three turns, and then another piece of β-strand [25].

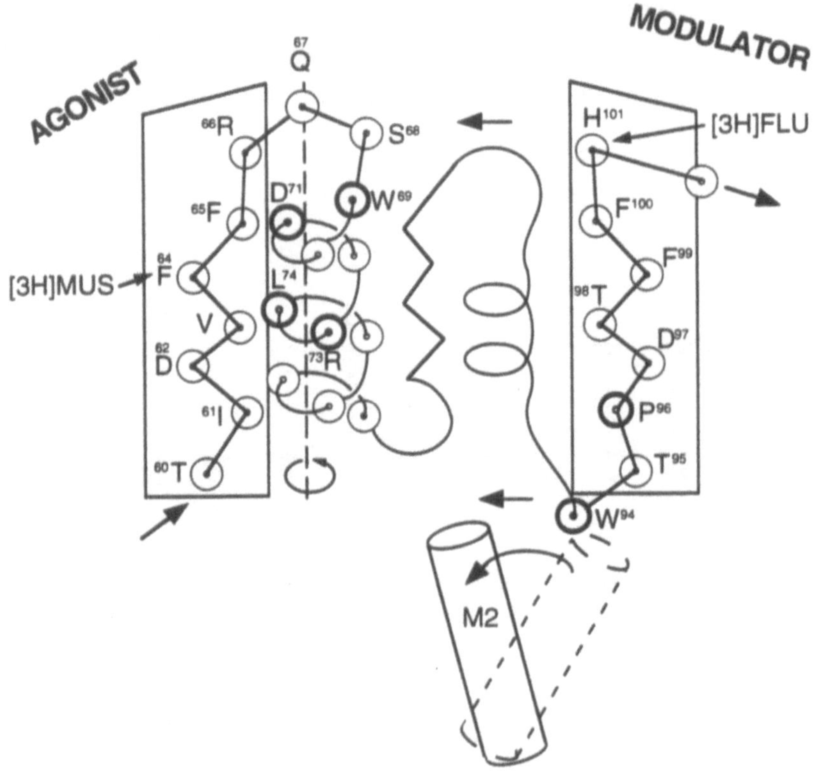

Figure 3. Speculative model of a functional domain of ligand-gated ion channel receptors based on the α1 subunit of the $GABA_A$ receptor. A functional unit is suggested which involves a portion of the ligand binding domains and translation of binding to a channel gating mechanism. The conserved motif (boxed) contains residues involved in ligand binding. The sequence beginning at T60 (rat numbering) contains the F64 residue ([^{3}H]MUS) shown to bind [^{3}H]muscimol and the region commencing at T95 contains the H101 residue ([3H]FLU) shown to bind benzodiazepine agonists. Highly conserved residues among ligand-gated ion channel members are indicated by highlighted circles. The figure shows how agonist binding to the muscimol-labeled region including the RQS(66-68) loop could result in a rotation of the helical portion which begins at W69. This rotational motion could be transduced to other regions of the receptor, such as the M2 channel lining helix, by interacting with the region between the two absolutely conserved W69 and W94. Motion of this region is indicated by left-pointng arrows. Ligands binding to the benzodiazepine allosteric modulatory site at H101 could influence the probability of complete transduction in a positive or negative manner. The observed homology between the boxed agonist-binding and benzodiazepine-binding segments suggests an evolutionary connection and also a possible structural correlate for this allosteric interaction. Reprinted with permission from Smith & Olsen, 1995 [24].

Similar $\beta\alpha\beta$ motifs (Fig. 3) form structural units commonly found in proteins of diverse function and cellular environments, and residues which contribute to ligand binding or enzyme active sites are often found within a hairpin turn connecting the C-terminus of the first β-strand and the N-terminus of the helix [27]. Fig. 3, upper left, shows how the residue photolabeled with muscimol "[^{3}H]MUS" near the top of the putative β-strand leads into a binding loop ("agonist site") with the helix which begins at W69. Such a helix-based binding motif in ligand-gated ion channel receptors would agree with recent electron micrographs of the nicotinic acetylcholine receptor showing rod-shaped α-helices in the extracellular domain, speculated to be associated with the agonist binding site [4].

This region formed by conserved residues between the absolutely conserved W69 and W94 residues could correspond to a conserved structural unit among members of this superfamily of receptors, and may also correspond to a functional domain which performs a similar role among members of the superfamily. We suggest that this segment, as depicted in Fig. 3, may participate in the complex transduction of ligand binding to channel activation. Agonist binding is suggested to cause a rotation of the helix and concerted movement of the larger motif (arrows in Fig. 3), a conformational change associated with channel gating in the M2 segment, with reorientation of the residues that provide steric impediment to ion flow. In this way, the energy of ligand binding would be magnified as a lever operating on the fulcrum of the two conserved W residues in this area. Positive and negative allosteric modulatory actions resulting from binding at the site on the C-terminal end of this structural unit could influence the extent or probability of ligand-induced rotation, conformational change of the structural unit, and channel gating.

Acknowledgements

Supported by NIH grant NS28772.

References

1. Schofield PR, Darlison MG, Fujita N, Burt DR, Stephenson FA, Rodriguez H, Rhee LM, Ramachandran J, Reale V, Glencorse TA, Seeburg PH, Barnard EA. Sequence and functional expression of the GABA-A receptor shows a ligand-gated receptor superfamily. Nature 1987;328: 221-227.

2. Olsen RW, Tobin AJ. Molecular biology of $GABA_A$ receptors. FASEB J 1990;4:1469-1480.

3. Macdonald RL, Olsen RW. $GABA_A$ receptor channels. Ann Rev Neurosci 1994;17:569-602.

4. Unwin N. Acetylcholine receptor channel imaged in the open state. Nature 1995;373:37-43.

5. Pritchett D, Sontheimer H, Shivers BD, Ymer S, Kettenmann H, Schofield PR, Seeburg P. Importance of a novel $GABA_A$ receptor subunit for benzodiazepine pharmacology. Nature 1989;338:582-585.

6. DeLorey TM, Olsen RW. γ-Aminobutyric $acid_A$ receptor structure and function. J Biol Chem 1992;267:16747-16750.

7. Sigel E, Stephenson FA, Mamalaki C, Barnard EA. A γ-aminobutyric acid/benzodiazepine receptor complex of bovine cerebral cortex: Purification and partial characterization. J Biol Chem 1983;258:6965-6971.

8. Stauber GB, Ransom RW, Dilber AI, Olsen RW. The γ-aminobutyric acid-benzodiazepine receptor protein from rat brain: large-scale purification and preparation of antibodies. Eur J Biochem 1987;167:125-133.

9. Bureau MH, Khrestchatisky M, Heeren MA, Zambrowicz EB, Kim H, Grisar TM, Colombini M, Tobin AJ, Olsen RW. Isolation and cloning of a voltage-dependent anion channel-like 36,000 Dalton polypeptide co-purified with the $GABA_A$ receptor from mammalian brain. J Biol Chem 1992;267:8679-8684.

10. Deng L, Ransom RW, Olsen RW. [^{3}H]Muscimol photolabels the γ-aminobutyric acid receptor binding site on a peptide subunit distinct from that labeled with benzodiazepines. Biochem Biophys Res Comm 1986;138:1308-1314.

11. Fuchs K, Sieghart W (1989) Evidence for the existence of several differenct α- and β-subunits of the GABA/benzodiazepine receptor complex from rat brain. Neurosci Lett 1989;97:329-333.

12. Bureau M, Olsen RW. Multiple distinct subunits of the γ-aminobutyric acid-A receptor protein show different ligand-binding affinities. Mol Pharmacol 1990;37:497-502.

13. Stephenson FA, Duggan MJ, Pollard S. The γ2 subunit is an integral component of the γ-aminobutyric $acid_A$ receptor but the α1 polypeptide is the principal site of the agonist benzodiazepine photoaffinity labeling reaction. J Biol Chem 1990;265:21160-21165.

14. Endo S, Olsen RW. Preparation of antibodies to beta subunits of $GABA_A$ receptors. J Neurochem 1992;59:1444-1451.

15. Endo S, Olsen RW. Antibodies specific for α subunit subtypes of $GABA_A$ receptors reveal brain regional heterogeneity. J Neurochem 1993;61:

16. Smith GB, Olsen RW. Identification of a [^{3}H]muscimol photoaffinity substrate in the bovine γ-aminobutyric $acid_A$ receptor α subunit. J Biol Chem 1994;269:20380-20387.

17. Sigel E, Baur R, Kellenberger S, Malherbe P. Point mutations affecting antagonist and agonist dependent gating of $GABA_A$ receptor channels. EMBO J 1992;11:2017-2023.

18. Weiss DS, Amin J. $GABA_A$ receptor needs two homologous domains of the beta-subunit for activation by GABA but not by pentobarbital. Nature 1993;366:565-569.

19. Wieland HA, Luddens H. Four amino acid exchanges convert a diazepam-insensitive inverse agonist-preferring $GABA_A$ receptor into a diazepam-preferring $GABA_A$ receptor. J Med Chem 1994;37:4576-4580.

20. Smith GB, Olsen RW. Identification of amino acid residues in the $GABA_A$ receptor α subunits photoaffinity labeled with the benzodiazepine flunitrazepam. Submitted.

21. Stephenson FA, Duggan MJ. Mapping the benzodiazepine photoaffinity-labelling site with sequence-specific γ-aminobutyric $acid_A$-receptor antibodies. Biochem J 1989;264:199-206.

22. Sieghart W, Eichinger A, Richards JG, Mohler H. Photoaffinity labeling of benzodiazepine receptor proteins with the partial inverse agonist [^{3}H]Ro15-4513: a biochemical and autoradiographic study. J Neurochem 1987;48:46-52.

23. Korpi ER, Kleingoor C, Kettenmann H, Seeburg PH. Benzodiazepine-induced motor impairment linked to point mutation in cerebellar $GABA_A$ receptor. Nature 1993;361:356-359.

24. Smith GB, Olsen RW. Functional domains of $GABA_A$ receptors. Trends Pharmacol Sci 1995;16:162-168.

25. Chou PY, Fasman GD. Empirical predictions of protein conformation. Ann Rev Biochem 1978;47:251-276.

26. Kyte J, Doolittle RF. A simple method for displaying hydropathic character of a protein. J Mol Biol 1982;157:105-132.

27. Branden C, Tooze J. *Introduction to Protein Structure*. Garland Press New York 1991.

GABA: Receptors, Transporters and Metabolism
ed. by C. Tanaka & N.G. Bowery

GABA$_A$-RECEPTOR SUBTYPES: PHARMACOLOGICAL SIGNIFICANCE AND MUTATIONAL ANALYSIS *IN VIVO*

H. Mohler, J.M. Fritschy, D. Benke, J. Benson, U. Rudolph, and B. Lüscher

Institute of Pharmacology, ETH and University of Zürich, Winterthurerstr, 190, CH 8057 Zürich, Switzerland

Summary: The subunit composition of the major GABA$_A$-receptors *in vivo* has been identified. They comprise about half a dozen major subtypes (e.g., $\alpha1\beta2\gamma2$, $\alpha2\beta3\gamma2$, $\alpha3\beta3\gamma2$) and one or two dozen minor subtypes, some of which contain two α-subunit variants. The functional significance of GABA$_A$-receptor subtypes was assessed by mutational analysis *in vivo*. Benzodiazepine-insensitive mice were generated by targeted disruption of the $\gamma2$-subunit gene. For drug development, GABA$_A$-receptor heterogeneity offers new opportunities to advance the therapy of anxiety, insomnia, and epilepsy with ligands acting at selected modulatory sites.

Introduction

GABAergic transmission, one of the most widespread signal transduction systems in the brain, operates via a multitude of structurally distinct GABA$_A$-receptors. This receptor diversity subserves various functions in adapting GABAergic transmission to biological requirements, in particular, developmental maturation, cell type-specific signal transmission, regulatory control at the protein level, and subcellular compartmentation. Receptor heterogeneity arises from a repertoire of at least 15 subunits which can be grouped by their degree of sequence homology into six α, three β, three γ, one δ, and two ρ subunits. The combinatorial assembly of these subunits (and splice variants of several of them) in a presumably pentameric heterooligomeric structure results in diverse receptor subtypes (for review, see refs. 1-4). The most prevalent types of GABA$_A$-receptors have recently been identified. Their allocation to particular neuronal circuits has opened the possibility of investigating their functional role *in situ*. Since GABA$_A$-receptors are the targets for many important neuroactive drugs including benzodiazepines (BZ), barbiturates, and steroids, the possibility exists of developing selective drugs with improved therapeutic profile for the treatment of anxiety syndromes, epilepsy, and insomnias. In the following, an overview of the major types of $GABA_A$-receptors in the brain, the pharmacology of receptor subtypes, and the analysis of receptor function by mutational analysis is given.

Major $GABA_A$-receptor subtypes in vivo

In view of the relevance of receptor heterogeneity to the pharmacology of BZ site ligands (see below), the major $GABA_A$-receptor subtypes that are sensitive to BZ site ligands are outlined first.

$GABA_A$-receptors containing α1β2γ2 subunits

The subunit combination α1β2γ2 constitutes the major $GABA_A$-receptor subtype in the brain, as demonstrated by immunoprecipitation and immunohistochemical localization of the three subunits in the same neurons (5-7). This receptor type is not only the main component in GABAergic signal transduction, but also mediates the basic pharmacological spectrum of the classical BZ site ligands. The α1β2γ2-receptors display high nanomolar affinity for all classical BZ site ligands, except CL 218872 (Tab. 1) (8, 9).

With regard to identified neurons, high levels of receptors containing the subunits α1β2γ2 are expressed in numerous populations of GABAergic neurons at all levels of the neuraxis (Tab. 2). In particular, interneurons in cerebral cortex and hippocampus, GABAergic neurons in the cerebellum, brainstem reticular formation, pallidum, substantia nigra, and basal forebrain are intensely immunoreactive for the α1-subunit (6, 10-12). The α1β2γ2 subunit combination has also been allocated to non-GABAergic neurons, such as olfactory bulb mitral cells, in association with the α3-subunit, and relay neurons in the thalamus, in combination with the δ-subunit (Tab. 2).

Only a few regions are devoid of α1-subunit staining, notably the striatal complex, the granule cell layer of the olfactory bulb, and the reticular nucleus of the thalamus (7).

$GABA_A$-receptors containing α2β3γ2 or α3β3γ2 subunits

Receptors containing the α2-subunit are most abundant in regions where the α1-subunit is absent or expressed at low levels (7, 9, 13), such as striatum, hippocampal formation, and olfactory bulb. As for the α2-subunit, receptors containing the α3-subunit are frequently distributed in regions expressing only low or moderate levels of the α1-subunit, including the lateral septum, the reticular nucleus of the thalamus, and several brainstem nuclei (7, 9, 10). The co-expression of α2- or α3-subunits with the β3- and γ2-subunits (9, 12) is evident on the cellular level, e.g., in hippocampal pyramidal cells (α2β3γ2) and in cholinergic neurons of the basal forebrain (α3β3γ2) (Tab. 2). The profile of BZ site ligands interacting with these receptors differs from that of the α1β2γ2 receptors in that βCCM displays a slightly lower displacing potency (4-5 fold) and zolpidem and CL 218872 a considerably lower displacing potency (9-14 fold) (Tab. 1). Thus, the neuronal circuits expressing α2β3γ2 and α3β3γ2 receptors are expected to be less prominently involved in the pharmacological responses of the latter drugs.

GABA$_A$-receptors containing the α5-subunit

Receptors containing the α5-subunit are of minor abundance in the brain and are concentrated mainly in the hippocampus, olfactory bulb, hypothalamus, and trigeminal sensory nucleus (7). They comprise various subtypes differentiated by the affinity of zolpidem. For instance, in the hippocampus and spinal cord, the receptor population immunoprecipitated by the α5-antiserum displays micromolar affinity for zolpidem, while the receptor population in the thalamus/hypothalamus, striatum, and brainstem shows nanomolar affinity for zolpidem (Tab. 1) (8). Differential affinities for zolpidem have also been observed in radioligand binding studies on cell membranes and tissue sections prepared from

Tab. 1 Drug binding profile of native GABA$_A$-receptor populations

	K_i [nM]; [^{3}H]Flumazenil binding					
	Receptor population immunoprecipitated by subunit-specific antisera					
	α1β2γ2	α2β3γ2	α3β3γ2	α5βxγx		
Displacer	Whole brain	Whole brain	Whole brain	Whole brain	Striatum	Hippocampus
Flumazenil*	0.9	1.2	1.1	0.5	–	–
Flunitrazepam	7	7	8	3	–	–
βCCM	2	7	7	2	–	–
Zolpidem	13	100	83	17	19	1220
CL 218'872	190	955	660	270	–	–

Displacement potencies of various BZ site ligands were determined in [^{3}H]flumazenil binding to GABA$_A$-receptors immunoprecipitated from membrane extracts of rat whole brain or brain areas indicated. The values in columns 1-3 indicate mean values obtained with immunoprecipitation with either the α- or the β-subunit-specific antisera. The presence of the γ2-subunit was inferred from the presence of high affinity BZ binding sites. For the receptors containing the α5-subunit only the α5-antiserum was used for immunoprecipitation. The data are taken from ref. (2). *K_D-values derived from Scatchard analysis.

hippocampus and spinal cord (14, 15). On recombinant receptors containing the α5-subunit, zolpidem is practically without effect (K_i>10μM; α5βxγx) (16-19), suggesting that the receptors immunopurified from hippocampus and spinal cord by the α5-subunit antiserum might largely be represented by these zolpidem-insensitive subunit combinations. In contrast, in brain areas where native α5-receptors display nanomolar affinities for zolpidem, the α5-subunit might be associated with an additional α-subunit, the α1- and α3-subunits being prime candidates (8).

Tab. 2 GABA$_A$-receptor subtypes in specified neurons

Neurons	Subunits
Olfactory bulb	
Mitral cells	α1 α3 β2 γ2
Granule cells	α2 β3 γ2
	α5 β3 γ2
Short-axon cells	α1 β2 γ2
Periglomerular cells	α2 α5 δ
Hippocampus	
Pyramidal cells	α2 β3 γ2
	α5 β3 γ2
Dentate gyrus granule cells	α2 β3 γ2
Most interneurons	α1 β2 γ2
Thalamus	
Relay neurons	α1 β2 γ2 δ
Reticular nucleus neurons	α3 γ2
Hypothalamus	
Supraoptic nucleus	α1 α2 β2,3 γ2
Ventromedial, arcuate nu.	α2 β3 γ2
	α5 β3 γ2
Cerebellum	
Purkinje cells	α1 β2,3 γ2
Granule cells	α1 α6 β2,3 γ2 δ
Golgi type II cells	α2 α3 γ2
Motoneurons (cranial nerve nuclei)	
Facial motor nu., Hypoglossal nu.	α1 α2 γ2
Trigeminal motor nu., Ambiguus nu.	α2 γ2
Transmitter markers	
GABA (most areas)	α1 β2 γ2
(basal forebrain, raphe)	α1 α3 β2 γ2
5-HT, ACh	α3 β3 γ2

Co-expressed subunits were visualized immunohistochemically. The subunits analyzed include α1, α2, α3, α5, α6, β2,3, γ2, and δ. In each combination the subunits not detected are not indicated.

Pharmacology of GABA$_A$-receptors subtypes

Ligands of the benzodiazepine site

Novel ligands of the BZ site, the medically most relevant modulatory site of GABA$_A$-receptors, are being developed with the aim of retaining the therapeutic effectiveness of classical benzodiazepines while reducing unwanted side-effects such as tolerance, dependence liability, memory impairment, and ataxia. Two strategies are being followed to achieve this goal: 1) reduction of the efficacy of the ligands at all or some receptor subtypes (partial agonists), 2) targeting of the ligands to particular receptor subtypes by selective affinity. Obviously, a combination of both approaches is also feasible. In the following, a brief account of the application of these principles in drug development is given.

Although there are individual variations (20), classical benzodiazepines such as diazepam interact with nearly all receptors with high affinity and high efficacy, as demonstrated for both native and recombinant receptors. In contrast, partial agonists, typified by bretazenil or imidazenil, display high affinity with most if not all receptors, but compared to classical BZ agonists act with reduced efficacy at practically all receptors (21-24). Partial agonistic activity is also exerted by the novel ligands DN 2327 (25) and Y 23684 (26), the latter being in phase II clinical trials as an anxiolytic. In addition, there are BZ site ligands which display efficacies that vary depending on the GABA$_A$-receptor subtype. For example, abecarnil acts as a partial agonist on recombinant receptors containing α2- and α5-subunits but acts as a full agonist on receptors containing the α1- and α3-subunits (27, 28). Abecarnil is currently undergoing clinical trials as an anxiolytic. In addition, the novel ligand U 101017 interacts with GABA$_A$-receptors containing α1-, α2-, or α5-subunits with reduced efficacy and displays potent anti-stress activity in rodents, but only limited anti-conflict activity (von Voigtlander, in preparation). Finally, there are ligands which discriminate among receptor subtypes exclusively by their affinity. The most prominent example is the hypnotic zolpidem. It shows high affinity for receptors containing the α1-subunit while displaying lower affinity for receptors containing the α2- and α3-subunits and even lower affinity for most receptors containing the α5-subunit (Tab. 1) (8, 16, 29). These examples indicate that favorable profiles of BZ site ligands can be achieved either by reducing the efficacy of the ligand at all receptors (bretazenil, imidazenil) or by avoiding a full activation of receptors containing α2- and α5-subunits (abecarnil, zolpidem). The findings concerning abecarnil and zolpidem are of particular interest with regard to receptor heterogeneity. Receptors containing α2- and α5-subunits might be expected to be located in neuronal circuits which are involved in mediating unwanted rather than desirable drug effects. Indeed, it is conspicuous that receptors containing the α2-subunit are strongly expressed in brain areas mediating reward (e.g., nucleus accumbens), while receptors containing the α5-subunit are concentrated in certain areas linked to memory functions (e.g., hippocampus) (7). In addition, the preliminary results on the prominent anti-stress activity of the novel ligand U 101017, which acts with reduced efficacy at receptors containing α1-, α2- or α5-subunits, indicate that novel pharmaceutical profiles can be generated by selective receptor subtype interactions.

Novel modulatory sites

Recently, novel modulatory drug binding sites located on $GABA_A$-receptors have been identified. Pyrrolopyrimidines such as U 89843A represent a novel class of modulators acting at a site distinct from the sites for benzodiazepines, barbiturates, neurosteroids, or loreclezole. In recombinant receptors, U 89843A enhanced the GABA-induced currents irrespective of the types of α-subunit (α1β2γ2, α3β2γ2, and α6β2γ2 receptors, but not α1β2 and α1γ2 receptors). In mice, U89843A induced sedation without loss of righting reflex (30). Furthermore, pyrazinones such as U 92813 appear to interact with yet another drug modulatory site of $GABA_A$-receptors. The unique characteristics of this site are demonstrated by the lack of flumazenil-sensitivity and the lack of requirement for the γ2-subunit in recombinant receptors, as well by the additive nature of its agonistic activity with that of barbiturates and neurosteroids. Since the binding site is present on an αβ-subunit combination (α1β2), it appears to be distinct from that of the pyrrolopyrimidines (31). These novel modulatory sites may hold the potential for further therapeutic advances.

Genetic alterations and disease

In view of the ubiquitous presence of GABAergic circuits in the brain, various neurologic or mental disorders might be linked to genetic alterations of $GABA_A$-receptors. However, up to now, only a few naturally-occurring mutations of $GABA_A$-receptor subunits have been found in animals and only a few tentative links to human diseases have been proposed.

Point mutations

In a selectively bred alcohol-sensitive rat strain (ANT), a single amino acid mutation was discovered in $GABA_A$-receptors located almost exclusively on cerebellar granule cells. The replacement of arginine in position 100 of the α6-subunit by glutamine converts α6-subunit-containing receptors from diazepam-insensitive to diazepam-sensitive (32). Correspondingly, the mutant animals were abnormally sensitive to the motor-impairing effects of BZ agonists (33) while the GABA sensitivity of the receptor remained unaffected (34). The enhanced motor impairment by ethanol was attributed to the point mutation. However, the effects of ethanol have recently been linked to the phosphorylation state of the receptor (35). Nevertheless, the results underline the role of granule cell activity in motor control by the cerebellum.

The widespread resistance of insects to cyclodiene pesticides appears to be due to a point mutation of $GABA_A$-receptors. In Drosophila the resistance to dieldrine and the GABA antagonist picrotoxin has been traced to a point mutation located in the second transmembrane region of the Rdl subunit (Ser 302 Ala) (36, 37). Mutations in equivalent positions in mammalian α-, β-, or γ-subunits likewise result in picrotoxin-resistant GABA$_A$-receptors (38). These observations underlines the view that, similarly to the nicotinic acetylcholine receptor (39, 40), the second transmembrane domain lines the channel wall.

Subunit deletions

Various mutants have been studied to identify potential deletions of GABA$_A$-receptor subunit genes as underlying cause of neurological defects. In particular, p^{cp} mice with a 95% penetrant, recessive, neonatally-lethal cleft palate, were investigated. They contain a deletion of the cleft palate 1 (cp1) locus on chromosome 7, closely linked to the pink-eyed dilution (p) locus (41). This deletion has recently been found to include the genes encoding the GABA$_A$-receptor subunits α5, γ3, and β3 (42, 43). A corresponding loss of certain GABA$_A$-receptor subtypes was demonstrated by radioligand binding, particularly in hippocampus where receptors containing the α5-subunit are prevalent (43; Benke, unpublished). The cp1 locus was localized to an interval beginning distally to the α5-subunit gene and ending within the coding region of the β3-subunit gene. Interestingly, mice containing a deletion which included the α5-and the γ3-subunit-genes, but not the β3-subunit gene, developed a normal palate (44). It was proposed that deletion of the β3-subunit gene, and not another gene in the deletion interval, causes cleft palate. To test this hypothesis, an experiment was performed to rescue the cleft palate phenotype by introducing a β3-subunit transgene under the control of a β-actin promoter into the mice homozygous for the cleft palate deletion. The rescued animals displayed no obvious neurological defects (45). Thus, abnormalities of GABA$_A$-receptors containing the β3-subunit contribute to the clefting defect observed in the p^{cp} mice. This result is consistent with earlier teratological observations that GABA or diazepam can interfere with normal palate development in mice (46). The human counterpart of the region deleted in p^{cp} is the locus 15q11-q13. Aberrations in this locus, especially those with a deletion or mutation of the β3-subunit gene, might contribute to familial or sporadic craniofacial abnormalities in man.

Alterations in chromosome 15q11-q13 are associated with Angelman (AS) syndrome, which is characterized by severe mental retardation, microencephaly, seizures, ataxia, and craniofacial abnormalities (47, 48). The smallest known maternal deletion resulting in AS involves the β3- but not the α5-subunit gene (49, 50), although a single AS patient containing a translocation was found with an intact β3-subunit gene (51). An AS-like paternal imprinting has not been detected for the corresponding region of the mouse chromosome 7 (52). It remains to be determined whether GABA$_A$-receptor genes play a role in the aetiology of AS.

A possible relationship to human disease has also been proposed for the α3-subunit gene. Its locus was mapped to a site on the X chromosome, Xq28, a location that makes it a potential candidate gene for the X-linked form of manic depression (53, 54).

Targeted disruption of the γ2-subunit gene

Apart from the analysis of naturally occurring mutants, GABA-receptor function can be analyzed by the targeted disruption of specific subunit genes. This has recently been accomplished for the γ2-subunit (55), which – in combination with α- and β-subunits – occurs in most $GABA_A$-receptors. While α- and β-subunits are sufficient to form GABA-

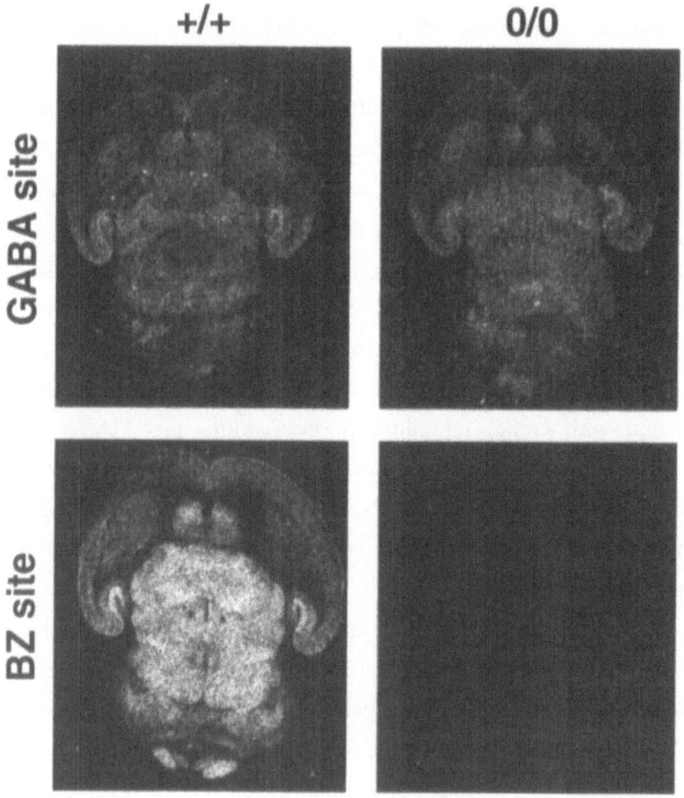

Figure 1. Distribution of GABA and benzodiazepine (BZ) binding sites in $\gamma2^{+/+}$ and $\gamma2^{0/0}$ mice: Autoradiographs of brain sections from newborn mice incubated with [^{3}H] SR 95531 (GABA site) and [^{3}H] flumazenil (BZ site), (x4.5). From ref. (55).

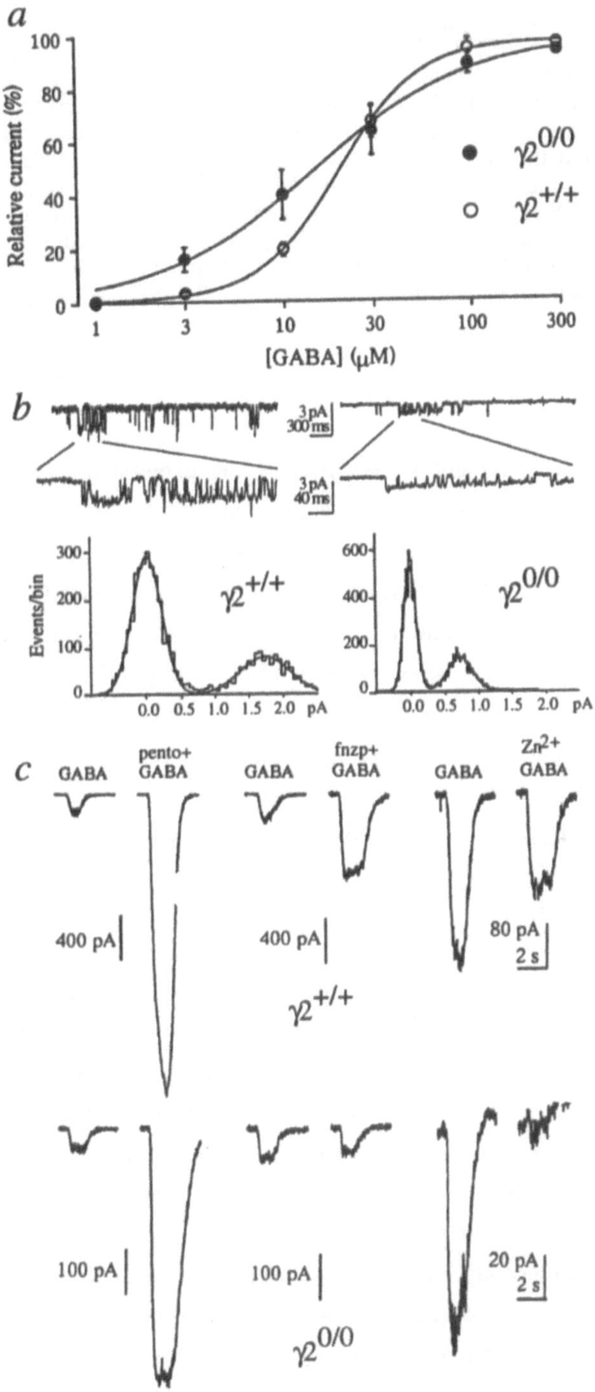
a
Relative current (%)
100
80
60
40
20
0
1
3
10
30
100
300
[GABA] (μM)
γ2^0/0
γ2^+/+
b
3pA
300ms
3pA
40ms
Events/bin
300
200
100
0
0.0 0.5 1.0 1.5 2.0 pA
γ2^+/+
600
400
200
0
0.0 0.5 1.0 1.5 2.0 pA
γ2^0/0
c
GABA
pento+ GABA
GABA
fnzp+ GABA
GABA
Zn2+ GABA
400 pA
400 pA
80 pA
2 s
γ2^+/+
100 pA
100 pA
20 pA
2 s
γ2^0/0

Figure 2

Figure 2. GABA-evoked whole cell and single channel currents from dorsal root ganglion (DRG) neurons of newborn $\gamma 2^{+/+}$ and $\gamma 2^{0/0}$ mice. (a) GABA dose-response curves, yielding EC50 values of 20.0±0.5 μM (n=7, $\gamma 2^{+/+}$) and 15.0±3.5 μM (n=5, $\gamma 2^{0/0}$). The Hill coefficients were 1.98±0.07 ($\gamma 2^{+/+}$) and 1.05±0.30 ($\gamma 2^{0/0}$) (b) Elementary currents evoked by GABA at low and high time resolution, illustrating the main conductance states. Single channel current amplitude histograms were fitted to the sum of two Gaussians representing the baseline current and the amplitude of the openings. From patches clamped at -60mV, current amplitudes were derived (5 μM GABA) corresponding to single channel main conductance states of 28.3±0.9 pS (n=5, $\gamma 2^{+/+}$) and 11.3±1.0 pS (n=4, $\gamma 2^{0/0}$). The traces given correspond to 28.6 and 12.3 pS, respectively. (c) Current traces showing the effects on the currents evoked by GABA pulses (2s, 5μM) in $\gamma 2^{+/+}$ and $\gamma 2^{0/0}$ DRG neurons of the test substances indicated (pento, pentobarbital, 100 μM; fnzp, flunitrazepam, 1 μM; Zn^{2+}, $ZnSO_4$, 10 μM). From ref. (55).

gated ion channels, the γ2-subunit is required for the expression of $GABA_A$-receptors containing the BZ site. The physiological relevance of the BZ sites for brain development and function is unresolved. This is particularly striking, since the BZ site is an evolutionary-conserved modulatory element of $GABA_A$-receptors. Following targeted disruption of the γ2-subunit gene, the number and distribution of GABA binding sites was practically unaltered in neonatal brain, while the BZ sites were almost entirely (94%) absent (Fig. 1). Correspondingly, diazepam was inactive behaviorally in neonates. Electrophysiological analysis showed that functional $GABA_A$-receptors had been formed from the remaining α- and β-subunits. The Hill coefficient and the single channel conductance of the GABA response corresponded to those of recombinant receptors composed of α- and β-subunits (Fig. 2). In keeping with this subunit composition, the GABA response was potentiated by pentobarbital but not by flunitrazepam (Fig. 2). These results demonstrate that functional $GABA_A$-receptors are formed *in vivo* in the absence of the γ2-subunit. Thus, the γ2-subunit serves no essential function for the assembly, transport, and insertion into the cell membrane of $GABA_A$-receptors. However, the γ2-subunit is essential for the establishment of normal signal transduction characteristics, as shown by the reduction of the Hill coefficient and the single channel conductance in the homozygous mutant mice compared to wild-type. In addition, the γ2-subunit is required for the formation of BZ sites on the normal receptor (55).

Embryonic development of the γ2-subunit homozygous mutant mice appeared to be normal according to body weight and histological analysis of the brain and various other organs. No impairment in feeding or lack of anterior pituitary hormones for endocrine control was apparent. Thus, there was no indication for an essential role of an endogenous BZ-site ligand in governing embryonic development. Postnatally, however, the reduced $GABA_A$-receptor function was associated with retarded growth, sensorimotor dysfunction, and drastically reduced life span. Most likely, the γ2-deficient $GABA_A$-receptors do not provide the precise control of GABAergic tone required in inhibitory synaptic transmission in order to control neuronal excitability. Thus, the time by which most GABAergic synapses are expected to become operative coincides with the maximal life-span of the mice lacking the γ2-subunit. Possibly, the lack of GABA$_A$-receptor regulation by a postnatally relevant endogenous BZ-site ligand might contribute to this phenotype.

References

1. Luddens H, Korpi ER, Seeburg PH. GABA$_A$/Benzodiazepine receptor heterogeneity: neurophysiological implications. Neuropharmacol 1995; 34: 245-254.

2. Mohler H, Fritschy JM, Luscher B, Rudolph U, Benson J, Benke D. The GABA$_A$-receptors: from subunits to diverse functions. In: Ion channels Vol. IV. Narahashi T, editor. New York: Plenum Publishing Corporation, in press.

3. Sieghart W. Structure and pharmacology of γ-aminobutyric acid$_A$ receptor subtypes. Pharmacol Rev 1995; 47: 181-234.

4. Smith GB, Olsen RW. Functional domains of GABA$_A$-receptors. Trends Pharmacol Sci 1995; 16: 162-168.

5. Benke D, Mertens S, Trzeciak A, Gillessen D, Mohler H. GABA$_A$ receptors display association of γ2-subunit with α1- and β2/3 subunits. J Biol Chem 1991; 266: 4478-4483.

6. Fritschy JM, Benke D, Mertens S, Oertel WH, Bachi T, Mohler H. Five subtypes of type A γ-aminobutyric acid receptors identified in neurons by double and triple immunofluorescence staining with subunit-specific antibodies. Proc Natl Acad Sci USA 1992; 89: 6726-6730.

7. Fritschy JM, Mohler H. GABA$_A$-receptor heterogeneity in the adult rat brain: differential regional and cellular distribution of seven major subunits. J Comp Neurol 1995; 359: 154-194.

8. Mertens S, Benke D, Mohler H. GABA$_A$ receptor populations with novel subunit combinations and drug binding profiles identified in brain by α5- and δ-subunit-specific immunopurification. J Biol Chem 1993; 268: 5965-5973.

9. Benke D, Fritschy JM, Trzeciak A, Bannwarth W, Mohler H. Distribution, prevalence and drug-binding profile of GABA$_A$-receptors subtypes differing in β-subunit isoform. J Biol Chem 1994; 269: 27100-27107.

10. Gao B, Fritschy JM, Benke D, Mohler H. Neuron-specific expression of GABA$_A$-receptor subtypes: differential associations of the α1- and α3-subunits with serotonergic and GABAergic neurons. Neurosci 1993; 54: 881-892.

11. Gao B, Fritschy JM. Selective allocation of GABA$_A$-receptors containing the α1-subunit to neurochemically distinct subpopulations of hippocampal interneurons. Eur J Neurosci 1994; 6: 837-853.

12. Gao B, Hornung JP, Fritschy JM. Identification of distinct $GABA_A$-receptor subtypes in cholinergic and parvalbumin-positive neurons of the rat and marmoset medial-septum-diagonal band complex. Neurosci 1995; 65: 101-117.

13. Marksitzer R, Benke D, Fritschy JM, Mohler H. $GABA_A$-receptors: Drug binding profile and distribution of receptors containing the α2-subunit *in situ*. J Recept Res 1993; 13: 467-477.

14. Ruano D, Vizuete M, Cano J, Machado A, Vitorica J. Heterogeneity in the allosteric interaction between the γ-aminobutyric acid (GABA) binding site and three different benzodiazepine binding sites of the $GABA_A$/benzodiazepine receptor complex in the rat nervous system. J Neurochem 1992; 58: 485-493.

15. Benavides J, Peny B, Ruano D, Vitorica J, Scatton B. Comparative autoradiographic distribution of central omega (benzodiazepine) modulatory site subtypes with high, intermediate and low affinity for zolpidem and alpidem. Brain Res 1993; 604: 240-250.

16. Pritchett DB, Seeburg PH. γ-Aminobutyric $acid_A$ receptor α5-subunit creates novel type II benzodiazepine receptor pharmacology. J Neurochem 1990; 54: 1802-1804.

17. Puia G, Vicini S, Seeburg PH, Costa E. Influence of recombinant γ-aminobutyric $acid_A$ receptor subunit composition on the action of allosteric modulators of γ-aminobutyric acid-gated Cl^- currents. Mol Pharmacol 1991; 39: 691-696.

18. Hadingham KL, Wingrove PB, Wafford KA, Bain C, Kemp JA, Palmer KJ, Wilson AW, Wilcox AS, Sikela JM, Whiting PJ. Role of the β subunit in determining the pharmacology of human γ-aminobutyric acid type A receptors. Mol Pharmacol 1993; 44: 1211-1218.

19. Luddens H, Seeburg PH, Korpi ER. Impact of β and γ variants on ligand-binding properties of γ-aminobutyric acid type A receptors. Mol Pharmacol 1994; 45: 810-814.

20. Ducic I, Puia G, Vicini S, Costa E. Triazolam is more efficacious than diazepam in a broad spectrum of recombinant receptors. Eur J Pharmacol 1993; 244: 29-35.

21. Haefely W, Martin JR, Schoch P. Novel anxiolytics that act as partial agonists at benzodiazepine receptors. Trends Pharmacol Sci 1990; 11: 452-456.

22. Puia G, Dudic I, Vicini S, Costa E. Molecular mechanisms of the partial allosteric modulatory effects of bretazenil at γ-aminobutyric acid type A receptor. Proc Natl Acad Sci USA 1992; 89: 3620-3624.

23. Wafford KA, Whiting PJ, Kemp JA. Differences in affinity and efficacy of benzodiazepine receptor ligands at recombinant γ-aminobutyric acid receptor subtypes. Mol Pharmacol 1993; 43: 240-244.

24. Auta J, Giusti P, Guidotti A, Costa E. Imidazenil, a partial positive allosteric modulator of $GABA_A$ receptors, exhibits low tolerance and dependence liabilities in the rat. J Pharmacol Exp Ther 1994; 270: 1262-1269.

25. Yasumatsu H, Morimoto Y, Yamamoto Y, Takehara S, Fukuda T, Nakao T, Setoguchi M. The pharmacological properties of Y-23684, a benzodiazepine receptor partial agonist. Brit J Pharmacol 1994; 111: 1170-1178.

26. Wada T, Fukuda N. Pharmacologic profile of a new anxiolytic, DN-2327: effect of Ro-15 788 and interaction with diazepam in rodents. Psychopharmacol 1991; 103: 314-322.

27. Knoflach F, Drescher U, Scheurer L, Malherbe P, Mohler H. Full and partial agonism displayed by benzodiazepine receptor ligands at different recombinant GABA$_A$ receptor subtypes. J Pharmacol Exp Ther 1993; 266: 385-391.

28. Pribilla I, Neuhaus R, Huba R, Hillmann M, Turner JD, Stephens DN, Schneider HH. Abercanil is a full agonist at some, and a partial agonist at other recombinant GABA$_A$ receptor subtypes. In: Anxiolytic β-carbolines. Stephens DN, editor. Berlin: Springer-Verlag, 1993: 50-61.

29. McKernan RM, Quirk K, Prince R, Cox PA, Gillard NP, Ragan CI, Whiting P. GABA$_A$ receptor subtypes immunopurified from rat brain with α subunit-specific antibodies have unique pharmacological properties. Neuron 1991; 7: 667-676.

30. Im HK, Im WB, Pregenzer JF, Carter DB, Hamilton BJ. U 89843 A represents a novel class of GABA$_A$-receptor agonists. J Pharmacol Exp Ther, in press.

31. Im HK, Im WB, Judge TM, Gamill RB, Hamilton BJ, Carter DB, Pregenzer JF. Substituted pyrazinones, a new class of allosteric modulators for γ-aminobutyric acid A receptors. Mol Pharmacol 1993; 44: 468-472.

32. Korpi ER, Kleingoor C, Kettenmann H, Seeburg PH. Benzodiazepine-induced motor impairment linked to point mutation in cerebellar GABA$_A$ receptor. Nature 1993; 361: 356-359.

33. Hellevuo K, Kiianmaa K, Korpi ER. Effect of GABAergic drugs on motor impairment from ethanol, barbital and lorazepam in rat lines selected for differential sensitivity to ethanol. Pharmacol Biochem Behav 1989; 34: 399-404.

34. Kleingoor C, Wieland HA, Korpi ER, Seeburg PH, Kettenmann H. Current potentiation by diazepam but not GABA sensitivity is determined by a single histidine residue. Neuroreport 1993; 4: 187-190.

35. Harris RA, McQuilkin SJ, Paylor R, Abeliovich A, Tonegawa S, Wehner JM. Mutant mice lacking the γ isoform of protein kinase C show decreased bevioral actions of ethanol and altered function of γ-aminobutyrate type A receptors. Proc Natl Acad Sci USA 1995; 92: 3658-3662.

36. ffrench-Constant RH, Rocheleau TA, Steichen JC, Chalmers AE. A point mutation in a Drosophila GABA receptor confers insecticide resistance. Nature 1993; 363: 449-451.

37. ffrench-Constant RH, Steichen JC, Rocheleau TA, Aronstein K, Roush RT. A single-amino acid substitution in a γ-aminobutyric acid subtype A receptor locus is associated with cyclodiene insecticide resistance in Drosophila populations. Proc Natl Acad Sci USA 1993; 90: 1957-1961.

38. Gurley D, Amin J, Ross PC, Weiss DS, White G. Point mutations in the M2 region of the α, β, or γ subunit of the GABA$_A$ channel that abolish block by picrotoxin. Recept Channel 1995; 3: 13-20.

39. Karlin A. Structure of nicotinic acetylcholine receptors. Curr Opin Neurobiol 1993; 3: 299-309.

40. Akabas M, Kaufmann C, Archdeacon P, Karlin A. Identification of acetylcholine receptor channel-lining residues in the entire M2 segment of the α subunit. Neuron 1994; 13: 919-927.

41. Lyon MF, King TR, Gondo Y, Gardner JM, Nakatsu Y, Eicher EM, Brilliant MH. Genetic and molecular analysis of recessive alleles at the pink-eyed dilution (p) locus of the mouse. Proc Natl Acad Sci USA 1992; 89: 6968-6972.

42. Culiat CT, Stubbs L, Nicholls RD, Montgomery CS, Russell LB, Johnson DK, Rinchik EM. Concordance between isolated cleft palate in mice and alterations within a region including the gene encoding the β3 subunit of the type A γ-aminobutyric acid receptor. Proc Natl Acad Sci USA 1993; 90: 5105-5109.

43. Nakatsu Y, Tyndale RF, DeLorey TM, Durham-Pierre D, Gardner JM, McDanel HJ, Nguyen Q, Wagstaff J, Lalande M, Sikela JM, Olsen RW, Tobin AJ, Brilliant MH. A cluster of three $GABA_A$ receptor subunit genes is deleted in a neurological mutant of the mouse p locus. Nature 1993; 364: 448-450.

44. Culiat CT, Stubbs LJ, Montgomery CS, Russel LB, Rinchik EM. Phenotypic consequences of deletion of the γ3, α5, or β3 subunit of the type A γ-aminobutyric acid receptor in mice. Proc Natl Acad Sci USA 1994; 91: 2815-2818.

45. Culiat CT, Stubbs L, Woychik RP, Russel LB, Johnson DK, Riuchik EM. Deficiency of the β3-subunit of the type A γ-aminobutyric acid receptor causes cleft palate in mice. Nature Genetics, in press.

46. Wee EL, Zimmerman EF. Involvement of GABA in palate morphogenesis and its relation to diazepam teratogenesis in two mouse strains. Teratol 1983; 28: 15-22.

47. Angelmann H. "Puppet" children. A report on three cases. Dev Med Child Neurol 1965; 7: 681-683.

48. Magenis RE, Toth-Fejel S, Allen LJ, Balck M, Brown MG, Budden S, Cohen R, Friedman JM, Kalousek D, Zonana J, et al. Comparison of the 15q deletions in Prader-Willi and Angelman syndromes: specific regions, extent of deletions, parental origin, and clinical consequences. Amer J Med Gen 1990; 35: 333-349.

49. Knoll JH, Sinnett D, Wagstaff J, Glatt K, Wilcox AS, Whiting PM, Wingrove P, Sikela JM, Lalande M. FISH ordering of reference markers and of the gene for the α5 subunit of the γ-aminobutyric acid receptor (GABAR5) within the Angelman and Prader-Willi syndrome chromosomal regions. Hum Mol Gen 1993; 2: 183-189.

50. Sinnett C, Wagstaff J, Glatt K, Woolf E, Kirkness EJ, Lalande M. High-resolution mapping of the γ-aminobutyric acid receptor subunit β3 and α5 gene cluster on chromosome 15q11-q13, and localization of breakpoint in two Angelman syndrome patients. Amer J Hum Genet 1993; 52: 1216-1229.

51. Reis A, Kunze J, Ladanyi L, Enders H, Klein-Vogler U, Niemann G. Exclusion of the $GABA_A$-receptor β3 subunit gene as the Angelman's syndrome gene. Lancet 1993; 341: 122-123.

52. Nicholls RD, Gottlieb W, Russell LB, Davda M, Horsthemke B, Rinchik EM. Evaluation of potential models for imprinted and nonimprinted components of human chromosome 15q11-q13 syndromes by fine-structure homology mapping in the mouse. Proc Natl Acad Sci USA 1993; 90: 2050-2054.

53. Bell MV, Bloomfield J, McKinley M, Patterson MN, Darlison MG, Barnard EA, Davies KE. Physical linkage of a $GABA_A$ receptor subunit gene to the DXS374 locus in human Xq28. Amer J Hum Genet 1989; 45: 883-888.

54. Buckle VJ, Fujita N, Ryder-Cook A, Derry JMJ, Barnard PJ, Lebo RV, Schofield PR, Seeburg PH, Bateson AN, Darlison MG, Barnard EA. Chromosomal localization of $GABA_A$ receptor subunit genes: relationship to human genetic disease. Neuron 1989; 3: 647-654.

55. Günther U, Benson J, Benke D, Fritschy JM, Reyes GH, Knoflach F, Crestani F, Aguzzi A, Arigoni M, Lang Y, Blüthmann H, Mohler H, Lüscher B. Benzodiazepine-insensitive mice generated by targeted disruption of the γ2-subunit gene of γ-aminobutyric acid type A receptors. Proc Natl Acad Sci USA 1995; 92: 7749-7753.

GABA: Receptors, Transporters and Metabolism
ed. by C. Tanaka & N.G. Bowery

REGULATION OF GABA$_A$ RECEPTORS BY MULTIPLE PROTEIN KINASES

Stephen J. Moss, Bernard McDonald, George H. Gorrie, Belinda K Krishek[1], and Trevor G. Smart[1].

The MRC Laboratory of Molecular Cell Biology, and Department of Pharmacology, University College London, Gordon Street, London WC1E 6BT.
[1]The Department of Pharmacology, The School of Pharmacy, 29-39 Brunswick Square London WC1N 1AX.

Summary
The role of protein phosphorylation in modulating the functional properties of recombinant and neuronal GABA$_A$ receptors was investigated using a combination of molecular, and physiological approaches. Direct phosphorylation by either PKA, or PKC inhibited receptor function, whilst tyrosine phosphorylation enhanced receptor function. Together these results suggest that GABA$_A$ receptors are under the control of multiple cell signalling pathways that can selectively up and down regulate function. Such "cross talk" may have profound effects on inhibitory synaptic transmission mediated by GABA$_A$ receptors.

Introduction

Molecular cloning has revealed a multiplicity of $GABA_A$ receptor subunits which can be divided based on homology into subunit classes with multiple members (1, 2): α (1-6), β (1-4), γ (1-4), and δ (1). This diversity is further enhanced by the alternative splicing of some subunit mRNA's, including the γ2, β2, and β4 subunits (3, 6). The subunits of $GABA_A$ receptors show extensive sequence homology to those of both nicotinic acetylcholine and glycine receptors (7, 8). Subunits of these ion channels are proposed to share many structural features, including a common transmembrane topology, (Fig. 1A) that comprises a large N-terminal extracellular domain, 4 transmembrane domains, with a major intracellular domain between transmembrane domains 3 and 4 (7, 8). *In situ* hybridisation and immunological studies have demonstrated that there is a large temporal and spatial heterogeneity of $GABA_A$ subunit expression in the CNS (1, 2, 9). $GABA_A$ receptors are believed to be pentameric hetrooligomers (10), however the exact subunit composition of a single population of $GABA_A$ receptors remains to be determined. Consensus opinion derived from a multiplicity of differing experimental approaches suggests that they consist of α, β, and γsubunits (1, 2, 7, 11).

Protein phosphorylation is widely regarded as a primary mechanism for regulating cellular processes including neurotransmission (12-14). The high levels of expression of many kinases, and phosphatases in the CNS (12, 13) suggest a role for this process in controlling neuronal function. The presence of consensus sites for protein kinases (7, 11) within the predicted intracellular domains of $GABA_A$ receptor subunits, suggests that this process may play an important role in regulating the function of $GABA_A$ receptors (14). Given the critical role of these receptors in mediating synaptic inhibition, such regulation may be of central importance in modulating neuronal excitability.

Fig 1: Proposed membrane topology of a $GABA_A$ receptor subunit

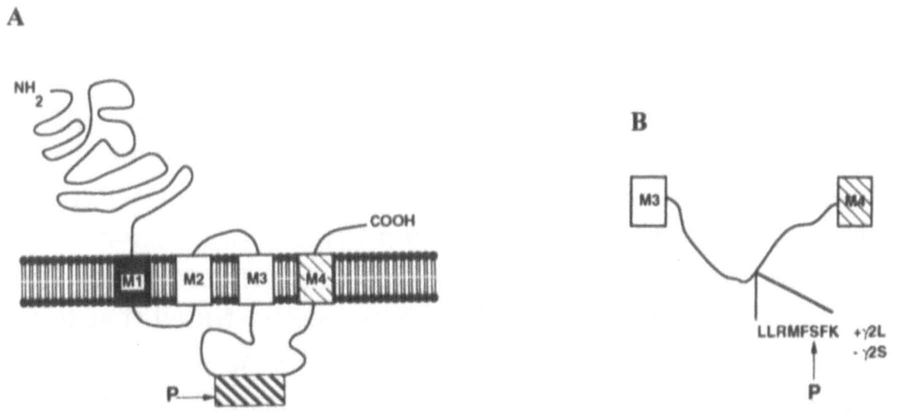

Legend The proposed membrane topology of a $GABA_A$ receptor subunit is shown in (A). The presence of consensus sites for phosphorylation (P) within the predicted major intracellular domain is shown. (B) shows the differing forms of the γ2 subunit intracellular domain generated by alternative splicing. The additional phosphorylation site present only in the γ2L subunit is indicated.

Identification of subunit phosphorylation sites.

Within the major intracellular domains of many $GABA_A$ receptor subunits, there are consensus sites for a range of both serine/threonine and tyrosine protein kinases. This is highlighted in the case of the γ2 subunit where alternative splicing produces 2 forms of this subunit (γ2L and γ2S) that differ by 8 amino acids; L-L-R-M-F-S-F-K (single letter code, Fig.1B, 3, 4, 11). This insertion, contains a serine residue, which conforms to the consensus for phosphorylation by a number of protein kinases including protein kinase C (PKC) (3, 4). In addition, alternative splicing of the chick β2, β4 and the human β2 subunit mRNAs generates multiple forms of these subunits, which all differ in the predicted major intracellular domain (3-6). In the case of the β2 splice variants, in both human and the chick, the added exons contain consensus sites for a number of serine/threonine protein kinases (5, 16).

The existence of an apparent consensus site for phosphorylation does not assure that a protein can be phosphorylated, nor is it complete indicator of which kinase(s) phosphorylates the site (15). Therefore to study the role of phosphorylation in modulating $GABA_A$ receptor function, it is essential to determine which subunits can be phosphorylated, and then relating subunit phosphorylation to receptor function.

Studies utilising benzodiazepine affinity purified preparations of $GABA_A$ receptors, have demonstrated that these receptors are substrates of a number of differing protein kinases *in vitro.* Both cAMP dependent protein kinase (PKA), and PKC phosphorylate "β" subunits based on molecular mass (17, 18)). In addition an "α" subunit can be phosphorylated by an unidentified serine/threonine protein kinase (19). Purified receptor preparations can also be phosphorylated by vSRC within the β and γ subunits (20). Due to the heterogenous nature of purified receptor preparations, and the low abundance of $GABA_A$ receptors in the CNS (1, 2), the precise identification of the subunits phosphorylated in these studies has proven difficult.

To overcome these problems the expression of the intracellular domains of $GABA_A$ receptors has been utilised. The system of choice for these purposes is the expression as glutathione-S-transferase (GST) fusion proteins in, *E. coli* (4, 20, 21) The intracellular domains of $GABA_A$ receptor subunits when expressed as GST fusion's are highly soluble, and can be purified to homogeneity on immobilised glutathione. Using a combination of molecular and biochemical methodologies, it has been possible to identify phosphorylation sites for a number of protein kinases within the predicted major intracellular domains of a range of $GABA_A$ receptor subunits (see table 1). These studies have demonstrated that the β subunits of $GABA_A$ receptors are high affinity substrates of a number of well characterised protein kinases, widely expressed in the CNS. Significantly all murine β subunits (β1-3) are phosphorylated to high stoichiometry with high affinity on a conserved serine residue (S409 in the β1 and β3 subunits, S410 in the β2), by PKA, PKC, Ca^{+2}/calmodulin type 2 dependent protein kinase (CAM KII), and cGMP dependent protein (PKG) kinase *in vitro* (table 1) (22, 23 McDonald and Moss unpublished observations). This site, of consensus: R-R-R-A(β1,β2)/S (β3)-S̲-L-Q-K is found in all β subunits isolated from a variety of vertebrate species, in addition to a β subunit isolated from the invertebrate *Lymnea* (1, 2, 24). There are additional sites in the murine β1 and β3 subunits for CAM KII phosphorylation identified as S384 and S383 respectively (23, McDonald and Moss unpublished observations). These observations suggest an important role for receptor β subunits in conferring cellular regulation upon $GABA_A$ receptors via phosphorylation. In addition to the phosphorylation by serine/threonine protein kinases, the intracellular domain of the β1 subunit can also be phosphorylated by vSRC (20).

Table 1: Subunit phosphorylation sites identified by *invitro* phosphorylation.

Substrate	PKA	PKG	PKC	CAM KII
GST - α1	-	-	-	-
GST - α6	-	-	-	-
GST - β1	S409	S409	S409	S409/S384
GST - β2	S410	S410	S410	S410
GST - β3	S409	S409	S409	S409/S383
GST - γ2L	-	-	S343/S327	S343/ S348/T350
GST - γ2S	-	-	S327	S348/S350

Legend: Subunit M3/M4 intracellular domains were expressed as GST fusion proteins. Phosphorylation sites were identified utilising site specific mutagensis, candidate Serine (S) or Threonine (T) residues were converted to alanine residues. - indicates not phosphorylated.

Phosphorylation of the γ2 subunit has been examined using similar methodologies. These studies demonstrate that both the γ2L and γ2S subunits are high affinity substrates of both PKC and CAM KII, however the γ2L intracellular domain is phosphorylated to a higher stoichiometry. This is due to the presence of an additional site for both of these kinases (S343) within the 8 amino acid insertion that differentiates the 2 forms of the γ2 subunit (3, 22). Additional distinct sites for these 2 kinases have been identified in both forms of the γ2 intracellular domain as S327 for PKC, and S348 and T350 for CAM KII respectively (22, 23 Table 1). These forms of the γ2 subunit have differing spatial and temporal patterns of expression in the brain (25), suggesting an important role for phosphorylation in the differential regulation of $GABA_A$ receptor function. Neither form of the γ2 subunit is phosphorylated by either PKA, or PKG, however the γ2L subunit intracellular domain can be phosphorylated by vSRC (20). The intracellular domains of the α1, α2, and α6 subunits are not significantly phosphorylated by PKA, PKC, PKG, CAM KII, or SRC (21, 22 McDonald, Hanley, and Moss unpublished). Overall these *in vitro* studies suggest that the β and γ2 subunits will be the major sites of phosphorylation within $GABA_A$ receptors *in vivo*.

The phosphorylation of $GABA_A$ receptor subunits expressed in mammalian cell lines has also been analysed. Murine $GABA_A$ receptors composed of either α1β1 or α1β1γ2S, subunits expressed in A293 cells are phosphorylated specifically by PKA on S409 (26). The β2 and β3 subunits are also phosphorylated on S410 and S409 respectively by PKA when expressed with α1 and γ2 subunits in A293 cells (McDonald and Moss unpublished observation). The phosphorylation of receptors composed of α1β1, α1β1γ2S and α1β1γ2L subunits by PKC has also been investigated. The α isoform of PKC specifically phosphorylates S409 in the β1 subunit in receptors produced form these subunit combinations (27). Studies of the phosphorylation of the γ2 subunit have proven extremely due to problems in producuing high affinity antisera against this subunit, and its extreme susceptible to proteolysis (25, 27).

These problems have been overcome by epitope tagging of the γ2L subunit, this has facilitated the study of $GABA_A$ receptor phosphorylation by tyrosine kinases. Co-expression of $GABA_A$ receptors consisting of α1β1 and γ2L subunits with the well characterised protein tyrosine kinase vSRC, the cellular homologue of which is widely expressed in the brain (13, 14), results in phosphorylation of the γ2L subunit on residues Y365 and Y367.

The β1 subunit is also phosphorylated but to a much lower stoichiometry on Y384 and Y386 (29). As with the sites for serine phosphorylation, the sites for tyrosine phosphorylation are found within the predicted intracellular domain of these 2 subunits.

Differential functional effects of serine and tyrosine phosphorylation on $GABA_A$ receptor function

The functional effects of phosphorylation on $GABA_A$ receptor function have been examined in a number of differing neuronal preparations. The effects appear complex and are often contradictory (14, 30). This in part may be due to $GABA_A$ receptor heterogeneity in the brain (1, 2, 9). These observations are however complicated by the fact that many drugs used to activate protein kinases, such as forskolin, and cAMP derivatives, appear to have effects on $GABA_A$ receptors that are independent of phosphorylation, and most likely result from direct interaction with the receptor (30). The use of heterologous expression to produce $GABA_A$ receptors of defined subunit composition is a powerful method of examining receptor regulation by phosphorylation, as the state of receptor phosphorylation can be directly related to function, via site specific mutagensis.

Using this approach the functional effects of PKA mediated phosphorylation have been studied on $GABA_A$ receptors expressed in A293 cells, composed of either α1β1, or α1β1γ2S subunits. Phosphorylation of S409 in the β1 subunit mediated by either intracellular dialysis with cAMP, or coexpression with the catalytic subunit of PKA, resulted in a time dependent decrease of GABA induced currents (26). The magnitude of this effect is however, dependent on the subunit composition of the expressed receptor. In the case of receptors composed of α1β1 subunits the rate of rapid desensitisation is also modulated (26). All of these functional effects could be abolished by mutation of S409, in the β1 subunit the sole site of PKA phosphorylation in these receptors (26). Similar modulation of $GABA_A$ function has been seen in cultured superior cervical ganglia (SCG) and spinal cord neurones, using intracellular dialysis with the catalytic subunit of PKA (26, 28). In the retina and cerebellum however, PKA activation appears to enhance $GABA_A$ receptor function (2, 14). The reasons for these differing functional effects of PKA phosphorylation on receptor function may be due to heterogeneity of $GABA_A$ receptor structure or the differential expression of specific receptor isoforms during development (1, 2, 9, 14). Alternatively the effects in these differing neurone types may not be due to direct receptor phosphorylation by PKA, but may be mediated by the activation of other unknown PKA stimulated signalling pathways. In addition PKA phosphorylation has also been suggested to play a role in controlling receptor assembly, via phosphorylation of the β1 subunit (31).

The effects of PKC phosphorylation have been examined using similar methodologies, utilising receptors composed of α1 β1, α1β1γ2S and α1β1γ2L, subunits. PKC phosphorylation reduces GABA mediated currents via phosphorylation of S409 (β1), S327 (γ2L /γ2S) and S343 (γ2L) as determined by site specific mutagenesis (Fig 2A-D, 27, 32). The effects of phosphorylating these residues do not sum in a linear fashion, however the biggest functional effect was seen on phosphorylating S343 in the γ2L subunit (27). Similar regulation of $GABA_A$ receptor function has been observed in a number of neuronal systems (14, 27, 33). Single channel recordings suggest that serine phosphorylation mediated either by PKA, or PKC decrease the frequency of channel opening (Smart and Moss unpublished observations). Intracellular dialysis of L929 cells expressing $GABA_A$ receptors composed of, α1β1 and γ2L subunits with trypsin activated PKC has been reported to enhance receptor function (34). Whether these effects are mediated by direct phosphorylation by PKC however has not been reported.

The modulation of $GABA_A$ receptors composed of α1 β1 and γ2L subunits expressed in A293 cells by tyrosine phosphorylation has also been analysed. These cells have low steady state levels of phosphotyrosine indicating low endogenous kinases, and or high phosphatase activities. This allows $GABA_A$ receptor tyrosine phosphorylation to be controlled by co-expressing activated tyrosine kinases, or by introducing purified kinases by intracellular dialysis (29). Co-expressing receptors comprised of α1 β1 and γ2L subunits with vSRC, results in constitutive $GABA_A$ receptor tyrosine phosphorylation. The major sites of phosphorylation are Y365, and Y367 within the predicted

Fig 2. Differential functional effects of PKC and SRC mediated phosphorylation on GABA receptors expressed in A293 cells

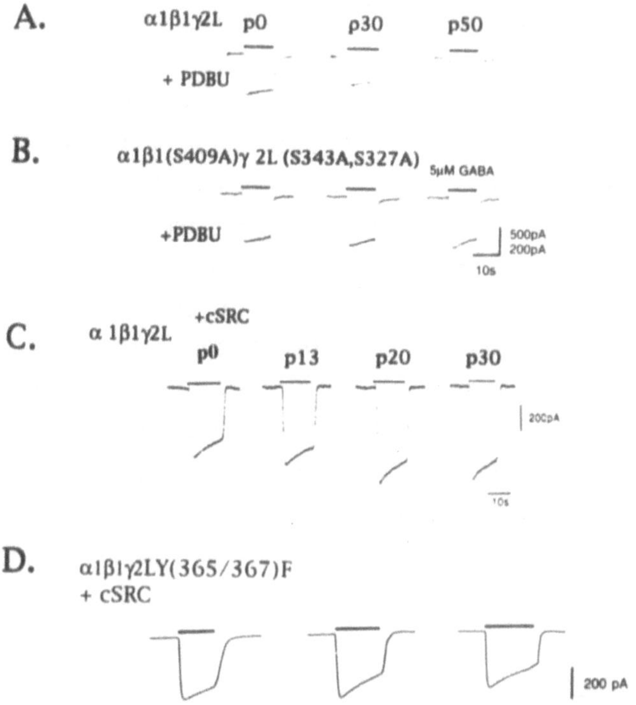

Legend: GABA activated currents analysed from transfected A293 cells expressing receptors composed of α1β1 and γ2L subunits, or mutated forms of the β1 and γ2L subunits in which biochemically defined receptor phosphorylation sites for either PKC or SRC had been mutated. Cells were exposed to phorbol dibutarate (PDBU), to test the effects of PKC phosphorylation on receptor function (A, B).

intracellular domain of the γ2L subunit (29). In addition the β1 subunit is also phosphorylated but to much lower stoichiometry on Y384 and Y386 (29). Tyrosine kinase inhibitors decrease GABA induced currents, an effect which can be blocked by mutating the sites for tyrosine phosphorylation identified on the γ2L subunit (29). These results suggest that tyrosine phosphorylation may be a means of maintaining or enhancing $GABA_A$ receptor function (14, 35). To confirm this, experiments were performed in A293 cells expressing $GABA_A$ receptors alone, which are not tyrosine phosphorylated. Intracellular dialysis with purified cSRC (Fig. 2 C-D)lead to a robust enhancement of GABA induced response (29). This effect was also abolished by mutating the sites for tyrosine phosphorylation within the γ2L subunit. Similar enhancements are also seen in $GABA_A$ responses recorded form SCG neurones. Single channel recording from SCG neurones suggest that tyrosine phosphorylation increases both mean open time and also the probability of channel opening (29).

The effects of tyrosine phosphorylation were analysed via intracellular dialysis with purified cSRC (C, D). The records were obtained at various times (p) after formation of the whole cell configuration, GABA was applied for the duration of the solid line (10 seconds). Holding potential in all cells was -50 mv

Conclusions and future directions.

$GABA_A$ receptor subunits can be phosphorylated by a multiplicity of protein kinases within the predicted intracellular domains of the β(1-3), γ2L and γ2S subunits (17-19, 26-30). Phosphorylation by PKC or PKA of sites in the β and γ2 subunits inhibit receptor function (14 26-28, 32-33 cf 31). In contrast to this, phosphorylation of tyrosine residues within the γ2 subunit enhance receptor function (29). The γ2 subunit is important in conveying benzodiazepine modulation upon $GABA_A$ receptors (1, 2, 11) and is therefore likely to be present in most neuronal $GABA_A$ receptor subtypes. Likewise $GABA_A$ receptor β subunits are believed to be crucial for receptor function(1, 2, 7). Together these results suggest that *in vivo*, $GABA_A$ receptors will be under the control of multiple cell signalling pathways that can selectively up and down regulate receptor function. Such regulation may have significant effects on neuronal excitability. The importance of this regulation is being assessed by using viral and transgenic methodologies, to express mutated receptor subunits, devoid of phosphorylation sites in neuronal networks. This will establish if receptor phosphorylation plays a significant role in controlling efficacy of inhibitory synaptic transmission mediated via $GABA_A$ receptors

Acknowledgements

This work was supported by the MRC and Welcome trust. We thank Dr Weasel for his help with solutions.

References

(1) Olsen, R.W. and Tobin, A.J. (1990) FASEB J. **4** 1469-1480.
(2) Macdonald R. L. and Olsen R.W. (1994) Ann. Rev. Neurosci **17** 569-602.
(3) Kofuji, P et al. (1991) J. Neurochem. **56** 713-715.
(4) Whiting, P. et al. (1990) P.N.A.S. (USA) **87** 9966-9970.
(5) Harvey, R.J. et al. (1994) J. Neurochem **62** 10-16.
(6) Bateson, A.N. et al (1991) J. Neurochem **56** 1437-1440.
(7) Schofield, P.R. et al. (1987) Nature **328** 221-227.
(8) Unwin N. (1993) Cell 72 Vol **10** 31-41.
(9) Wisden, W. and Seeburg, P.H. (1992) Curr. Opin. Neurobiol. **2**, 263-269.
(10) Nayeem, N. M. et al. (1994) J. Neurochem **62** 815-818.
(11) Pritchett, D.B. et al. (1989) Nature **338** 582-585.
(12) Hemmings, H. G., et al. (1989) FASEB J. **3** 1583-1592.
(13) Wagner, K.R. et al. (1991) Curr. Opin Neurobiol **1** 65-73.
(14) Swope, S.L. et al. (1992) FASEB J. **6** 2514-2523 (1992).
(15) Kennelly, P.J. and Krebs, E.G. (1991) J. Biol. Chem. **266** 15555-15557.
(16) McKinley, D.D. et al. (1995) Mol. Brain Res **28** 175-179
(17) Kirkness, E.F. et al (1989) Biochem. J. **259**, 613-616.
(18) Browning, M.D. et al. (1990) P.N.A.S. (USA). **87**, 1315-1318.
(19) Sweetnam, P.M. et al. (1988) J. Neurochem. **51**, 1274-1284.
(20) Valenzuela, et al. (1995) Mol. Brain Res. (In press).
(21) Smith, D.B. and Johnson K.S. (1989) Gene **67** 31-42.
(22) Moss, S. J. et al. (1992) J. Biol. Chem. **267** 14470-14476.
(23) McDonald, B. M. and Moss S.J. (1994) J. Biol Chem. **269**18111-18117.
(24) Harvey, R.J. et al. (1991) EMBO J. **10** 3239-3245.
(25) Glencourse, T.A. et al. (1993) E.J. Neurosci **4,** 271-278.
(26) Moss, S.J. et al. (1992) Science **257** 661-665.
(27) Krishek, B.J. et al. (1994) Neuron **12** 1081-1095.

(28) Porter, N.M. et al. (1990) Neuron **5** 789-796.
(29) Moss, S.J. et al. (1995) submitted for publication.
(30) Leidenheimer, N.J. et al. (1991) TIPS. **12**:84-87.
(31) Angelotti, T. P. et. al. (1993) Mol. Pharmacol **44** 1202-1210.
(32) Kellenberger, S. et al. (1992) J. Biol. Chem. **267** 25660-25663.
(33) Ledenheimer, N.J. et al. (1992.) Mol. Pharmacol. **41** 1116-1123.
(34) Lin, Y.F. et al. (1994) Neuron **13** 1421-1431.
(35) Stelzer, A. et al. (1988) Science **241** 339-341 (1988).

GABA: Receptors, Transporters and Metabolism
ed. by C. Tanaka & N.G. Bowery

ENHANCEMENT OF GABA-INDUCED CURRENT BY 20-HYDROXY-ECDYSONE IN CULTURED CORTICAL NEURONS

M. Sasa, S. Tsujiyama, K. Ishihara, [1]R. Hanaya, [2]M. Fujita, [1]K. Kurisu, [2]K. Yajin and [3]T. Serikawa

Department of Pharmacology, [1]Neurosurgery and [2]Oto-Rhino-Laryngology, Hiroshima University School of Medicine, Hiroshima 734, and [3]Institute of Laboratory Animals, Faculty of Medicine, Kyoto University, Kyoto 606, Japan

Summary: Effects of a neurosteroid, 20-hydroxyecdysone (20-HE), on epileptic seizures of spontaneously epileptic rat (SER: zi/zi, tm/tm), medial vestibular nucleus (MVN) neurons and GABA-induced currents in primarily cultured cortical neurons were examined. 20-HE (10-50mg/kg) inhibited tonic convulsions in SER without affecting the absence-like seizures. The inhibition was antagonized by pretreatment with bicuculline. The rotation-induced firing of the MVN neuron was inhibited by microiontophoretically applied 20-HE. Similarly this inhibition was antagonized by bicuculline. Using whole-cell patch clamp method application of 20-HE with the U-tube system did not affect the membrane potential, but potentiated GABA-induced hyperpolarization and current at all membrane potentials without affecting the reversal potential of approximately -18mV. These results suggest that 20-HE acts on $GABA_A$ receptor, probably at a modulatory site to inhibit the seizure and response of MVN neurons.

Introduction

GABAA receptors contain modulatory sites responsible for the action of benzodiazepine, barbiturate and neurosteroids (1). In contrast to the facts that benzodiazepine and barbiturate act on GABAA receptors as an agonist, neurosteroids show a complex action on the receptors (2). For instance, pregnenolene sulfate and dehydroepiandrosterone sulfate inhibit GABA-induced Cl^- influx, thereby showing excitatory action of neurons (3, 4, 5, 6). However, alfaxalone and tetrahydroproigesterone potentiate GABA-induced currents by directly opening GABAA receptor-coupled Cl^- channel (7, 8), although multiple recognition sites of GABAA receptors for neuroactive steroids have been demonstrated (2, 9).

One of the neuroactive steroids 20-hydroxyecdysone (20-HE) (Fig. 1), is known to be a biologically active endogenous ecdysteroid hormone related to metamorphosis and certain behaviors in some insects (10, 11). However, since the action of this neurosteroid on central nervous system remains unclear, the effects of 20-HE on epileptic seizures of spontaneously epileptic rat which is a double mutant (SER: zi/zi, tm/tm) (12), medial vestibular nucleus (MVN) neurons of anesthetized rats and primarily cultured cortical cells were examined in this study.

Materials and Methods

Effects on epileptic seizures in SER

SER shows both tonic convulsion and absence-like seizures characterized by a sudden appearance of 5-7Hz spike-wave complex in cortical and hippocampal EEG (12). The electrodes for recording EEG were chronically implanted in the left cerebral cortex and hippocampus of SER (aged: 9-10 weeks) under pentobarbital anesthesia. After one-week recovery period the effects of 20-HE on seizures of the animals were tested. After 30-min control recording of EEG, 20-HE (dissolved in polyethylene-glycol) was intraperitoneally administered at 10, 20, 30 and 50mg/kg. Blowing stimulus was given to the back of the animal every 5min to induce tonic convulsion and maintain alertness in the animal during continuous EEG recording for 2hr. The number and total duration of tonic convulsion and absence-like seizures of post-20-HE administration were measured with reference to the EEG and compared with the control data using paired Student's t-test.

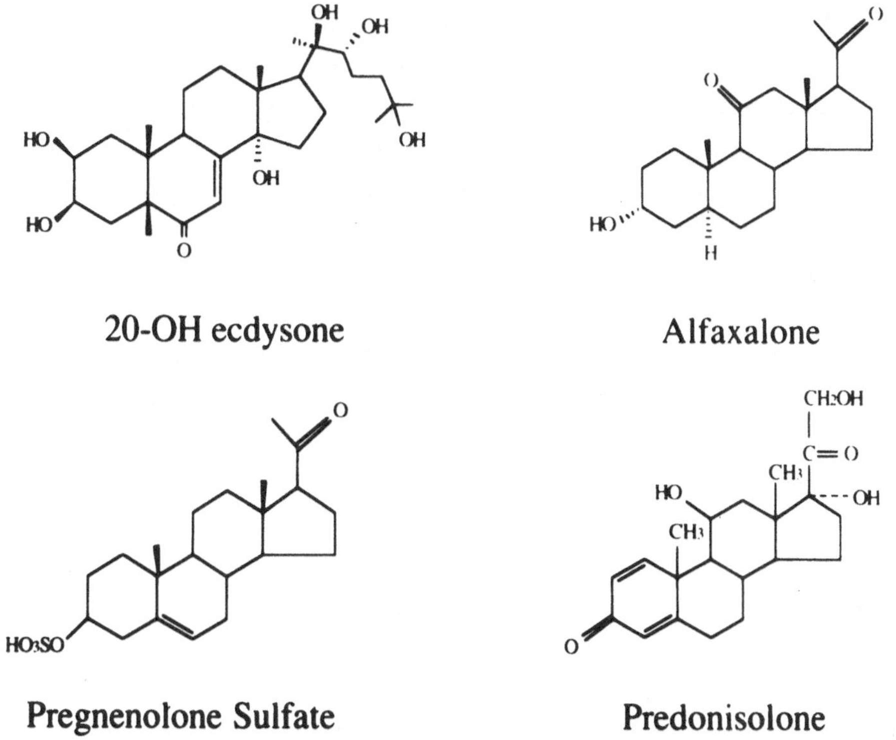

Figure 1 Chemical structures of neurosteroids

Effects on MVN neurons

Chloral hydrate-anesthetized Wistar rat was fixed in a stereotaxic instrument placed on the turn-table. Extracellular action potentials were discriminated and traced on the recticorder via a spike-counter using a silver-wire microelectrode attached along a seven-barreled micropipette. Each of the pipette was filled with 20-HE, bicuculline, glutamate, or 2M NaCl accordingly. These chemicals were microiontophoretically applied to the neurons recorded. The turn-table was manually and sinusoidally rotated in directions ipsilateral and contralateral to the recording site (13).

Effects on cultured cortical neurons

Primary cultured cortical neurons obtained from Wistar rat fetuses on pregnancy day 18 were used for the study (14). Cells cultured for 14-16 days were placed in a chamber continuously perfused with a solution containing: 165mM NaCl, 5mM KCl, 2mM $CaCl_2$, 10mM D-glucose and 5mM HEPES (pH: 7.3) with and without 0.3μM tetrodotoxin (Wako, Osaka, Japan) in the voltage and current clamp studies, respectively. Drugs such as 20-HE (Daicel, Tokyo, Japan), GABA (Katayama, Osaka, Japan) and (+)-bicuculline (Sigma, St. Louis, U.S.A.) were directly applied to target cells using the U-tube system (15, 16). The patch microelectrode (3-5MΩ) contained 135mM K-methanesulfonic acid, 5mM KCl, 1mM $CsSO_4$, 2mM $MgSO_4$, 11mM EGTA and 10mM HEPES (pH: 7.2) for the current clamp study, while 80mM CsCl, 80mM CsF, 10mM EGTA and 10mM HEPES (pH: 7.4) were used for the voltage clamp investigation.

The membrane potential and current amplified by Axopatch 200A (Axon Instrument, U.S.A.) were displayed on the oscilloscope and stored in a computer using DIGIDATA 1200 data acquisition board (Axon Instrument). The peak amplitudes of GABA-induced currents and hyperpolarizations were measured with the pCLAMP software (Axon Instrument) (14).

Results

Effects on epileptic seizures in SER

SER, which is obtained by mating heterozyte tremor rat (tm/+) with homozyte zitter rat, spontaneously shows both tonic convulsion and absence-like seizures characterized by sudden appearance of 5-7Hz spike-wave complex in cortical and hippocampal EEG (12). Light stimuli such as air-puff on the body easily induce tonic convulsion in SER. This convulsion is inhibited by phenytoin, phenobarbital, valproate, carbamazepine and diazepam, but not trimethadione or ethosuximide, which is effective against absence-like seizures in SER. Therefore, this animal is an useful model for evaluating antiepileptic drugs. When 20-HE (at 10, 20, 30 and 50mg/kg) was intraperitoneally administered to SER, the tonic convulsion was inhibited with 50mg/kg in all 4 animals examined (Fig. 2). This inhibition was antagonized by pretreatment with bicuculline (1mg/kg), a $GABA_A$ receptor antagonist. However, the absence-like seizures were not significantly affected by 20-HE up to 50mg/kg.

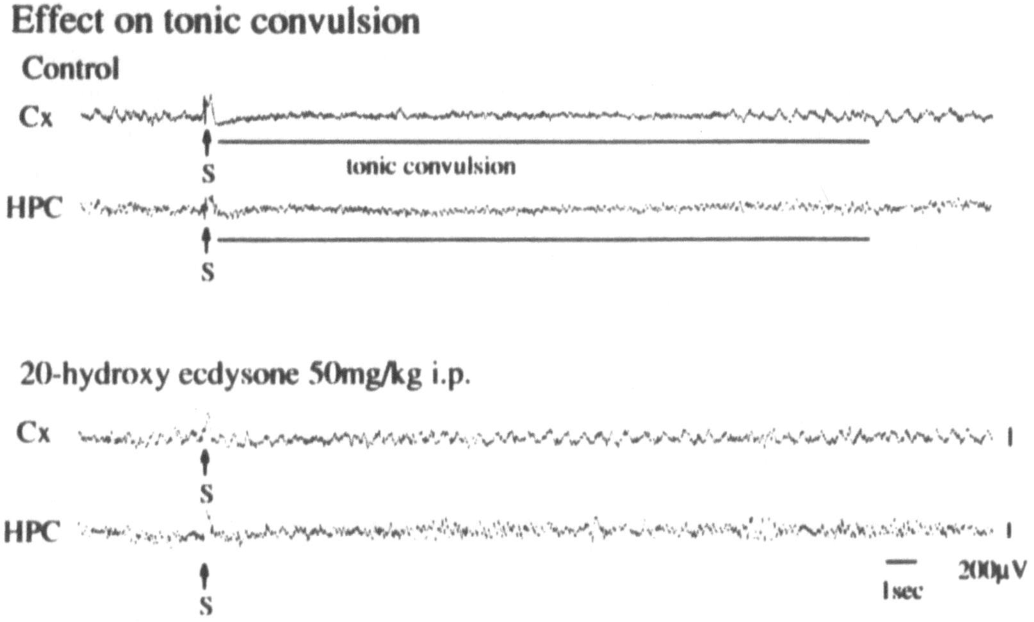

Figure 2 Effects of 20-hydroxyecdysone on tonic convulsion induced by blowing stimuli on the back of spontaneously epileptic rat (SER).
The animal shows low-voltage fast activity on cortical and hippocampal EEG during tonic convulsion. Cx: Cerebral cortex, HPC: hippocampus

Effects on MVN neurons

The MVN neurons are known to receive dense GABAergic innervations from cerebellar Purkinje cells and intranuclear GABA-containing neurons. Two type of MVN neurons receiving semicircular inputs were classified according to responses to the horizontal, sinusoidal rotation; type I showed an increase and decrease in firing when ipsilateral and contralateral to the recording site, respectively, and type II rotated responds to rotation in directions contradictory to those type I neurons (13) (Fig. 3). The effects of 20-HE were examined on type I neurons. Microiontophoretic applications of 20-HE at doses of 10-60nA for 60sec inhibited the rotation-induced increase in firing of the MVN type I neurons in a dose-dependent manner. This inhibition was abolished with concomitant application of bicuculline up to 60nA. The firing induced by ipsilateral rotation, which was significantly reduced to 65.5% (n=7) of the control value by 20-HE at 60nA, recovered to 103.1% (n=5) in the presence of bicuculline (60nA).

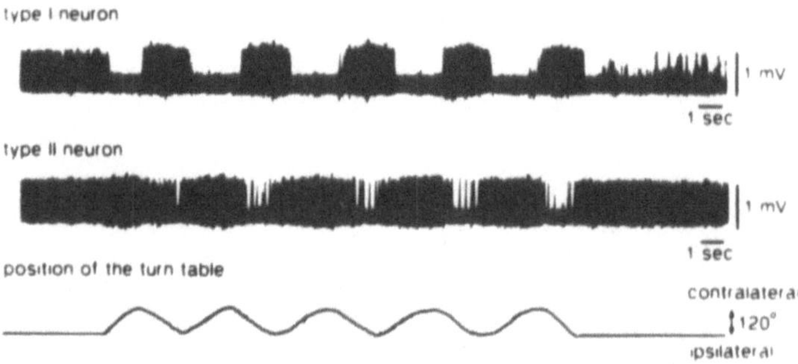

Figure 3 Responses of type I and II neurons in medial vestibular nucleus to horizontal, sinusoidal rotation.

Effects on primary cultured cortical cells

When GABA of 1, 10, 100, 300 and 1000μM was ejected to cells for 5sec by the U-tube system, a dose-dependent hyperpolarization was obtained with the current clamp. Similarly, a dose-dependent inward current was also observed with GABA under voltage clamp at -60mV (Fig. 4C). The estimated EC50 of GABA for inward current was 10.9μM. The reversal potential was -18mV (Fig. 4B). This coincided well to the theoretical value of -19.9mV when calculated with the Nernst equation. The GABA-induced hyperpolarizations and currents were completely blocked by concomitant bath application of bicuculline (100μM).

When 20-HE alone (up to 1000μM) was ejected on the cell for 5sec, neither obvious alteration of the membrane potentials nor induction of current in the current and voltage clamps was observed, respectively. However, concomitant application of 20-HE with GABA for 5sec potentiated GABA-induced hyperpolarization and produced a dose-dependent inward current (Fig. 4A). The potentiation by 20-HE (100μM) of the GABA (10μM)-induced inward current was observed at all remembrane potentials tested without changes in the reversal potential (Fig 4B).

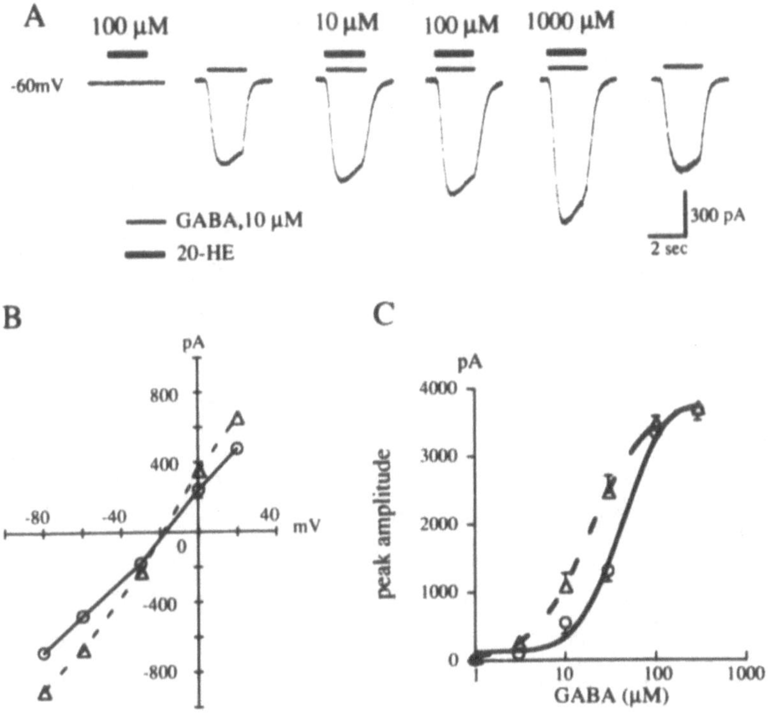

Figure 4 Effects of 20-hydroxyecdysone (20-HE) on GABA-induced current in cultured rat cortical neurons. (Ref. 14 with permission from Jpn. J. Pharmacol.)

A: Dose-dependent increase by 20-HE in GABA (10μM)-induced current, B: Voltage-current relationship of GABA (10μM) in the presence (Δ) and absence (O) of 20-HE (100μM), C: Concentration-response curves in the presence (Δ) and absence (O) of 20-HE (100μM) (mean ± S.E., n=3). The curves were obtained with the least-square fit to an equation of the form

$$R_{exp} = \frac{R_{max}}{1 + \left(\frac{EC_{50}}{[A]}\right)^n}$$

where [A] is the concentration of the agonist, n is the Hill coefficient, R_{exp} is the expected response and R_{max} is the maximum response.

Discussion

Alfaxalone potentiates GABA-induced current in cortical neurons at concentrations in the nM range and induces bicuculline-sensitive Cl^- current at higher concentrations in the μM range (8, 17). In contrast, 20-HE enhanced GABA-induced inward current in the cortical neurons in a dose-dependent manner without producing any current (14). These results are in line with the findings that steroids with 17β-substituents potentiate the GABAA receptor-gated Cl^- current (18). It is suggested, therefore, that 20-HE acts on the modulatory site of the GABAA receptor to potentiate the GABA-induced current.

The tonic convulsion of SER is inhibited by conventional antiepileptic drugs effective for grand mal epilepsy, such as phenytoin, and agents acting on the GABAA receptor such as diazepam (12, 19). In the present study, tonic convulsion of SER was inhibited by 20-HE. In addition, bicuculline at a dose that did not precipitate the seizure antagonized against 20-HE-induced inhibition of the seizure. Therefore, 20-HE may have acted on the modulatory site of GABAA receptors, thereby potentiating endogeneous GABA effects to inhibit the tonic convulsion. Actually, the SER hippocampal CA3 field neurons, which showed long-lasting depolarization shift accompanied by repetitive firing with a single stimulus given to the mossy fiber, were much more sensitive to GABA than those of normal Wistar rats. Furthermore, 20-HE produced bicuculline-reversible inhibition of the rotation-induced firing of MVN type I neurons. Type I neurons receive excitatory inputs from the peripheral vestibule and inhibitory inputs from the cerebellar Purkinje cells as well as the branch of contralateral type II neurons, according to the nucleus related to eye movement. The inhibition of rotation-induced firing of type I neurons by 20-HE was probably due to potentiation of endogeneous GABAergic inhibition, since GABA also inhibited firing of the same neuron in a manner similar to that of 20-HE. The 20-HE-induced inhibition was antagonized by bicuculline. Thus, 20-HE appeared to have modulated the vestibulo-ocular reflex.

In conclusion, 20-HE probably acts on the modulatory site of GABAA receptors to potentiate the GABA-induced Cl^- current, thereby inhibiting the epileptic seizures in SER and modulating the vestibulo-ocular reflex in normal rats.

Acknowledgment

The authors are grateful to Drs. Y. Yamaguchi and T. Matsumoto, Daicel Chemical Industries, for the gift of 20-hydroxyecdysone. We would like to thank Ms S. Makita for her assistance in preparing the manuscript.

References

1. Sieghart W. GABAA receptors: ligand-gated Cl^- ion channels modulated by multiple drug-binding sites. Trends Pharmacol Sci 1992; 13: 446-450.

2. Paul SM. Purdy RH. Neuroactive steroids. FASEB J 1992; 6: 2311-2322.

3. Majewska MD, Schwartz RD. Pregnenolone-sulfate: an endogenous antagonist of the γ-aminobutyric acid receptor complex in brain? Brain Res 1987; 404: 355-360.

4. Majewska MD, Mienville JM, Vicini S. Neurosteroid pregnenolone sulfate antagonizes electrophysiological responses to GABA in neurons. Neurosci Lett 1988; 90: 279-284.

5. Majewska MD, Demirgoren S, Spivak CHE, London ED. The neurosteroid dehydroepiandrosterone sulfate is an antagonist of the GABAA receptor. Brain Res 1990; 526: 143-146.

6. Carette B, Poulain P. Excitatory effect of dehydroepiandrosterone, its sulphate esther and pregnenolone sulphate, applied by iontophoresis and pressure, on single neurons in the septo-optic area of the guinea pig. Neurosci Lett 1984; 45: 205-210.

7. Harrison NL, Majewska MD, Harrington JW, Barker JL. Structure-activity relationships for steroid interaction with the γ-aminobutyric acidA receptor complex. J Pharmacol Exp Ther 1987; 241: 346-353.

8. Cottrell GA, Lambert JJ, Peters JA. Modulation of GABAA receptor activity by alphaxalone. Br J Pharmacol 1987; 90: 491-500.

9. Shingai R, Sutherland ML, Barnard EA. Effects of subunit types of the cloned GABAA receptor on the response to a neurosteroid. Eur J Pharmacol 1991; 206: 77-80.

10. Truman JW, Talbot WS, Fahrbach SE, Hogness DS. Ecdysone receptor expression in the CNS correlates with stage-specific responses to ecdysteroids during Drosophia and Manduca development. Development 1994; 120: 210-234.

11. Prugh J, Croce KD, Levine RB. Effects of the steroid hormone, 20-hydroxyecdysone, on the growth of neurites by identified insect motoneurons in vitro. Dev Biol 1992; 154: 331-347.

12. Sasa M, Ohno Y. Ujihara H, Fujita Y, Yoshimura M, Takaori S, Serikawa T, Yamada J. Effects of antiepileptic drugs on absence-like and tonic seizures in the spontaneously epileptic rat, a double mutant rat. Epilepsia 1987; 29: 503-513.

13. Kawabata A, Sasa M, Ujihara H, Takaori S. Inhibition by enkephalin of medial vestibular nucleus neurons responding to horizontal pendular rotation. Life Sci 1990; 47: 1355-1363.

14. Tsujiyama S, Ujihara H, Ishihara K, Sasa M. Potentiation of GABA-induced inhibition by 20-hydroxyecdysone, a neurosteroid, in cultured rat cortical neurons. Jpn J Pharmacol 1995; 68: 133-136.

15. Ujihara H, Albuquerque EX. Ontogeny of *N*-methyl-D-asparate-induced current in cultured hippocampal neurons. J Pharmacol Exp Ther 1992; 263: 859-867.

16. Ujihara H, Sasa M, Ban T. Selective blockade of P-type channels by lead in cultured hippocampal neurons. Jpn J Pharmacol 1995; 67: 267-269.

17. Mistry DK, Cottrell GA. Actions of steroids and bemegride on the GABAA receptor of mouse spinal neurons in culture. Exp Physiol 1990; 75: 199-209.

18. Hu Y, Zorumski CF, Covey DF. Neurosteroid analogues: Structure-activity studies of benzindene modulators of GABAA receptor function. 1. The effect of 6-methyl substitution on the electrophysiological activity of 7-substituted benzindene-3-carbonitriles. J Med Chem 1993; 36: 3956-3967.

19. Sasa M, Ujihara H, Ishihara K, Ohno Y, Fujita Y, Yoshimura M, Nakamura J, Serikawa T, Yamada J, Takaori S. Responsiveness of the spontaneously epileptic rat (SER), a double mutant, to antiepileptic drugs. Adv Neurol Sci 1989; 33: 909-918.

GABA: Receptors, Transporters and Metabolism
ed. by C. Tanaka & N.G. Bowery

DIFFERENTIAL SENSITIVITY DURING DEVELOPMENT OF GABA-MEDIATED SYNAPTIC TRANSMISSION TO A NEUROACTIVE STEROID IN RAT HIPPOCAMPUS AND CEREBELLUM

Elizabeth J. Cooper, Graham A. R. Johnston and Frances A. Edwards

Department of Pharmacology, The University of Sydney, NSW 2006 Australia

Summary: The neuroactive steroid 5α-pregnan-3α,21-diol-20-one (THDOC) at 1 μM reversibly potentiated GABA-mediated synaptic events recorded in dentate gyrus granule cells and cerebellar Purkinje cells in brain slices 18-21 day old male rats. Lower concentrations of THDOC were ineffective. This was also true for cerebellar Purkinje cells from 10-13 day old animals. In contrast, in dendate granule cells from these younger animals, GABA-mediated synaptic events were considerably more sensitive to THDOC with a 20% increase in the decay time constant of the ipscs induced by 50 nM THDOC. While THDOC levels of 1 μM are unlikely under normal physiological conditions, levels of 50 nM may be possible in discrete brain regions during stress.

Introduction

The CNS depressant actions of steroids has been known since 1927 when Cashin and Moravek (1) injected a colloidal suspension of cholesterol into cats causing deep anaesthesia, but it was not until the extensive investigations of Seyle (2) that it became apparent that a wide range of natural and synthetic steroids have anaesthetic actions. These studies led to the development of steroid anaesthetic agents, such as alfaxalone. Electrophysiological studies showed that alfaxalone selectively enhanced the activation of $GABA_A$ receptors by the inhibitory neurotransmitter GABA, thus providing a basis for the anaesthetic action of alfaxalone involving a specific receptor site (3).

Majewska and her colleagues (4) discovered that steroid hormone metabolites found in the brain are 'barbiturate-like modulators' of the $GABA_A$ receptor. This led to the concept of neuroactive steroids that can modulate $GABA_A$ receptor function in the brain by acting on cell surface receptors rather than on genomic receptors. Studies on the synthesis of pregnenolone, and metabolites, such as allopregnanolone (3α-hydroxy-5α-pregnan-20-one), from cholesterol in brain tissue led the term 'neurosteroid' (5). On the other hand, allotetrahydrodeoxycorticosterone (3α,21-dihydroxy-5α-pregnan-20-one; THDOC) is a 'neuroactive steroid' because the sole source of this steroid appears to be the adrenals. Nonetheless, THDOC is found in the brain where its concentration is increased during stress (6). Allopregnanolone and THDOC are among the most potent known steroid modulators of

$GABA_A$ receptors. Steroids produced in the adrenals influence the expression of $GABA_A$ receptor subunits in the brain as shown by adrenalectomy (7).

Neuroactive steroids appear to be able to modulate $GABA_A$ receptor activity only when applied extracellularly being inactive on intracellular administration (8). There are regional differences in the sensitivity of $GABA_A$ receptors to modulation by neuroactive steroids (9) and the effects of neuroactive steroids are dependent on the subunit composition of the $GABA_A$ receptors (10).

Although many of the actions of neuroactive steroids are similar to those of barbiturates on $GABA_A$ receptors, steroids and barbiturates interact with different sites on $GABA_A$ receptors (11). GABA autoreceptors are modulated by barbiturates but not by steroids (12). Insect GABA receptors are only weakly influenced by neuroactive steroids (13).

Various stressors have profound effects on $GABA_A$ receptors likely to be mediated via changes in neuroactive steroids. Foot shock causes a rapid decrease in $GABA_A$ receptors in handling habituated rats (14). A simple warm swim stress of female mice in only 3 minutes substantially increases the apparent numbers of cortical $GABA_A$ receptors (15, 16). Swim stress also increases the apparent number of $GABA_A$ receptors in the brains of male rats, an effect abolishes by adrenalectomy (17). These experiments show that $GABA_A$ receptors are rapidly regulated in the brain, in the main via neuroactive steroids, such as THDOC, produced in the adrenals (6).

The present study was designed to observe the effects of THDOC on GABAergic synaptic currents in rat brain slices in order to assess whether any changes are likely to occur at physiological levels of such a neuroactive steroid. Most effects of neuroactive steroids on the activation of $GABA_A$ receptors by GABA agonists have been reported at μM concentrations while the highest brain levels of these steroids which have been detected have been in the low nM range (6). Extremely potent effects of corticosteroids have been found in the guinea-pig ileum, where pM concentrations of cortisol and cortisone influence $GABA_A$ receptors (18, 19). Such potent actions of corticosteroids may be restricted to particular $GABA_A$ receptors since cortisol has little effect on $GABA_A$ responses in the rat cuneate nucleus (20) though there may well be species differences regarding the effects of corticosteroids since rats do not employ 17α-hydroxy-corticosteroids whereas guinea-pigs do (21).

Materials and Methods

Two different brain areas, hippocampal dentate gyrus and cerebellum, were compared at different developmental stages using tissue from 10-13 day old and 18-21 day old male rats. These areas were selected because differences in the time course of GABA-mediated synaptic currents in these regions suggest that different subtypes of $GABA_A$ receptors may underlie the synaptic response.

Using whole cell patch clamp methodology as previously described (22), miniature inhibitory postsynaptic currents (ipscs) were recorded from dendate granule cells and cerebellar Purkinje cells in brain slices (300 μm thick) from the two age groups of male rats. Recordings were made in the presence of tetrodotoxin (1 μM) and the glutamate antagonist 6-cyano-7-

nitroquinoxaline-2,3-dione (CNQX, 10 μM). Evoked and spontaneous ipscs were also measured in the absence of tetrodotoxin. These and the miniature ipscs were sensitive to the $GABA_A$ receptor antagonist bicuculline (10 μM). The extracellular bathing solution contained NaCl 125 mM, KCl 2.5 mM, $NaHCO_3$ 26 mM, NaH_2PO_4 1.25 mM, glucose 25 mM, $CaCl_2$ 2 mM and $MgCl_2$ 1 mM, gasses the 5%/95% CO_2/O_2. The intracellular solution contained CsCl 140 mM, EGTA 10 mM, HEPES, 10 mM, $CaCl_2$ 1 mM, and MgATP 2 mM, adjusted to pH 7.3 with CsOH. The electrodes (3-6 MΩ) were pulled from thick walled Borosilicate glass (O.D. 1.5 mm, I.D. 0.84 mm). Ipscs were recorded in the presence of various concentrations of THDOC (0, 0.05, 0.1, 1, 2 μM).

Results

In recordings made from Purkinje cells in cerebellar slices from both 10-13 day old rats and 18-21 day old male rats, 1 and 2 μM THDOC increased the decay time constant of the miniature ipscs was increased considerably without any measurable change in amplitude or rise time. This effect was washed out after about 30 minutes washing in control bathing solution. No effects were seen with 0.05 or 0.1 μM THDOC.

Similar results were obtained in recordings made from dentate granule cells in hippocampal slices made from 18-21 day old male rats. In contrast, in recordings made from dentate granule cells from 10-13 day old animals, the miniature ipscs were considerably more sensitive to THDOC with an approximately 20% increase in the decay time constant at 50 nM and 50% at 100 nM. Thus there is a distinct developmental change in the sensitivity of dentate granule cells to THDOC.

The decay time constants for the miniature ipscs differed in each age group between the two cell types with those for the dentate granule cells being considerably slower than those for the cerebellar Purkinje cells.

The developmental change seen in miniature currents was also investigated by recording of evoked ipscs in dentate granule cells. A similar change in the sensitivity of the evoked ipscs to THDOC was observed with the cells from the 10-13 day old rats being more sensitive than those from 18-21 day old rats.

Discussion

The ability of THDOC to enhance the activation of $GABA_A$ receptors at μM concentrations is in accord with many other findings in the literature. A recent study by Teschemacher et al. (23) showed that 1 μM THDOC increased $GABA_A$ receptor-mediated inhibitory postsynaptic potentials in rat neocortical neurones. These authors noted the possible difficulties of investigating lipophilic substances, such as steroids, in brain slices where neighbouring tissue may take up steroid molecules before they reach their target sites. They found that the effects of 10 μM THDOC were very difficult to wash out and that some 90 minutes were needed to observe recovery from the effects of 1 μM THDOC. This suggests that exogenous THDOC may be slowly released from lipophilic stores. In the present study, recovery from the effects of 1 μM THDOC was observed in about 30 minutes. Whole cell patch clamp studies on recombinant $GABA_A$ receptors expressed in HEK 293 cells, where lipophilic stores are unlikely to have a large influence on the potency of neuroactive steroids, have shown enhancement of GABA responses by 10 nM allopregnanolone (24).

Purdy et al. (6) have shown that the levels of THDOC in the brains of nonstressed male rats are less than 1 nM but rise quite rapidly in response to a mild swim stress to 10-20 nM. While it appears that THDOC levels of 1 μM will be reached under normal physiological conditions, levels of 50 nM may be possible in discrete brain regions during stress. The present study shows that 50 nM THDOC can clearly potentiate GABA-mediated synaptic events in dentate gyrus granule cells from 10-13 day old male rats. It will be important to measure the levels of THDOC and related neuroactive steroids in different brain regions during development.

The developmental changes in sensitivity to THDOC found in the hippocampus suggests that neuroactive steroids may well act as neuromodulators of $GABA_A$ receptors during development particularly under conditions of stress. It is interesting to speculate on the possible significance of an increased sensitivity to steroid modulation of $GABA_A$ receptor function in certain brain areas in younger animals. As animals get older and able to move effectively in escape or attack, the more appropriate response to a stressful situation would generally involve being as awake and acutely aware of surroundings as possible. Increasing central inhibition mediated by $GABA_A$ receptors would thus seem unlikely to enhance survival. In contrast in very young animals, dependent on being moved by their mothers, the more appropriate response would be to be able to be hidden effectively - thus increasing central inhibition may be appropriate, inhibiting movement and vocalisation.

The observed differences in the kinetics of the GABA-mediated ipscs between cerebellar Purkinje cells and dendate granule cells strongly suggests that different subtypes of $GABA_A$ receptors may underlie the synaptic response. Similar differences have been described within the cerebellum between Purkinje cells and cerebellar granule cells (25) and also attributed to likely subunit differences. Factors such as the phosphorylation states of the various $GABA_A$ receptor subunits may also be important. In the light of the greater sensitivity to THDOC of the dentate granule cells from younger compared to older rats, there is probably also a developmental change in the expressed $GABA_A$ receptor subunits or their phosphorylation state in these neurones. Dentate granule cells are likely to show a rich variety of functional $GABA_A$ receptor subtypes since *in situ* hybridisation studies have shown that RNA for all the known subunits of $GABA_A$ receptors except the α_6 subunit are present in this cell type (26, 27).

Conclusion

The $GABA_A$ receptors of dentate gyrus granule cells in the hippocampus of 10-13 day old male rats are at least 10 fold more sensitive to modulation by the neuroactive steroid THDOC than receptors of the same cells from 18-21 day old animals or receptors on cerebellar Purkinje cells from animals of both age groups. The levels of THDOC in male rat brain during acute mild stress could rise to be sufficient to influence GABA-mediated synaptic transmission in the hippocampus of the younger rats. Differences in $GABA_A$ receptor subunit composition could underlie differences in susceptibility to modulation by neuroactive steroids.

Acknowledgements

The authors are grateful to the National Health and Medical Research Council of Australia and the Australian Research Council for financial support.

References

1. Cashin F, Moravek V. The physiological action of cholesterol. Am. J. Physiol. 1927; 82:294-298.

2. Seyle H. Correlations between the chemical structure and the pharmacological actions of the steroids. Endocrinology 1942; 30:437-453.

3. Harrison NL, Simmonds MA. Modulation of the GABA receptor complex by a steroid anaesthetic. Brain Res. 1984; 323:287-292.

4. Majewska MD, Harrison NL, Schwartz RD, Barker JL, Paul SM. Steroid hormone metabolites are barbiturate-like modulators of the GABA receptor. Science 1986; 232:1004-1007.

5. Baulieu EE. Neurosteroids: a new function in the brain. Biol. Cell 1991; 71:3-10.

6. Purdy RH, Morris AL, Moore PH, Paul SM. Stress-induced elevations of γ-aminobutyric acid type A receptor-active steroids in the rat brain. Proc. Natl. Acad. Sci. U. S. A. 1991; 88:4553-4557.

7. Orchinik M, Weiland NG, McEwen BS. Adrenalectomy selectively regulates $GABA_A$ receptor subunit expression in the hippocampus. Mol. Cell. Neurosci. 1994; 5:451-458.

8. Lambert JJ, Peters JA, Sturgess NC, Hales TG. Steroid modulation of the $GABA_A$ receptor complex: electrophysiological studies. In: Steroids & Neuronal Activity. Chadwick, D. and Widdows, K. eds, Wiley, Chichester, UK, 1990: 56-82,.

9. Jussofie A. Brain region-specific effects of neuroactive steroids on the affinity and density of the GABA-binding site. Biol. Chem. Hoppe-Seyler 1993; 374:265-270.

10. Puia G, Ducic I, Vicini S, Costa E. Does neurosteroid modulatory efficacy depend on GABAA receptor subunit composition? Receptors & Channels 1993; 1:135-142.

11. Kerr DIB, Ong J. GABA agonists and antagonists. Med. Res Rev. 1992; 12:593-636.

12. Ennis C, Minchin MCW. Modulation of the $GABA_A$-like autoreceptor by barbiturates but not by steroids. Neuropharmacology 1993; 32:355-257.

13. Rauth JJ, Vassallo JG, Lummis SCR, Wafford KA, Sattelle DB. Steroids reveal differences between GABA-operated chloride channels of insects and vertebrates. Mol. Neuropharmac. 1993; 3:1-9.

14. Biggio G, Corda MG, Concas A, Demontis G, Rossetti Z, Gessa GL. Rapid changes in GABA binding induced by stress in different areas of the rat brain. Brain Res. 1981;229:363-9.

15. Skerritt JH, Trisdikoon P, Johnston GAR. Increased GABA binding in mouse brain following acute swim stress. Brain Res. 1981; 215:398-403.

16. Akinci MK, Johnston GAR. Sex differences in the effects of acute swim stress on binding to $GABA_A$ receptors in mouse brain. J. Neurochem. 1993; 60:2212-2216.

17. Schwartz RD, Wess MJ, Labarca R, Skolnick P, Paul SM. Acute stress enhances the activity of GABA receptor-gated chloride channels in brain. Brain Res. 1987; 411:151-155.

18. Ong J, Kerr DIB, Johnston GAR. Cortisol: a potent biphasic modulator at $GABA_A$-receptor-complexes in the guinea-pig isolated ileum. Neursci. Letters 1987; 82:101-106.

19. Ong J, Kerr DIB, Capper HR, Johnston GAR. Cortisone, a potent $GABA_A$ antagonist in the guinea-pig isolated ileum. J. Pharm. Pharmac. 1990; 42:662-664

20. Andres-Trelles F, Bibby V, Lustman S, Simmonds MA. Effects of cortisol on $GABA_A$ receptor-mediated responses compared in the guinea-pig ileum and rat cuneate nucleus. Neuropharmacology 1989; 28:705-708.

21. Kerr DIB, Ong J, Johnston GAR. Stress and cortisol modulation of GABA receptors. Stress & Anxiety 1990; 13:209-213.

22. Edwards FA, Konnerth A, Sakmann B, Takahashi T. A thin slices preparation for patch clamp recordings from neurones of the mammalian central nervous system. Pflugers Arch. 1989; 414:600-612.

23. Teschemacher A, Zeise ML, Holsboer F, Zeiglgänsberger W. The neuroactive steroid 5α-tetrahydrodexoycorticosterone increases GABAergic postsynaptic inhibition in rat neocortical neurons *in vitro*. J. Neuroendocrin. 1995; 7:233-240.

24. Hauser CAE, Chesnoy-Marchais D, Robel P, Baulieu EE. Modulation of recombinant $\alpha_6\beta_2\gamma_2$ $GABA_A$ receptors by neuroactive steroids. Eur. J. Pharmac. Mol. Pharmac Section 1995; 289:249-257.

25. Puia G, Costa E, Vicini S. Functional diversity of GABA-activated Cl^- currents in Purkinje versus granule neurons in rat cerebellar slices. Neuron 1994; 12:117-126.

26. Wisden W, Laurie DJ, Monyer H, Seeburg PH. The distribution of 13 $GABA_A$ receptor subunit mRNAs in the rat brain. I. Telencephalon, diencephalon and mesencephalon. J. Neurosci. 1992; 12:1040-1062.

27. Laurie DJ, Seeburg PH, Wisden W. The distribution of 13 $GABA_A$ receptor subunit mRNAs in the rat brain. II. Olfactory bulb and cerebellum. J. Neurosci. 1992: 12:1063-1076.

GABA: Receptors, Transporters and Metabolism
ed. by C. Tanaka & N.G. Bowery

ONTOGENIC CHANGES OF $GABA_A$ FUNCTION OF THE RAT MEYNERT NEURON

N. Akaike, J.S. Rhee, Y.H. Jin and K. Ono

Department of Physiology, Kyushu University Faculty of Medicine, Fukuoka 812-82, Japan

Summary : Functional and developmental changes in $GABA_A$ response were studied in the acutely dissociated rat Meynert neurons. The whole-cell GABA current displayed greater efficacy in the order of 0-2day > 6-month > 2-week-old rat neurons with changing the receptor affinity and cooperativity. Intracellular Cl^- concentration decreased through the developmental stages; i.e. 35 and 12 mM in 0-day and 6-month-old rat neurons, respectively. Intracellular ATP sensitive-GABA response was the most evident in 2-week-old rat neurons.

Introduction

The nucleus basalis of Meynert is populated by large cholinergic neurons and is major source of cholinergic input to cerebral cortex. When the cholinergic neurons were degenerated by either electrical or chemical lesions, the impairment in avoidance response or spatial cognition was observed. These behavioral studies suggest that cholinergic system has an important role in learning and memory. In addition, there are degeneration and reduction in the number of cholinergic neurons in the basal forebrain structures in senile dementia of Alzheimer's type and Alzheimer's disease. The Meynert neurons receive GABAergic glutamatergic, noradrenergic, serotonergic and dopaminergic fibers. Recent molecular biological and pharmacological studies revealed the existence of multiple kinds of $GABA_A$ receptor subtypes which differ during development and in brain regions, though less is known about the functional meanings of these developmental and regional differences. To understand the functional role of this neurotransmitter, the postnatal development of $GABA_A$ receptor-mediated Cl^- responses was studied in the freshly dissociated rat Meynert neurons. Pathological processes thought to be dependent, in part, on NMDA receptor activation are frequency depending on aging. The neuronal damage under hypoxic/ischemic conditions is more serious in the immature rats than in adult rats (1). Such immature rat brain is also more susceptible to the excitotoxic action induced by exogenous application of NMDA (2). Therefore, as the second aim of present study, it is of interest to know the developmental interactions between the GABAergic and glutamatergic systems in this brain region.

Materials and Methods

Preparation : The neurons acutely dissociated from the nucleus basalis of Meynert of the rats were used in the present study. Wistar rats of different age groups were decapitated under ether anesthesia. The brain was sliced at a thickness of 400μm with a microlicer. Following 30-60 min maintenance in incubation medium saturated with 95% O_2 - 5% CO_2 at room temperature, the slices were treated with dispase at 31°C. The treatment time and the concentration of enzyme were 25 min with 700 pu/ml, 40min with 1300 pu/ml, 90 min with 2500 pu/ml and 110 min with 3400 pu/ml for 0-2-day, 2-week, 6-month and 1.5-year-old rat neurons, respectively. The micropunched-out nucleus basalis of Meynert region was mechanically triturated with a fire-polished Pasteur pipette. The dissociated neurons adhered to the bottom of the dish within 20-30min and remained viable for up to 6 hours.

Electrical measurement : Electrical measurements were performed with the nystatin or gramicidin perforated patch recording modes (3). In some experiments, the conventional whole-cell patch recording mode was used. The current and voltage were measured with a patch-clamp amplifier, monitored on both a storage oscilloscope and a pen recorder, and stored on video tapes after digitization with a pulse-coded modulation processor. All experiments performed at room temperature (21-23°C).

Solution : The standard external solution had the composition (in mM): NaCl 150, KCl 5, $CaCl_2$ 2, $MgCl_2$ 1 glucose 10 and HEPES 10. The ionic composition of the internal (patch-pipette) solution for nystatin or gramicidin perforated patch recording was (in mM): KCl 150 and HEPES 10. The internal solution for the conventional whole-cell patch recording had the composition (in mM): NaCl 20, KCl 50, K-gluconate 70, $MgSO_4$ 6, $CaCl_2$ 0.25, EGTA 5, and HEPES 10. The pH of the external and internal solutions were adjusted to 7.4 and 7.2, respectively, with Tris-OH. Drugs were applied by a fast application method, termed the "Y-tube system" (4). By this technique, the solution surrounding a dissociated neuron could be exchanged within 10-20 ms.

Results

Age-related changes of GABA responses

Whole-cell current measurements were carried on the Meynert neurons freshly dissociated from the immature and mature rats at a holding potential (V_H) of -50mV using the nystatin-perforated patch recording configuration. The neurons from the 0-2-day, 2-week and 6-month rats exhibited a heterogeneity of GABA responses. Since the number of the receptor-channel complexes depend on the surface area of neuron, it is necessary to compare the current amplitude per unit cell membrane. The peak component of GABA-induced inward current (I_{GABA}) decreased rapidly within 2-week. But, I_{GABA} of 6-month-old rat neurons was greater than that of 2-week ones but smaller than that of 0-2-day ones. In the neurons dissociated from 0-2-day, 2-week and 6-month rats, GABA induced the inward currents in a sigmoidal fashion with increasing concentration (Fig. 1). The maximal current amplitudes were 111 (n=4), 48.4 (n=5) and 74 (n=4) pA/pF in the 0-2-day, 2-week and 6-month-old rat neurons, respectively. The EC_{50} values were 21.8, 8.2 and 9.0 μM for 0-2-day, 2-week and 6-month rat neurons, respectively. When all current amplitudes were normalized to the respective maximal one in each concentration-response curve, there were difference among the three curves. The least-squares fitting of the data points gave the apparent Hill coefficient (^nH) of 0.9, 1.5 and 1.2 for 0-2-day, 2-week and 6-month rat neurons, respectively. The data suggest a change in the maximum efficacy with changing in the receptor affinity or cooperativity with age.

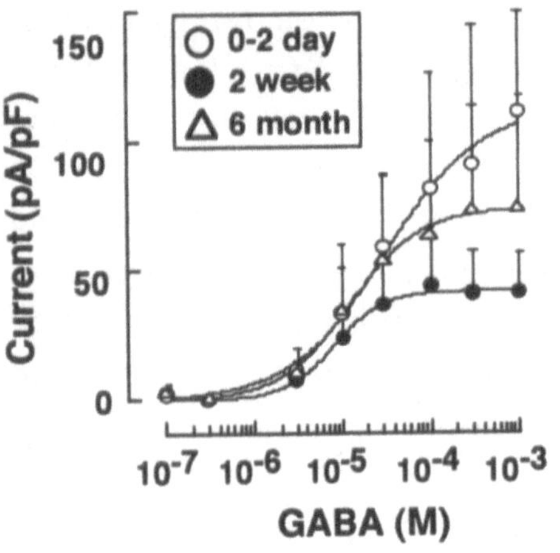

Fig. 1 GABA concentration-response relationships in Meynert neurons of different aging rats.

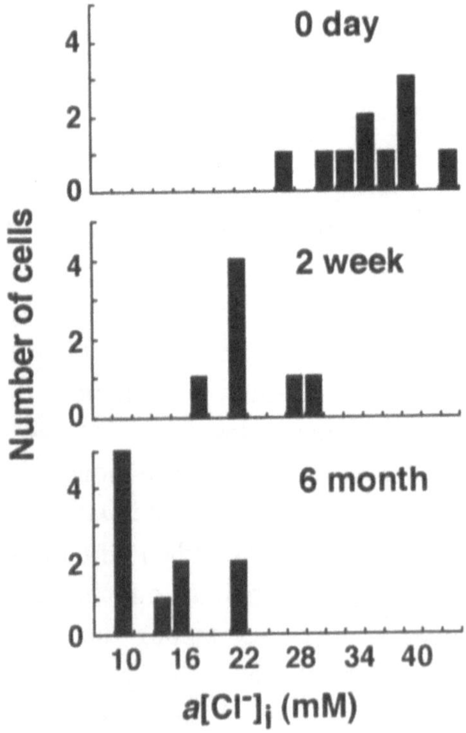

Fig. 2 Developmental change of intracellular Cl^- activities.

Developmental change of intracellular Cl^- activities ($a[Cl^-]_i$)

The electrical recording of neuronal responses with physiological $a[Cl^-]_i$ and the measurement of $a[Cl^-]_i$ in mammalian neurons have been technically hampered. Recently, we have overcome these problems by developing the 'gramicidin-perforated patch recording mode' which maintains the physiological composition of intracellular anions (3). By applying this method to Meynert neurons, we investigated I_{GABA} with physiologically intact $a[Cl^-]_i$ and estimated the $a[Cl^-]_i$ from the GABA reversal potential (E_{GABA}). The $a[Cl^-]_i$) calculated from the Nernst equation using both extracellular Cl^- activity and E_{GABA} ranged with a mean value of 34.8, 22.3 and 13.1 mM for 0-day, 2-week and 6-month rat neurons, respectively (Fig. 2).

Ontogenic changes of glutamatergic responses

We investigated the ontogenic changes in ionotropic and metabotropic glutamate receptor responses (iGluR and mGluR responses, respectively) in the rat Meynert neurons. With aging, NMDA-induced inward current dramatically decreased while KA response gradually increased over developmental stage. The AMPA response showed a bell-shaped increase with a peak at 2-month-old rat neurons. The activation of mGluR evoked a small inward current followed by a little age-related decline (Fig. 3). The Ca^{2+}-activated K^+ current (I_{KCa}) was elicited during an application of NMDA and KA and immediately after washing out these agonists, and the outward currents showed the age-related increase.

Discussion

Most of biochemical and electrophysiological data indicated that there was no change or a decrease in GABA receptor binding sites during development (5-7). In the present study, whole-cell current density was the greatest in the early postnatal rat neurons. In addition, our recent inside-out patch recording clearly showed the existence of a lot of GABA-operated single channels per unit cell membrane in 0-2-day rat neurons than in 2-week and 6-month ones, suggesting that the age-related decrease in the current density in 2-week rat neurons could be explained as the decrease of a number of GABA receptors on unit cell membrane surface. On the other hand, the whole-cell current density of 6-month rat neurons became greater than that of 2-week ones without changing EC_{50} values, and there was no difference in the number of single channels per unit cell membrane between 2-week and 6-month rat neurons. The result suggests the increase of single channel conductance with aging. In fact, the measured main conductance states were 33 and 38.5 pS in 2-week and 6-month rat neurons, respectively.

The resting potentials of the Meynert neurons ranged between -45 and -55 mV regardless of the difference of aging. The Cl^- equilibrium potentials calculated from the Nernst equation using $a[Cl^-]_i$ values (Fig. 2) were about -35, -42 and -62 mV for 0-day, 2-week and 6-month rat neurons, respectively. The results indicate that GABA depolarizes the immature rat neurons but hyperpolarizes the mature ones. Consequently, GABA behaves like as an excitatory transmitter rather than an inhibitory one in the immature rat neurons. Since the I_{KCa} during and after the application of NMDA and KA also developed with aging and GABA induced hyperpolarization in adult rat neurons, both I_{KCa} and GABA response may function as negative feedback systems, preventing the excess cell excitation and neuronal damages in the mature rat.

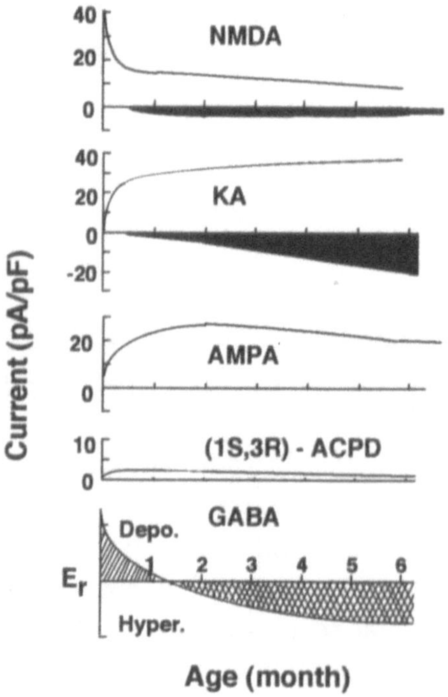

Fig. 3 Age-related changes of glutamatergic and GABAergic responses.
Excitatory response
I_{KCa} (inhibitory response)
Depolaryzation
Hyperpolaryzation
E_r Resting potential

References

I. Ikomomidou C, Mosinger JL, Shahid SK, Labruyere J, Olney JW. Sensitivity of the developing rat brain to hypobaric/ischemic damage parallels sensitivity to N-methyl-D-aspartate neurotoxicity. J Neurosci 1989; 9: 2809-2818.

2. McDonald JW, Silverstein FS, Johnston MV. Neurotoxicity of N-methyl-D-aspartate is markedly enhanced in developing rat central nervous system. Brain Res 1988; 459: 200-203.

3. Ebihara S, Shirato K, Harata N, Akaike N. Gramicidin-perforated patch recording : GABA response in mammalian neurones with intact intracellular chloride. J Physiol (Lond.) 1995; 484. 1: 77-86.

4. Murase K, Randic M, Shirsaki T, Nakagawa T, Akaike N. Serotonin suppresses N-methyl-D-aspartate responses in acutely isolated spinal dorsal horn neurons of the rat. Brain Res 1990; 525: 84-91.

5. Erdo SL, Wolff JR. Age-related loss of t-[^{35}S]-butylbicyclophosphorothionate binding to the gamma-aminobutyric $acid_A$ receptor-chloride ionophore in rat cerebral cortex. J Neurochem 1989; 53: 648-651.

6. Lippa AS, Critchett DJ, Ehlert F, Yamamura HI, Enna SJ, Bartus RT. Age-related alterations in neurotransmitter receptors: an electrophysiological and biochemical analysis. Neuobiol Aging 1981; 2: 3-8.

7. Mhatre MC, Ticku MK. Aging related alterations in $GABA_A$ receptor subunit mRNA levels in Fischer rat. Mol Brain Res 1992; 14: 71-78.

GABA: Receptors, Transporters and Metabolism
ed. by C. Tanaka & N.G. Bowery

GABAERGIC SYSTEM MODULATES THE FORMATION OF LTP (LONG-TERM POTENTIATION) IN THE SUPERIOR COLLICULUS

Yasuhiro OKADA and Hirokazu HIRAI

Department of Physiology, School of Medicine, Kobe University, Kusunoki-cho, Chuo-ku, Kobe 650, JAPAN

Summary: High amount of gamma-aminobutyric acid (GABA) is contained in the superior colliculus (SC) and the highest level of GABA in the CNS was found in the superficial gray layer (SGL) of SC in guinea pig, rat and cat averaging 37-40 mmol/kg dry weight. The distribution of activity of GAD, a GABA synthesizing enzyme, paralleled with that of GABA. Inhibitory interaction between corticotectal and retinotectal response within the SGL was blocked by application of picrotoxine(25 mg/kg i.p.) and methoxy-pyridoxine (100 mg/kg i.p.) suggesting this inhibitory interaction is mediated by activation of GABA interneuron within the SGL. On the other hand, tetanic stimulation (50 Hz, 20 sec) to optic nerve or optic layer induced long-term potentiation (LTP), a model of synaptic plasticity, in neurotransmission of SGL in SC slices. The application of GABA (300μM-1mM) inhibited the appearance of LTP in the SC slices of guinea pig and rat *in vitro*. In the *in vivo* preparation of rat, LTP was only elicited by tetainic stimulation when picrotoxine or methoxypyradoxine was administered prior to the tetanic stimulation and when the ipsilateral visual cortex was removed. These results indicate that GABAergic interneurons in the SGL activated by cortical input may modify the formation of LTP in the superior colliculus.

Introduction

The superior colliculus (SC) plays an important role in the integration of visual auditory and sensory-motor information, especially in relation to eye-movement. GABA must be involved in the integrative action of the SC because the SC contains neumerous GABAergic neurons and concentration of GABA and activity of glutamate decarboxylase (GAD), a GABA synthesizing enzyme, are the highest in the CNS (1).

Long-term potentiation (LTP), characterized as a long lasting enhancement of synaptic efficacy following afferent tetanic stimulation, was first reported in the hippocampus by Bliss and Lomo (2). The mechanism underlying the phenomenon and the significance of LTP have not yet been fully understood although the LTP in the hippocampus has been extensively studied in relation to memory formation. LTP is, however, considered to represent a unique form of synaptic plasticity, which raises the intriguing possibility that similar mechanisms may play a role in the aquisition of lasting modifications of neural integration.

In this paper we are reporting the appearance of LTP in the SC of *in vitro* and *in vivo* preparation from the guinea pig and rat and showing the possibility that GABAergic system in the SC may regulate the occurrence of LTP in the superficial gray layer of SC.

Results

(1) Fine distribution of GABA and GAD in the superior colliculus.

GABA is widely distributed in the nervous system of vertebrates and invertebrates. As regards the regional distribution of GABA in the central nervous system (CNS), the SC contains a high amount of GABA (3). Histologically, the SC is characterized by cells and fibers that are organized in a laminated pattern. Laminar analysis of the distribution of GABA as well as GAD, a GABA synthesizing enzyme, was performed within the SC of the rabbit, cat and guinea pig (1) using microassay methods (4). The regional distribution of GABA and GAD activity obtained for each layer of the superior colliculus is summarized in Table 1. The distribution pattern of GABA within the SC was similar in

TABLE 1

	Rabbit		Cat	Guinea pig
	GAD	GABA	GABA	GABA
SGL (U)	239.4	43.6 ± 2.0	(U) 40.3	(U) 37.4
(L)		34.8 ± 1.8	(L) 36.8	(L) 23.4
OL	108.8	22.3 ± 0.6	28.2	15.5
IG	95.5	22.1 ± 0.5	24.1	15.0
IW	85.3	17.7 ± 0.5	24.0	10.8
DG	82.4	19.1 ± 0.5	25.2	13.1
DW	72.1	17.1 ± 0.4	24.5	10.9

TABLE 2

Dissected layer	mmol/kg (dry) ± S.E.M.
1	44.3 ± 0.9
2	44.8 ± 1.0
3	40.7 ± 2.2
4	34.5 ± 1.6
5	33.3 ± 1.7
6	26.4 ± 1.3
OL	21.0 ± 1.5

Table 1. GAD activity and GABA levels in the layers of the superior colliculus of rabbit, cat and guinea pig. GABA, mmol/kg dry; GAD, ~mmol produced GABA/kg dry/h; SGL, superficial gray layer ((U) upper half of SGL) ((L) lower half of SGL); OL, optic layer; IG, intermediate gray layer; IW, intermediate white layer; DG, deep gray layer; DW, deep white layer.

Table 2. GABA concentration within the superficial gray layer of the rabbit superior colliculus. The dessected SGL was further cut into 6 thin laminated pieces (50-80 µm width), and GABA content in each tissue piece was determined.

each species studied. The highest level of GABA was found in the superficial gray layer (SGL) averaging 37-40mmol/kg dry weight. The GABA levels in the optic and intermediate gray layers were each only half that of the concentration in the SGL. GABA content in the intermediate white, deep gray and deep white layers was lower than the concentration in the optic layer, ranging from 10-22 mmol/kg dry weight. GABA concentration of the whole SC was 22.9 mmol/kg dry weight in the rabbit, 29.0 in the cat, and 18.0 in the guinea pig. These GABA values based upon dry weight are in good agreement with those by wet weight if the dry weight of the tissue is assumed to be 20-25% of the wet weight.

The SGL, which contained the highest level of GABA, was further dissected into 6 thin laminated layers (50-80μm in width). Table 2 shows the GABA distribution within the SGL of the rabbit SC. The GABA concentration of the superficial half was in the range of 40-44 mmol/kg dry weight, while the GABA content in the deep layers was 26-35 mmol/kg. This was true for the cat and guinea pig as shown in Table 1. A dry weight level of GABA of 44 mmol/kg is the same as that in the substantia nigra (SN) and the madial forebrain bundle which have the highest amount of GABA in the mammalian brain (1, 5). GAD activity parallels the GABA levels in each layer of the SC and is highest in the SGL, where the highest level of GABA was also found. The GAD activity of other layers ranged from 30 to 49% of that in the SGL. Thus the distribution of GAD activity in each layer agrees well with that of GABA.

TABLE 3

Treatment	GABA concentration (mmol/kg (dry))	
	right-SGL	left-SGL
No surgical operation (control)	40.6 ± 1.5 (20)	40.9 ± 1.3 (20)
Ablation of right visual cortex	39.8 ± 1.1 (20)	39.8 ± 1.3 (20)
Transection of left optic nerve	38.2 ± 1.1 (18)	39.4 ± 1.2 (20)
Sections of superior collicular commissure	39.5 ± 1.0 (18)	40.5 ± 0.9 (18)

Table 3. GABA concentrations in the SGL after denervation of main input pathways to SC of the rabbit.

The SC receives a substantial input from the visual cortex and retina as well as from other nuclei in the brain stem (6, 7). Fibers from the retina and visual cortex terminate in the SGL and optic layers in an orderly and precise fashion. Physiological studies have revealed that both retina and cortex exert and early facilitation and later inhibition on collicular neurons (8). The SC on one side also exerts an inhibition on the contralateral tectum. To investigate whether the large amount of GABA in the SGL is contained in the afferent fibers terminating in the SC or originates intrinsically within interneurons, three major inputs to the SC of the rabbit were destroyed. In one group, the left visual cortex was ablated; in another group, the left optic nerve was transected just behind the eyeball; in a third group, the SC commissure was cut by knife. The GABA level and GAD activity in the SGL were determined in each animal at 12 days after these surgical operations. No decrease in GABA content in the SGL was found by comparison with that of unoperated controls (Table 3). These results indicate that the GABA concentrated in the SGL is intrinsic to the layer and likely contained within interneurons in the SGL.

Numerous histological and immunohistochemical studies have indicated the existence of GABAergic interneurons in the SGL (9). Cajal (10) designated the upper SGL the "zone of horizontal cells" and the lower SGL the "zone of vertical fusiform cells". Mize showed that 45% of the SGL neurons and 30% of the intermediate grey neurons in the cat SC are GABA-immunoreactive (9). Electron microscopic studies have also shown that there exist many nerve terminals with flattened vesicles in the SGL of the rat (6), suggesting that GABAergic neurons also are located within the SGL of this species. In this connection, it is to be noted that the secondary inhibition evoked from the optic tract and the visual cortex is mediated by a single mechaism intrinsic to the SGL as shown in following section.

(2) LTP formation in the superior colliculus and the role of GABA

In, 1973, Bliss and Lomo (2) discovered the phenomenon of long-term potentiation (LTP) in the hippocampus of the rabbit which was maintained for long periods after tetanic stimulation. LTP formation is interpreted to be a substantial increase in synaptic efficacy. The phenomenon has attracted great interst because of the possibility that LTP might underlie some aspect of memory storage. For this reason, research findings on the formation of LTP in the mammalian brain have mainly come from studies of the hippocampus (11, 12).

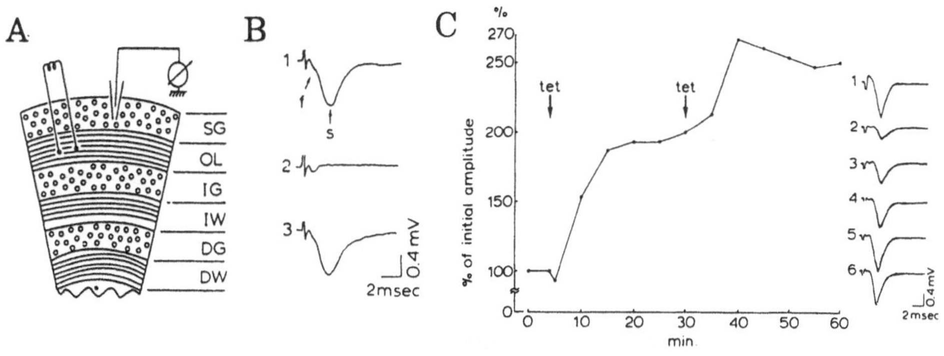

Fig. 1-A, B. A schematic drawing of SC slices showing the placement of the stimulating and recording electrodes [A] and postsynaptic potentials in SGL elicited by stimulation to optic layer. [B] B-1; Two kinds of negative potenials in the control response. Note the earlier deflection (f) in the declining phase of the large potential(s). B-2. 10 min after removal of Ca^{2+} from the standard medium, the large potential(s) was abolished but not the earlier response (f). The early response (deflection) can now be clearly seen. B-3. The recovery of the later potential 10 min after reintroduction of Ca^{2+} into the standard medium. In A, SG, superficial gray layer; OL, optic layer; IG, intermediate gray leyer; IW, intermediate white layer; DG, deep gray layer; DW, deep white layer.

Fig. 1-C. The appearance of LTP in SGL of SC slice and the time course of a typical example of the LTP formation. Right panel shows the PSPs elicited in the SGL of the SC slice after the stimulation to OL. (1) indicates the PSP of maximum amplitude with one test stimulus. In (2) the stimulus intensity was adjusted to evoke PSP for the amplitude to be about 1/3 of the maximum amplitude. (3) and (4) show potentialed PSPs 5 and 15 min after the tetanic stimulation (50 Hz, 20 sec), respectively. Furthermore (5) and (6) show more potentialed PSPs 10 and 20 min after the second tetanic stimulation, respectively. Left panel indicates the time course of LTP formation of the slice shown in the right panel. In the figure the adjusted amplitude of PSP in (2) of right panel was taken as 100%. At tet↓, the tetanic stimulation was applied to OL.

On the other hand, it has been suggested that LTP might represent a general synaptic plasticity for modifying synapses throughout the brain. If this is so, it would be expected that LTP could be reliably recorded in many parts of the central and peripheral nervous system. Besides the hippocampus, the LTP phenomenon has been observed in several areas of cerebral cortex (13), the limbic forebrain (14), the medial geniculate body (15). LTP also has been observed in non-mammalian neural tissue such as goldfish tectum (16).

We are reporting here the LTP formation in the SGL of the SC in *in vitro* (17) and *in vivo* (18., 19) preparations and showing that LTP formation can be modified by GABAergic interneurons within the SGL.

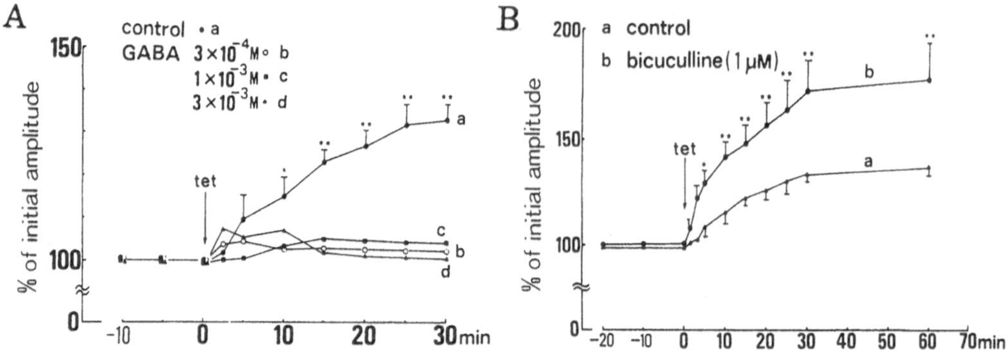

Fig. 2-A. Effect of GABA on LTP formation in the SC slices. a: time course of the change in the PSP amplitude after tetanic stimulation in the absence of GABA. b-d: time course of the change in the amplitude after tetanic stimulation in the presence of GABA at the concentration of 0.3 mM (b), 1 mM (c) and 3 mM (d). In the figure, the PSP amplitude just before the application of tetanic stimulation (tet↓) was taken as 100%. Each plot indicates the average of the results from 5-7 slices and vertical bers show the S.E.M. ANOVA with Fischer's least-significant difference test shows significant difference (*P <0.005, **P<0.001) among (a) and (b) or (c) or (d).

Fig. 2-B. Effect of bicuculline methiodide (BM) on the LTP formation. a: time course of the change in the PSP amplitude after tetanic stimulation in the control slices. b: time course of the change in the PSP amplitude after tetanic stimulation in the presence of BM (1 μM). Each plot indicates the average of the results from 7 slices and vertical bars show the S.E.M. ANOVA with Fischer's least-significant test shows significant difference (*P< 0.05, **P<0.01) between (a) and (b).

In SC slices, electrical stimulation to optic layer elicited the postsynaptic potential (population spike, PS) in the SGL of SC from the gunea pig and rat. Tetanic stimulation (optimal parameter; 50 Hz, 20 sec in duration) to optic layer induced distinct LTP in the PS elicited in the SGL (Fig.1-C). After application of GABA (300μM-3mM) in the perfusion medium, tetanic stimulation failed to induce LTP in the SC slice (Fig.2-A). On the other hand after application of bicuculline methiodide (1μM), a GABAA receptor antagonist, facilitated the formation of LTP as shown in Fig.2-B.

In the *in vivo* preparation, a sharp negative field potential 4-6ms latency was induced on the surface of the superficial gray layer in the rat anesthetized with urethane (Fig.3). This negative potential reversed in polarity, about 200 μm to the optic layer. At this depth, unitary discharges on the negative wave were frequently superimposed on the negative field potential after optic nerve stimulation. This negative potential agreed well with the C2

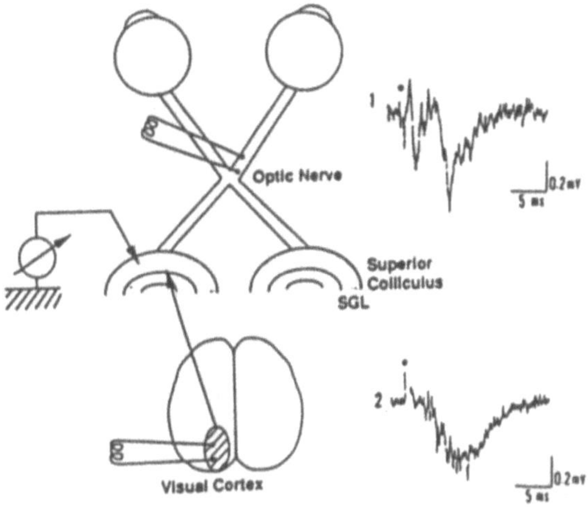

Fig. 3. Schematic drawing of the arrangement of the recording and stimulating electrodes in the *in vivo* experiment. A recording glass electrode (10 M Ω) was inserted into the superficial gray layer of the left superior colliculus stereotactically, using electrophysiological guidance. The reference electrode was placed on the dural surface of the left frantal lobe. Bipolar stimulating electrodes (200 μm diameter) were placed stereotaxically in the right optic nerve anterior to the optic chiasma and another bipolar stimulating electrode was placed in the visual cortex (area 18). The stimulus was applied with square wave pulses of 100 μs in duration. The right inset traces show the postsynaptic field potenitial with unitary discharges induced in the superficial gray layer after stimulation of the optic nerve (1) or the visual cortex (2). The stimulation was applied at the dot.

wave of Sefton and that recorded in the SGL of SC slices as mentioned above. This postsynaptic field potential produced in the layer 150-200μm below the SC surface was used to test the induction of LTP in the superior colliculus. 20 min after a stable response in the postsynaptic field potential from the SC, tetanic stimulation (50 Hz, 20 sec) was applied to the optic nerve. However, LTP could not be induced under these intact conditions (Fig.4).

McIlwain and Fields (8) showed that conditioning stimuli to the ipsilateral visual cortex of the cat exerted an inhibitory action on the retinotectal response in the SGL. We looked for this phenomenon in the rat. The visual cortex was stimulated and a negative field potential was also recorded from the same electrode from which the postsynaptic field potential was recorded. The cortical response had a latency of 4-10 ms (Fig.3). Unit responses were frequently superimposed on the negative wave when visual cortex about 2

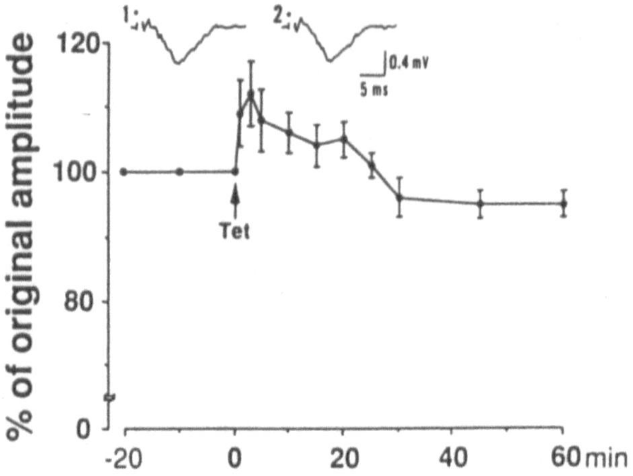

Fig. 4. Failure to form LTP in the intact animal. The postsynaptic field potential was evoked in the superficial gray layer after stimulation to the optic nerve. Tetanic stimulation (50 Hz, 20 sec) was applied to the optic nerve. The inset traces at the top show the postsynaptic field potential records before (1) and 30 min after (2) tetanic stimulation. The graph on the bottom shows the time course of the change in postsynaptic field potential. Tetanic stimulation was applied at the arrow. Each point shows the mean of the results from 10 animals. The vertical bars indicate the S.E. of the mean.

mm below the surface was stimulated. Fig.3-1 shows the postsynaptic field potential produced after optic nerve stimulation and Fig.3-2 shows the postsynaptic field potential produced after ipsilateral visual cortical stimulation. The optic nerve was stimulated after the conditioning stimulus to the visual cortex, and the postsynaptic field potential amplitude produced in the SGL after optic nerve stimulation was dramatically reduced and unit responses were also depressed. When the visual cortex was stimulated after the conditioning stimulus to the optic nerve, the postsynaptic field potential amplitude in the SGL was reduced and unit responses dissapeared. This inhibitory interaction between optic nerve— superficial gray layer and visual cortex— superficial gray layer was observed when the interval between conditioning and test stimulus was approximately 20-350 ms.

This inhibitory interaction between optic nerve — superficial gray layer and visual cortex— superficial gray layer was completely blocked by application of picrotoxine (25 mg/kg i.p.) or by the pretreatment with methoxypyridoxine (100 mg/kg i.p.), an anti-glutamate decarboxylase agent which reduced GABA concentration in the superficial gray layer to 58% of original level in 40 min. These results suggested that the inhibitory

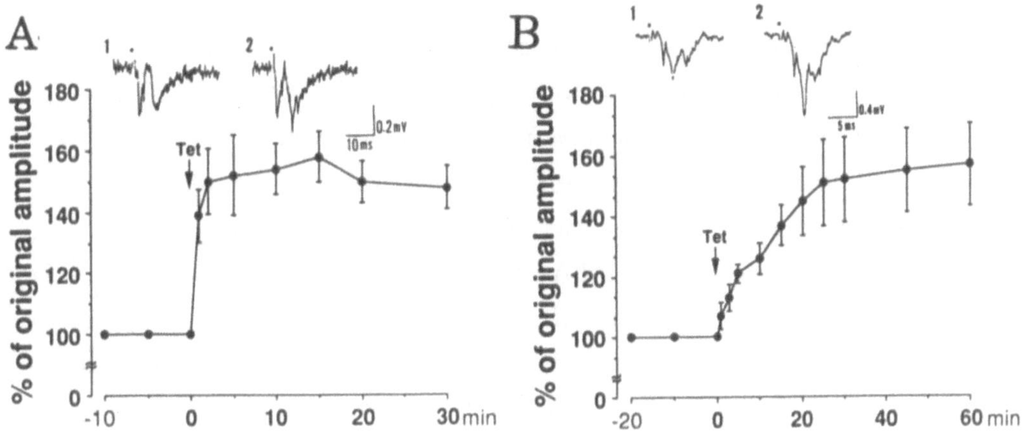

Fig. 5-A. Time course of the appearance of LTP in the superior colliculus of the rat treated with picrotoxin. The upper inset traces show the responses of the superficial gray layer before (1) and 20 min after (2) tetanic stimulation. The graph shows the time course of LTP up to 40 min after an i.p. application of picrotoxin (2.5 mg/kg). Tetanic stimulation (50 Hz for 20 sec) of the optic nerve is shown at the arrow. Each point indicates the average of the postsynaptic field potential amplitude from 6 animals. The vertical bars indicate the S.E. of the mean.

Fig. 5-B. Time course of the appearance of LTP in superior colliculus in the rat after ablation of the ipsilateral visual cortical area (area 18). Tetanic stimulation was applied to optic nerve at the arrow. Each point indicates the average of the postsynaptic field potential amplitude from 6 animals. The vertical bars indicate the S.E. of the mean. The inset traces at the top show the postsynaptic field potentials recorded before (1) and 20 min after (2) tetanic stimulation.

interaction between cortical and retinal response is mediated by GABAergic interneuron within the SGL to which both cortical and retinal projections are converged through glutamatergic pathway. In this connection we tested the effects of picrotoxine, methoxypyridoxine and of ablation of ipsilateral visual cortical area on the appearance of LTP elicited by tetanic stimulation to optic nerve. LTP in SGL was only elicited by tetanic optic nerve stimulation when picrotoxine (25 mg/kg i.p.) and methoxypyridoxine (100 mg/kg i.v.) was administered to the animal and when the ipsilateral visual cortex was removed. These results indicate that GABAergic interneurons in the SGL activated by corticotectal input may modify the formation of LTP in the SGL.

Discussion

In the present experiment, tetanic stimulation to optic layer induced easily the LTP in the isolated SC slices *in vitro* whereas in *in vivo* preparation, we could not induce LTP in the PS by tetanic stimulation to the optic nerve of the intact rat. However, LTP was elicited by tetanic stimulation either when the ipsilateral visual cortex was removed or when picrotoxin, a GABA antagonist, was administered to the animal before tetanic stimulation. In the SC slices, the application of GABA to the pefusion medium inhibited LTP formation and application of bicuculline facilitated the induction of LTP. These results indicate that GABAergic activity, whether it is extrinsic or intrinsic in the SC, can modulate the induction of LTP in the SGL.

Concerning the involvement of GABAergic neurons in modifying LTP formation, application of bicuculline and picrotoxin facilitate the induction of LTP in hippocampal slices (20). In slices of visual cortex, application of low doses of bicuculline induces long-term depression by tetanic stimulation whereas bicuculline at high doses elicits LTP (13). The induction of LTP may thus be influenced by the excitability or the level of membrane potential of postsynaptic neurons which is modulated by GABAergic input.

The involvement of extrinsic GABAergic afferent to SC can not be completely excluded as sources of modulation of LTP formation in the SGL. However, application of picrotoxin in animals or the ablation of ipsilateral visual cortical areas induced LTP after tetanic stimulation *in vivo* preparations. The ipsilateral corticotectal pathway has been reported to exert an inhibitory action on the neural activity evoked by the retinotectal pathway (8). This inhibition is found to be mediated by GABAergic interneurons located in the SGL because the corticotectal pathway is glutamatergic (21). In the isolated slice preparation of the SC, LTP can be easily induced by tetanic stimulation and the formation of LTP is modified by GABA agonists and antagonists. These results strongly suggest that corticotectal efferents tonically inhibit the induction of LTP that is elicited by tetanic stimulation of the optic nerve, probably by activating GABAergic interneurons within SGL. This ability of neurons to induce LTP in the SC may depend upon the delicate balance between excitatory and inhibitory inputs through the retinotectal, corticotectal, or other extrinsic pathways. GABAergic systems thus may have an important role in maintaining a delicate balance of neural activity within SC. The true mechanism and function of LTP formation in the SC in connection with GABAergic inhibitory processes remains to be investigated in further studies.

References

1. Okada Y. The distribution and function of gamma-aminobutyric acid (GABA) in the superior colliculus. In: Mize RR, Marc RE, Sillito AM, editors. GABA in the retina and central nervous system. Progress in Brain Research, 1992; 90:249-262.

2. Bliss TVP, Lomo T. Long-lasting potentiation of synaptic transmission in the dentate area of the anaesthetized rabbit following stimulation of the perforant path. J Physiol 1973; 232:331-356.

3. Okada Y, Nitsch-Hassler C, Kim JS, Bak IJ, Hassler R. Role of γ-aminobutyric acid (GABA) in the extrapyramidal motor system, I Regional distribution of GABA in rabbit, rat, guinea pig and baboon CNS. Exp Brain Res 1971; 13:514-518.

4. Okada Y, Taniguchi H, Chimada Ch. High concentration of GABA and high glutamate decarboxylase activity in rat pancreatic islets and human insulinoma. Science 1976; 194:620-622.

5. Okada Y. Regional distribution of GABA (γ-aminobutyric acid), GAD (glutamate decarboxylase), GABA-T (GABA-transaminase) and glutamate in the rat central nervous system. In: Ito M, editor. Integrative Control Function of Brain III. Tokyo: Kodansha, 1980; 26-28.

6. Lund RD, Lund JS. Modifications of synaptic patterns in the superior colliculus of the rat during development and following deafferentation. Vision Res, Suppl 1971; 3:281-298.

7. Sprague JM. Mammalian tectum: intrinsic organization, afferent inputs, and integrative mechanism. Neurosci Res Prog Bull 1975; 13:204-213.

8. McIlwain JT, Fields HL. Interactions of cortical and retinal projections on single neurons of the cat's superior colliculus. J Neurophysiol 1971; 34:763-772.

9. Mize RR. Immunocytochemical localization of gamma-aminobutyric acid (GABA) in the cat superior colliculus. J Comp Neurol 1988; 276:169-187.

10. Cajal SRY. Histologie du System Nerveux de l'Homme et des Vertebres. Consejo, Superior de Investigacions Cientificas, Madrid. 1955; Vol.2.

11. Teyler TJ, Discenna P. Long-term potentiation. Annu Rev Neurosci 1987; 10:131-161.

12. Lynch G, Markus K, Arai A, Larson J. The nature and causes of hippocampal long-term potentiation. In: Storm-Mathisen J, Zimner J, Ottersen OP, editors. Progress in Brain Research, Amsterdam: Elsevier, 1990; 233-250.

13. Artola A, Brocher S, Singer W. Different voltage-dependent thresholds for inducing long-term depression and long-term potentiation in slices of rat visual cortex. Nature 1990; 347:69-72.

14. Racine RJ, Milgram NW, Hafner S. Long-term potentiation phenomena in the rat limbic forebrain. Brain Res 1983; 260:217-231.

15. Gerren RA, Weinberger NM. Long-term potentiation in the magnocellular medial geniculate nucleus of the anesthetized cat. Brain Res 1983; 265:138-142.

16. Lewis D, Teyler TJ. Long-term potentiation in the goldfish optic tectum. Brain Res 1986; 375:246-250.

17. Okada Y. Long-term potentiation in the superior colliculus slices. In: Rahmann H. editor. Fortschritte der Zoologie. Progress in Zoology - Fundamentals of Memory Formation: Neuronal Plasticity and Brain Function, Gustav Fischer Verlag, Stuttgart, New York, 1989; 37:190-196.

18. Shibata Y, Tomita H, Okada Y. The effects of ablation of the visual cortical area on the formation of LTP in the superior colliculus of the rat. Brain Res 1990; 537:345-348.

19. Hirai H, Okada Y. Ipsilateral corticotectal pathway inhibits the formation of long-term potentiation (LTP) in the rat superior colliculus through GABAergic mechanism. Brain Res 1993; 629:23-30.

20. Wigstrom H, Gustafsson B. Facillitation of hippocampal long-term potentiation by GABA antagonists. Acta Physiol Scand 1985; 125:159-172.

21. Sakurai T, Miyamoto T, Okada Y. Reduction of glutamate content in rat superior colliculus after retinotectal denervation. Neurosci Lett 1990; 109:299-303.

GABA: Receptors, Transporters and Metabolism
ed. by C. Tanaka & N.G. Bowery

GABA RECEPTOR-CHLORIDE CHANNEL COMPLEX AS A TARGET SITE OF ALCOHOL

Masanobu Nakahiro [1], Osamu Arakawa [2], Toshio Narahashi [3] and Tsuyoshi Nishimura [4]

[1]National Hospital Shou-rai-soh, Nara, Japan; [2]Faculty of Fisheries, Kagoshima University, Kagoshima, Japan; [3]Department of Molecular Pharmacology and Biological Chemistry, Northwestern University Medical School, Chicago, U.S.A.; and [4]Department of Psychiatry, Osaka University Medical School, Osaka, Japan

Summary: The effects of n-alcohols on the γ-aminobutyric acid (GABA) receptor-Cl^- channel complex of rat dorsal root ganglion neurons in primary culture were studied by the whole-cell patch-clamp technique. n-Alcohols (C_1 - C_{11}) enhanced the GABA-induced current, their effects increasing with their carbon chain lengths. n-Alcohols (C_2, C_8) accelerated desensitization of the GABA-induced current. At higher concentrations, n-alcohols (C_2, C_4, C_6 and C_8) alone gated the Cl^- channel of the receptor-channel complex to generate a Cl^- current.

Introduction

The mechanisms underlying the behavioral effects of ethanol have been extensively studied. At the cellular level, ethanol is reported to interact with a wide variety of neurotransmitter receptors and ion channels (1). Recently, much attention has been focused on the γ-aminobutyric acid type A ($GABA_A$) receptor channel as a target site of ethanol (2). Other straight-chain primary alcohols (n-alcohols) have similar pharmacological effects to ethanol (3, 4), and these n-alcohols probably have a common mode of action. Accordingly, a series of n-alcohols may be useful for studying the mechanisms of action of alcohol. In this study, we examined the effects of n-alcohols on the $GABA_A$ receptor channel.

Materials and Methods

The methods used were previously described (5, 6). Neurons were dissociated from dorsal root ganglia of 1 - 2 day old Sprague-Dawley rats and maintained in culture for 1 - 2 weeks before use in experiments. The whole-cell patch-clamp technique (7) was used to record ionic currents induced by bath application of GABA and alcohols at a membrane potential of -60 mV. The internal (pipette) solution contained (in mM): CsCl, 140; $MgCl_2$, 1; EGTA, 5; and HEPES, 10. The external (bath) solution contained (in mM): choline chloride, 136; $CaCl_2$, 2; $MgCl_2$, 1; and HEPES 10. Both solutions were adjusted to pH 7.3 with Tris base. In experiments for determination of the reversal potential of an alcohol-induced current, the internal concentration of Cl and the membrane potential were altered. Experiments were performed at room temperature (22 ± 2℃). Data are presented as means ± S.D.

Results and Discussion

Bath application of GABA (1 - 1000 μM) induced an inward current at a membrane potential of -60 mV with external and internal solutions both containing 142 mM Cl^-. This current was due to outward movement of Cl^- ions as previously described (8). GABA at low concentrations of 1 - 3 μM induced an inward steady-state current; and at higher concentrations of 10 - 1000 μM, it evoked a peak current which decayed to a reduced steady-state level due to desensitization. The peak current amplitude reached a maximum at concentrations of 300 - 1000 μM GABA. The dose-response curve for the nondesensitized and peak current showed a K_d value of 55 μM and Hill coefficient of 1.9. Thus, two GABA molecules were bound to one receptor.

Our initial study of the effect of alcohol (5) was performed with ethanol (C_2) and n-octanol (C_8). Ethanol (300 mM) and n-octanol (30 μM) increased the peak current induced by 30 μM GABA alone, 127 ± 10 % (n = 3) and 130 ± 20 % (n = 3), respectively, when added to the bath simultaneously with GABA. Both alcohols increased the current to the same extent when added before or with GABA, indicating very rapid reactions. Accordingly, we applied alcohols simultaneously with GABA in subsequent experiments.

Modulation of desensitization may be a novel mechanism for regulation of ion channel activity. For example, glycine reduces desensitization at N-methyl-D-aspartate (NMDA) receptors and thereby potentiates responses to NMDA (9). While the alcohols enhanced the nondesensitized and peak current, they accelerated decay of the current due to desensitization. Therefore, we measured the time constant of decay of the 30 μM GABA-induced current in the absence and presence of alcohol. The time constant of the current decay in the absence of alcohol was 17.9 ± 6.7 sec (n = 16). Ethanol (300 mM) and n-octanol (30 and 100 μM) shortened the decay time constant to 11.8 ± 6.2 sec (n = 4), 13.3 ± 1.9 sec (n = 3) and 9.3 ±2.4 sec (n = 4), respectively. Volatile general anesthetics and benzodiazepines are reported to accelerate desensitization of the $GABA_A$ receptor channel (8, 10). In this study, we observed similar effects of the alcohols on the receptor channel. Since these three groups of drugs all enhance GABA-induced current prior to desensitization, the mechanism for current enhancement appears to be closely related to acceleration of desensitization at the $GABA_A$ receptor channel.

n-Alcohols have general anesthetic effects and their anesthetic potencies increase with their carbon chain length (3, 4). To determine whether the GABA system is responsible for the anesthetic activity of alcohol, we studied the effects of various n-alcohols on the GABA-induced current. Each n-alcohol (C_1 - C_{11}) enhanced the peak current induced by 30 μM GABA, dose-dependently. The concentrations of the alcohols tested were 100 mM - 1 M for C_1, 30 - 300 mM for C_2, 10 - 300 mM for C_3, 1 - 30 mM for C_4, 100 μM - 3 mM for C_5, 30 μM - 1 mM for C_6, 10 - 300 μM for C_7, 3 - 100 μM for C_8, and 3 - 30 μM for C_{9-11}. To compare the potency of the alcohols, we estimated the concentrations required to enhance the current to 125 % of the control from the dose-response relationships. Table 1 summarizes the concentrations of the alcohols inducing 125 % current enhancement. Their effects increased progressively with their carbon chain length up to C_8, but not at higher carbon numbers (C_9 - C_{11}). n-Dodecanol (C_{12}) at up to 30 μM did not enhance the current (unpublished observation). This indicates a cutoff of enhancement of the GABA-induced current in the series of n-alcohols. The anesthetic potencies of n-alcohols are known to exhibit a cutoff (3, 4), so a cutoff for current enhancement might underlie the cutoff for anesthetic potencies of the alcohols. Therefore, the findings are compatible with the hypothesis that potentiation of the GABA response is a mechanism of general anesthesia (11).

Table 1. Concentrations of n-alcohols required to enhance the GABA-induced current to 125 % of the control.

n-Alcohol	C_n	Concentration (M)
Methanol	1	5.33×10^{-1}
Ethanol	2	2.47×10^{-1}
Propanol	3	5.00×10^{-2}
Butanol	4	4.55×10^{-3}
Pentanol	5	1.10×10^{-3}
Hexanol	6	1.89×10^{-4}
Heptanol	7	6.03×10^{-5}
Octanol	8	1.79×10^{-5}
Nonanol	9	1.31×10^{-5}
Decanol	10	1.41×10^{-5}
Undecanol	11	1.03×10^{-5}

n-Alcohols at the concentrations tested in the preceding section induced no current in the absence of GABA. At higher concentrations, however, n-alcohols by themselves generated an inward current (6). Ethanol (1 - 3 M), n-butanol (100 - 300 mM), n-hexanol (3 - 10 mM) and n-octanol (300 μM - 3 mM) generated an inward current at a membrane potential of - 60 mV with external and internal Cl^- concentrations of 142 mM. The alcohol-induced current had similar characteristics to the GABA-induced current: low concentrations of the alcohols evoked a sustained current, and higher concentrations produced peak currents decaying to a lower level due to desensitization. However, the observed amplitudes of the alcohol-induced currents were smaller than those of the GABA-induced current. Since alcohol concentrations higher than those tested damaged the neurons, we could not determine maximum responses to the alcohols. The potencies of the alcohols to generate current increased with their carbon chain length, as did those for enhancing the GABA-induced current.

In order to identify the ions that carry the alcohol-induced current, the reversal potentials for the peak current induced by 1 mM n-octanol were measured at three different concentrations of internal Cl^- (20, 50 and 142 mM). The reversal potentials for the current changed with the internal Cl^- concentrations in a manner predicted by the Nernst equation, although the observed reversal potentials were slightly more positive than the theoretical values. These data indicate that the current is carried largely by Cl^-.

To determine whether the $GABA_A$ receptor channel is the site of action of the alcohols in generating the Cl^- current, we examined the effects of drugs known to act on the $GABA_A$ receptor channel. Bicuculline (10 μM) and picrotoxin (30 μM), which are an antagonist and a channel blocker of the $GABA_A$ receptor channel, respectively, suppressed the current induced by 1 mM n-octanol to 32.9 ± 1.8 % (n = 3) and 15.8 ± 3.7 % (n = 4) of the control, respectively. Benzodiazepines and barbiturates are assumed to bind to their specific regulatory sites of the $GABA_A$ receptor channel and to enhance responses to GABA. Chlordiazepoxide (10 μM) and pentobarbital (10 μM) enhanced the current induced by 300 μM n-octanol to 164 ± 19 % (n = 3) and 153 ± 17 % (n = 3), respectively, of the control. These observations suggest that n-octanol acts on the $GABA_A$ receptor channel to generate the Cl^- current. Probably other alcohols also generate a current through the Cl^- channel of the $GABA_A$ receptor channel. Accordingly, n-alcohols mimic barbiturates, enhancing a GABA-induced Cl^- current and by themselves gate the Cl^- channel of the $GABA_A$ receptor channel to generate a Cl^- current (12).

Recent cloning studies have demonstrated that there are several subunits and subunit subtypes of the $GABA_A$ receptor channel and that enhancement of GABA responses by ethanol is dependent on phosphorylation of a splice-variant long form of $\gamma 2$ subunit (γ2L) by protein kinase C (2). As all n-alcohols (C_1 - C_{11}) probably act on the $GABA_A$ receptor channel by a common mechanism, enhancement of GABA-induced current by n-alcohols other than ethanol would also require the phosphorylated γ2L subunit. It is of particular interest whether the other actions of the alcohols observed in this study, that is, acceleration of desensitization and generation of Cl^- current by themselves, depend on the phosphorylation of the γ2L subunit or on mechanism(s) requiring other subunit(s).

Acknowledgments

We wish to thank Dr. Kihachi Saito for his critical reading of the manuscript and Mrs. Mieko Nakamura for her secretarial assistance.

References

1. Deitrich RA. Dunwiddie TV, Harris RA, Erwin VG. Mechanism of action of ethanol: initial central nervous system actions. Pharmacol Rev 1989;41:489-537.

2. Macdonald RL. Ethanol, γ-aminobutyric acid type A receptors, and protein kinase C phosphorylation. Proc Natl Acad Sci USA 1995;92:3633-3635.

3. Pringle MJ. Brown KB, Miller KW. Can the lipid theories of anesthesia account for the cutoff in anesthetic potency in homologous series of alcohols? Mol Pharmacol 1981;19:49-55.

4. Alifimoff JK. Firestone LL, Miller KW. Anaesthetic potencies of primary alkanols: implications for the molecular dimensions of the anaesthetic site. Br J Pharmacol 1989;96: 9-16.

5. Nakahiro M, Arakawa O, Narahashi T . Modulation of γ-aminobutyric acid receptor-channel complex by alcohols. J Pharmacol Exp Ther 1991;259:235-240.

6. Arakawa O. Nakahiro M, Narahashi T. Chloride current induced by alcohols in rat dorsal root ganglion neurons. Brain Res 1992;578:275-281.

7. Hamill OP, Marty A, Neher E, Sakmann B, Sigworth FJ. Improved patch-clamp technique for high resolution current recordings from cells and cell-free membrane patches, Pflugers Arch1981;391:85-100.

8. Nakahiro M, Yeh JZ, Brunner E, Narahashi T. General anesthetics modulate GABA recepto channel complex in rat dorsal root ganglion neurons, FASEB J 1989;3:1850-1854.

9. Mayer ML. Vyklicky Jr L, Clements J. Regulation of NMDA receptor desensitization in mouse hippocampal neurons by glycine. Nature 1989;338:425-427.

10. Mierlak D. Farb DH. Modulation of neurotransmitter receptor desensitization: chlordiazepoxide stimulates fading of the GABA response. J Neurosci 1988;8:814-820.

11. Franks NP, Lieb WR. Molecular and cellular mechanisms of general anaesthesia. Nature 1994;367:607-614.

12. Akaike N, Maruyama T, Tokutomi N. Kinetic properties of the pentobarbitone-gated chloride current in frog sensory neurones. J Physiol 1987;394:85-98.

GABA: Receptors, Transporters and Metabolism
ed. by C. Tanaka & N.G. Bowery

FUNCTIONAL ANALYSIS ON GABAB RECEPTOR USING A RECONSTITUTED SYSTEM WITH PURIFIED GABAB RECEPTOR, Gi/Go PROTEIN AND ADENYLYL CYCLASE

M. Hirouchi, H. Mizutani, M. Nishikawa, H. Nakayasu and K. Kuriyama

Department of Pharmacology, Kyoto Prefectural University of Medicine,
Kawaramachi-Hirokoji, Kamikyo-Ku, Kyoto 602, Japan

Summary: GABAB receptor was purified from the bovine cerebral cortex using baclofen-affinity gel. The eluate from the baclofen-affinity gel indicated possible presence of multiplicity in GABAB receptor. The Gi/Go protein and adenylyl cyclase prepared from the bovine cerebral cortex were reconstituted with the eluate from baclofen-affinity gel. This reconstituted system showed the GABAB receptor-mediated inhibition on the forskolin-stimulated cAMP accumulation. Subsequently, it was demonstrated that the inhibition of GABAB receptor binding by ethanol was accompanied with the decrease of GABAB receptor-mediated inhibition on forskolin-stimulated cAMP accumulation in the reconstituted systems.

Introduction

GABAB receptor has been introduced as a new GABA receptor subtype, which is insensitive to bicuculline, a classical GABA (GABAA) receptor antagonist, and sensitive to baclofen (1). The GABAB receptor binding was found to be inhibited by GTP analogues and the treatment of membranes by islet-activating protein (IAP) prevented this inhibition (2,3). Based on these data, it was postulated that the GABAB receptor was negatively coupled to adenylyl cyclase through IAP-sensitive GTP-binding protein (4,5).

Moreover, it was reported that GABAB receptor directly hyperpolarized neuronal cells, due to the increase of K^+ conductance (6). In contrast, the voltage-dependent Ca^{2+} channels was shown to be inhibited by the activation of GABAB receptor (7). It has been also suggested that these actions induced by GABAB receptor stimulation may be involved in the modulation of release of various neurotransmitters (8,9).

On the other hand, the molecular characteristics of GABAB receptor have not been well clarified as compared with those on GABAA receptors. The GABAB receptor was solubilized from crude synaptic membrane of the bovine cerebral cortex (10) and the ligand-affinity gel using baclofen as an immobilized ligand was developed to isolate the GABAB receptor (11,12) in our laboratory. Using the baclofen-gel, GABAB receptor was purified approximately 11,000-fold as compared with that in solubilized fraction from the bovine cerebral cortex (12). Furthermore, the monoclonal antibody against GABAB receptor was raised and applied it to immunoaffinity column for further purification (13,14). The purification by this method showed the presence of 80 kDa protein as the actual GABAB receptor.

In this study, we have examined whether or not the purified sample from baclofen-affinity gel has the same protein as found in the immunoaffinity purification. Moreover, the action of ethanol on the GABAB receptor-mediated system was analyzed using the reconstituted GABAB receptor with Gi/Go protein and adenylyl cyclase.

Materials and Methods

Materials

Baclofen was a kind gift from Daiichi Pharmaceuticals (Tokyo, Japan). GABA and 2-hydroxy saclofen were obtained from Sigma Chemicals (St. Louis, MO, USA) and Tocris Neuramin (Bristol, UK), respectively. [2,3-^{3}H]GABA (3.5 TBq/mmol) and [^{35}S]-guanosine 5'-[γ-thio]triphosphate (GTPγS: 43.7 TBq/mmol) were purchased from New England Nuclear (Boston, MA, USA). cAMP[^{125}I] assay system was obtained from Amersham (Des plaines, IL, USA). Other reagents used were of analytical grade and commercially available.

Preparation of baclofen-Toyopearl gel

The ligand-affinity gel was prepared using baclofen and AF-Epoxy Toyopearl 650M (TOSOH, Japan) (12). In brief, baclofen (about 2.5 g) was reacted with swelling gel (about 6 g as a dry gel) at 45 °C for 16 hours in pH 11.5 according to the manual of TOSOH.

Preparation of GABAB receptor fraction

Crude synaptic membranes were prepared from the bovine cerebral cortex according to the method of Zukin *et al.* (15) in the presence of various protease inhibitors as described below. These membrane fractions were solubilized by the incubation with 50 mM Tris-HCl (pH 7.4), 3.0% CHAPS/0.01% asolectin, 1 mM phenylmethylsulphonyl fluoride (PMSF), 2 mM EDTA, 5 μg/ml antipain, 5 μg/ml leupeptin and 5 μg/ml pepstatin A for 60 min at 4 °C and then centrifuged at 100,000 xg for 120 min. The supernatant obtained was dialyzed overnight against 50 mM Tris-HCl (pH 7.4) containing 0.15% CHAPS, 0.01% asolectin, 0.3 mM PMSF, 1 μg/ml antipain and

1 µg/ml leupeptin. The solubilized fraction, which was added $CaCl_2$ (final 2 mM) after dialysis, was applied to baclofen-affinity column (1 x 30 cm) equilibrated with dialysis buffer containing 2 mM $CaCl_2$. After washing with the same buffer, the dialysis buffer containing 5 mM baclofen was applied. The eluate was dialyzed overnight against the dialysis buffer and then concentrated by the ultrafiltration using Millipore Molcut II TGC.

The eluted protein was used for SDS-polyacrylamide gel electrophoresis and the gel was transferred onto a nitrocellulose filter (BA85, Schleicher & Schuell). This blot was analyzed using the monoclonal antibody against $GABA_B$ receptor (13,14).

Preparations of Gi/Go protein and adenylyl cyclase

The partially purified Gi/Go protein and adenylyl cyclase were prepared for the reconstitution and examination of the function of purified $GABA_B$ receptor protein. The partial purification of Gi/Go protein from the bovine cerebral cortex was performed using DEAE-sephacel and Ultragel AcA34 by the method of Sternweis and Robishaw (16). The confirmation of the function of Gi/Go protein was carried out by the determination of $[^{35}S]GTP\gamma S$ binding (16). Similarly, adenylyl cyclase from the bovine cerebral cortex was partially purified using DEAE-Toyopearl 650(S) and Toyopearl HW-55(S) by the modification method of Katada *et al.* (17).

The characteristics of these proteins were examined by the assay of cAMP formation and the immunoblot using antibodies against $GABA_B$ receptor, Gs protein and Gi/Go protein as previously described (14). It was confirmed that neither $GABA_B$ receptor nor adenylyl cyclase fractions contained Gi/Go protein. Also, the contamination of Gs protein in these fractions was not detected, although both Gi and Go proteins were mixed in this method. Moreover, the preparation of adenylyl cyclase had no $GABA_B$ receptor and Gi/Go proteins.

Reconstitution into phospholipid vesicles

The eluate from baclofen-affinity gel was reconstituted into phospholipid vesicles using the method reported previously for the reconstitution of various receptor-adenylyl cyclase systems (18,19). Namely, these three components were incubated with lipid solution, which was consisted of the Folch I fraction from bovine brain extract (final 1.4 mg/ml), dimyristoyl phosphatidylcholine (final 0.6 mg/ml) and CHAPS (final 7 mM), for 60 min at 4 °C. Thereafter, the mixture (total volume: 1.4 ml) was gel-filtrated using Sephadex G-50 (1.2 x 25 cm) equilibrated with 50 mM Tris-HCl (pH 7.4), 1 mM EDTA, 2 mM $MgSO_4$ and 100 mM NaCl. The eluted void fraction, which contained the reconstituted vesicles, was used for various assays such as $GABA_B$ receptor binding and adenylyl cyclase activities.

Measurement of GABAB receptor binding

The GABAB receptor binding was performed by the filtration method using polyethyleneglycol 6000 (11). The eluate from baclofen-affinity column or reconstituted vesicles was incubated with 12.5 nM [^{3}H]GABA in the presence or absence of 1 mM baclofen for 60 min at 4 °C.

Assay of cAMP formation

The adenylyl cyclase activity was determined under various conditions such as the basal, 10 μM forskolin-stimulated and 10 μM forskolin + GABAB receptor-stimulation (20). Namely, the reaction mixture (total volume: 160 μl) of reconstituted vesicles containing various drugs were incubated with 50 mM Tris-HCl (pH 7.4), 2 mM $CaCl_2$, 2 mM $MgSO_4$ and 1 mM EDTA (basal buffer) containing 5 mM theophylline, 0.5 mM isobutylmethylxanthine, 0.5 mM ATP, 6 mM creatine phosphate, 50 units/ml creatinephosphokinase, 5 μM GTP and 5 μM GDP. After incubating for 30 min at 37 °C, the reaction mixture was boiled for 2 min and then placed on ice. It was then centrifuged and the supernatant obtained was used to determine the cAMP content by radioimmunoassay using cAMP[^{125}I] assay system.

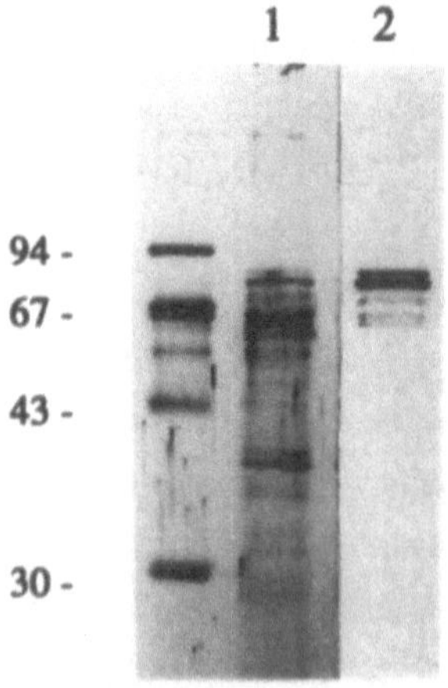

Figure 1. Eluate from baclofen-affinity column chromatography.
1: Protein staining; 2: Immunoblot

Results and Discussion

The GABAB receptor from bovine cerebral cortex was obtained using the baclofen-affinity column chromatography. The eluate from baclofen gel was examined on SDS-polyacrylamide gel electrophoresis and its immunoblot using a monoclonal antibody against GABAB receptor obtained previously (13,14). The immunoblot showed the presence of five protein bands which predominantly represented 79 kDa and 80 kDa proteins, although the protein staining by colloidal gold indicated the presence of many proteins (Fig. 1). In contrast, the immunoaffinity purification of GABAB receptor usually indicated the presence of 80 kDa protein (14). Occasionally, the minor bands of 65, 61 and 21 kDa were detected with 80 kDa protein, although these minor proteins having lower molecular weight than 80 kDa may be degradation products of 80 kDa protein. Recently, it has been proposed, however, that GABAB receptor is multiple from the various distinct pharmacological properties (21). Therefore, it may be possible that major proteins in the eluate from the baclofen-affinity column represent the presence of multiplicity in GABAB receptors. These points should be clarified in future studies including the cDNA cloning of GABAB receptors.

Subsequently, the eluate from baclofen-affinity gel was examined whether or not it had the pharmacological properties as GABAB receptor using the reconstituted system with partially purified Gi/Go protein and adenylyl cyclase. The application of 10 μM forskolin induced an increase of cAMP accumlation in the reconstituted vesicles (Fig. 2). The forskolin-stimulated cAMP accumulation was inhibited invariably by GABA and baclofen and this inhibition was eliminated by 2-hydroxy saclofen, a GABAB receptor antagonist. These results indicate that the purified GABAB receptor has a pharmacological characteristics as the receptor and this reconstituted system is useful for the analysis on the pharmacological characteristics of GABAB receptor systems in the brain. Therefore, it has been evaluated that the GABAB receptor prepared from baclofen-affinity gel has similar properties as compared to those in immunoaffinity purified GABAB receptor (14). In fact, the presence of the same 80 kDa protein in the eluate from baclofen-affinity gel was recognized by the monoclonal antibody.

It is well known that ethanol affects on various receptor-mediated systems (22). For example, the GABAA receptor-mediated Cl^- channel activity was depressed in the presence of ethanol (23). Therefore, we have examined whether or not ethanol has direct effects on GABAB receptor-mediated system using the reconstituted system. The binding activity to partially purified GABAB receptor from baclofen-affinity gel was inhibited by 500 mM ethanol. The forskolin-stimulated cAMP accumulation also decreased in the presence of 500 mM ethanol, although [^{35}S]GTPγS binding to Gi/Go protein preparation was not affected by ethanol. Similarly, it was found that 500 mM ethanol induced the suppression of GABAB receptor-mediated inhibition on the forskolin-

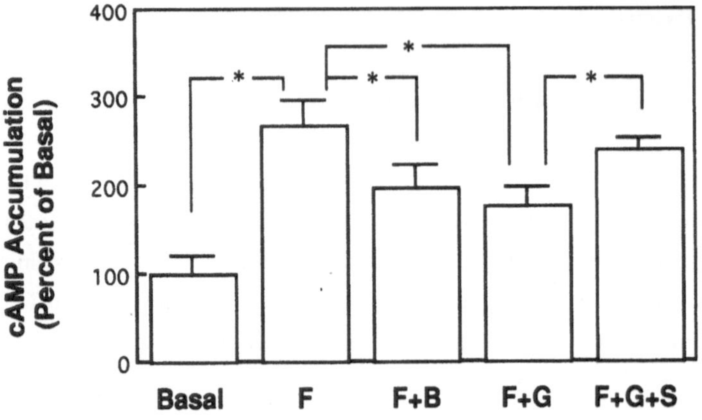

Figure 2. Effect of various drugs on cAMP accumulation in reconstituted vesicles with GABAB receptor, Gi/Go protein and adenylyl cyclase.

F: 10 μM forskolin; B: 100 μM baclofen; G: 10 μM GABA;

S: 100 μM 2-hydroxy saclofen

*$p<0.05$ (Bartlett's test; N=4-5)

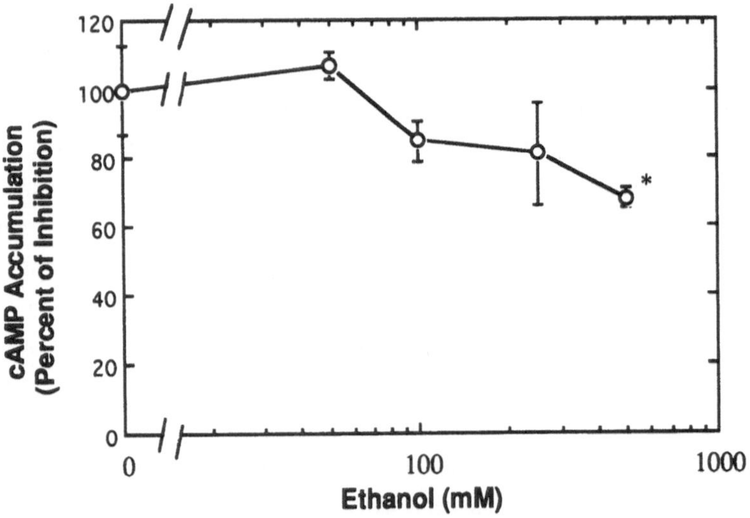

Figure 3. Effect of ethanol on GABA-induced suppresion of forskolin-stimulated cAMP accumulation in reconstituted vesicles.

*$p<0.05$; compared with the value in the absence of ethanol.

(Bartlett's test; N=3)

stimulated cAMP accumulation in the reconstituted vesicles (Fig. 3). These results indicate that the decreased binding activity of $GABA_B$ receptor by ethanol leads to the suppression of the response transduced by $GABA_B$ receptor, possibly due to the decreased inhibition to adenylyl cyclase. In contrast, the decrease of forskolin-stimulated cAMP accumulation was not detected, if these three components were reconstituted following the respective preincubation with 500 mM ethanol. These results also suggest that the inhibition of $GABA_B$ receptor systems by a high concentration of ethanol may be transient and reversible.

Conclusion

The $GABA_B$ receptor was isolated from the bovine cerebral cortex using baclofen-affinity gel. This receptor protein exhibited a typical characteristics of $GABA_B$ receptor such as the inhibition of adenylyl cyclase in the presence of Gi/Go protein. Furthermore, the eluate from baclofen-affinity gel indicated a possible presence of heterogeneity in $GABA_B$ receptors by the immunoblot. Possible inhibition of $GABA_B$ receptor functions by a high concentration of ethanol was also suggested.

Acknowledgment

This work was supported, in part, by Grant-in-Aid for Scientific Research and by International Scientific Research Program of the Japanese Ministry of Education, Science and Culture.

References

1. Hill DR, Bowery NG. ^{3}H-baclofen and ^{3}H-GABA bind to bicuculline-insensitive $GABA_B$ sites in rat brain. Nature 1981; 290: 149-152.

2. Hill DR, Bowery NG, Hudson AL. Inhibition of $GABA_B$ receptor binding by guanyl nucleotides. J Neurochem 1984; 42: 652-657.

3. Asano T, Ui M, Ogasawara N. Prevention of the agonist binding to γ-aminobutyric acid B receptors by guanine nucleotides and islet-activating protein, pertussis toxin, in bovine cerebral cortex. J Biol Chem 1985; 260: 12653-12658.

4. Wojcik WJ, Neff NH. γ-Aminobutyric acid B receptors are negatively coupled to adenylate cyclase in brain, and in the cerebellum these receptors may be associated with granule cells. Mol Pharmacol 1984; 25: 24-28.

5. Xu J, Wojcik WJ. Gamma aminobutyric acid B receptor-mediated inhibition of adenylate cyclase in cultured cerebellar granule cells: blockade by islet-activating protein. J Pharmacol Exp Ther 1986; 239: 568-573.

6. Newberry NR, Nicoll RA. Direct hyperpolarizing action of baclofen on hippocampal pyramidal cells. Nature 1984; 308: 450-452.

7. Holz GG, Rane SG, Dunlap K. GTP-binding proteins mediate transmitter inhibition of voltage-dependent calcium channels. Nature 1986; 319: 670-672.

8. Bowery NG, Hill DR, Hudson AL, Doble A, Middlemiss DN, Shaw J, Turnbull M. (-)Baclofen decreases neurotransmitter release in the mammalian CNS by an action at a novel GABA receptor. Nature 1980; 283: 92-94.

9. Bonanno G, Raiteri M. γ-Aminobutyric acid (GABA) autoreceptors in rat cerebral cortex and spinal cord represent pharmacologically distinct subtypes of the GABAB receptor. J Pharmacol Exp Ther 1993; 265: 765-770.

10. Ohmori Y, Hirouchi M, Taguchi J, Kuriyama K. Functional coupling of the γ-aminobutyric acidB receptor with calcium ion channel and GTP-binding protein and its alteration following solubilization of the γ-aminobutyric acidB receptor. J Neurochem 1990; 54: 80-85.

11. Ohmori Y, Kuriyama K. Solubilization and partial purification of GABAB receptor from bovine brain. Biochem Biophys Res Commun 1990; 172: 22-27.

12. Kuriyama K, Mizutani H, Nakayasu H. Purification and identification of 61 kilodalton GABA (γ-aminobutyric acid)B receptor from bovine brain. Mol Neuropharmacol 1992; 2: 155-157.

13. Nakayasu H, Mizutani H, Hanai K, Kimura H, Kuriyama K. Monoclonal antibody to GABA binding protein, a possible GABAB receptor. Biochem Biophys Res Commun 1992; 182: 722-726.

14. Nakayasu H, Nishikawa M, Mizutani H, Kimura H, Kuriyama K. Immunoaffinity purification and characterization of γ-aminobutyric acid (GABA)B receptor from bovine cerebral cortex. J Biol Chem 1993; 268: 8658-8664.

15. Zukin SR, Young AB, Snyder SH. Gamma-aminobutyric acid binding to receptor sites in the rat central nervous system. Proc Natl Acad Sci USA 1974; 71: 4802-4807.

16. Sternweis PC, Robishaw JD. Isolation of two proteins with high affinity for guanine nucleotides from membranes of bovine brain. J Biol Chem 1984; 259: 13806-13813.

17. Katada T, Oinuma M, Ui M. Mechanisms for inhibition of the catalytic activity of adenylate cyclase by the guanine nucleotide-binding proteins serving as the substrate of islet-activating protein, pertussis toxin. J Biol Chem 1986; 261: 5215-5221.

18. May DC, Ross EM, Gilman AG, Smigel MD. Reconstitution of catecholamine-stimulated adenylate cyclase activity using three purified protein. J Biol Chem 1985; 260: 15829-15833.

19. Haga K, Haga T, Ichiyama A, Katada T, Kurose H, Ui M. Functional reconstitution of purified muscarinic receptors and inhibitory guanine nucleotide regulatory protein. Nature 1985; 316: 731-733.

20. Nishikawa M, Kuriyama K. Functional coupling of cerebral γ-aminobutyric acid $(GABA)_B$ receptor with adenylate cyclase system: effect of phaclofen. Neurochem Int 1989; 14: 85-90.

21. Bonanno G, Raiteri M. Multiple $GABA_B$ receptors. Trends Pharmacol Sci 1993; 14: 259-261.

22. Kuriyama K, Ohkuma, S. Alteration in the function of cerebral neurotransmitter receptors during the establishment of alcohol dependence: neurochemical aspects. Alcohol & Alcoholism 1990; 25: 239-249.

23. Ueha T, Kuriyama K. Direct action of ethanol on cerebral $GABA_A$ receptor complex: analysis using purified and reconstituted $GABA_A$ receptor complex. Neurochem Int 1991; 19: 319-325.

GABA: Receptors, Transporters and Metabolism
ed. by C. Tanaka & N.G. Bowery

PARADOXICAL FACILITATION OF THE VOLTAGE-DEPENDENT CALCIUM CURRENT FOLLOWING ACTIVATION OF GABA$_B$ RECEPTORS

N. Ogata[1], S. Fujikawa[2] and H. Motomura[2]

[1]Second Department of Physiology, Faculty of Medicine, Hiroshima University, Hiroshima 734 and [2]Department of Pharmacology, Faculty of Medicine, Kyushu University, Fukuoka 812, Japan

Summary: The mechanisms underlying the enhancement of the calcium current (I_{Ca}) after application of baclofen, a GABA$_B$ agonist, were studied in neurons of the rat dorsal root ganglia using nystatin perforated patch clamp recording. Baclofen (50 μM) decreased I_{Ca} and slowed the onset of I_{Ca}. However, when baclofen was rapidly washed out from the medium, the amplitude of I_{Ca} was paradoxically augmented exceeding the control value measured before application of the drug. This enhancement of I_{Ca} by baclofen was not due to desensitization of GABA$_B$ receptors or a liberation from tonic G protein-mediated inhibition of I_{Ca}. From its extremely prolonged time course, an involvement of some intracellular signal transduction system was strongly suggested.

Introduction

It is well known that an activation of GABA$_B$ receptors causes inhibition of the voltage-dependent calcium current (I_{Ca}) through a G protein-mediated mechanism in a variety of tissues (1). In addition to this well-known effect we found that the amplitude of I_{Ca} is paradoxically increased following washout of baclofen in rat dorsal root ganglia (DRG). We now examined the mechanisms underlying this paradoxical enhancement of I_{Ca} by baclofen. Our results suggest that the calcium channel mediating the baclofen-induced facilitation of I_{Ca} may be distinct from the calcium channels which are linked to G protein-mediated inhibition.

Materials and Methods

Dissociated cultures of rat DRG neurones were prepared as described elsewhere (1). The methods for electrical recording used in the present study were similar to those previously described (1). Membrane currents were recorded with a nystatin perforated patch recording. The pipette solution contained (in mM): 120 Cs-glutamate, 10 NaCl, 2.5 $MgCl_2$, 5 glucose, 5 Hepes, and 5 EGTA. The pH of the pipette solution was adjusted to 7.0 with CsOH. Nystatin (450μg/ml) was added to the internal solution just belore sealing. The external solution contained (in mM): 120 NaCl, 5 CsCl, 1.8 $CaCl_2$, 1 $MgCl_2$, 5 Hepes, 25 glucose. The pH of the external solution was adjusted to 7.4 with NaOH. Tetrodotoxin (TTX) in a concentration of 0.2 mM was added to block TTX-sensitive Na current.

All the experiments were performed with an on-line system using a personal computer. In some cases, capacitative and leakage currents were subtracted digitally by the P-P/4 procedure (1). All experiments were done at room temperature (22 - 24℃). Drugs were applied through a rapid micro-superfusion system (2). This system uses a fine polyethylene needle (20μM tip internal diameter) placed near (about 50 μm) the cell. Electromagnetic valves are used to apply test solutions. This micro-superfusion system enabled us a rapid (latency, less than 0.1 s) and localized application of test solutions and their rapid washout (less than 0.5 s). (-)-Baclofen and CGP-35348 were generous gifts from Ciba-Geigy (Japan).

Results and Discussion

Fig. 1 shows a typical example of the effect of baclofen on I_{Ca} recorded in neurons of the cultured rat dorsal root ganglia. An application of baclofen produced typical G protein-mediated inhibition of I_{Ca} accompanied by a slowed activation (compare b with a). When baclofen was rapidly removed from the medium (c), the amplitude of I_{Ca} was paradoxically increased, exceeding the control value measured before application of the drug. Since this increase was reversible (d) and reproducible, the augmentation of I_{Ca} was not due to technical problems.

Next, we checked whether or not the paradoxical increase of calcium current following washout of baclofen (I_{BR}) was truly mediated by $GABA_B$ receptors. A combined application of CGP-35348, a $GABA_B$ antagonist, (50-100μM) with baclofen blocked (50 μM) both G protein-mediated inhibition and I_{BR} to a similar extent, indicating that I_{BR} was surely mediated by $GABA_B$ receptors.

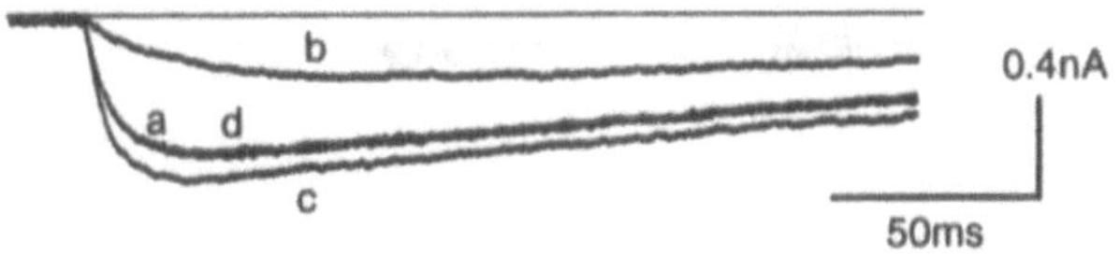

Figure 1. Rebound enhancement of the I_{Ca} following washout of baclofen. In this and subsequent figures: I_{Ca}s were recorded from rat dorsal root ganglion neuron; a step depolarization (200 ms in duration) to -10 mV from a holding potential of -80 mV was delivered; downward and upward deflections represent inward and outward currents, respectively. Step pulses were applied before (a), during (b), 30 s (c) or 10 min (d) after application of baclofen (50 μM). Four successive current traces are shown superimposed.

Fig. 2 shows the current-voltage curves obtained before, during and after application of baclofen. Baclofen did not cause any detectable shift of the voltage-axis. Thus, I_{BR} was not due to a shift of the voltage axis.

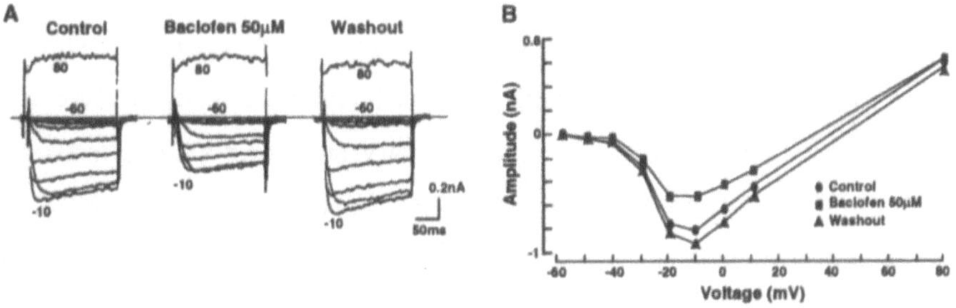

Figure 2. Current-voltage curves for I_{Ca} obtained before, during and after application of baclofen. In A, step depolarizations (200 ms in duration) to various potential levels (-60 - +80 mV) from a holding potential of -80 mV were delivered, and successive current traces were shown superimposed. The graph in B plots the peak amplitude of I_{Ca} against the test potential level.

There may be a possibility that I_{BR} is due to conductances for other ions. For example, a decreased permeability to potassium or chloride ions or an increased permeability to sodium ions may contribute to I_{BR}. Such possibilities should be taken into consideration, since baclofen affects potassium currents in other type of neurons and an influx of calcium activates chloride currents. However, these possibilities were excluded by ion substitution experiments where these ions were replaced by impermeant ions.

Fig. 3 shows the time course of I_{BR}. Baclofen was applied for 10s. I_{BR} decayed exponentially with an extremely prolonged time constant of the order of mins, strongly suggesting an involvement of an intracellular second messenger system.

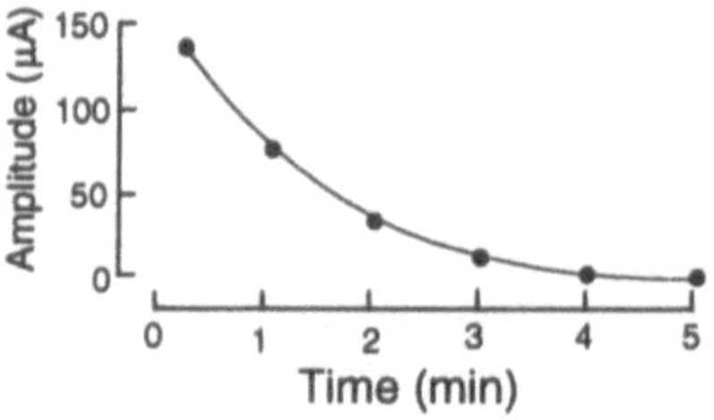

Figure 3. Time course of the baclofen-induced enhancement of I_{Ca}. I_{Ca}s were recorded after application of baclofen (50 μM). The amplitude of I_{BR} was plotted against the time elapsed after removal of baclofen from the medium.

I_{BR} was already apparent at 50nM of baclofen, whereas the G protein-mediated inhibition was still minimal at this concentration, and both responses increased in parallel with the increase in concentration of baclofen. One interesting feature of I_{BR} was its dependence on the holding potential. In Fig. 4, I_{BR} was induced with different holding potentials of -60 and -80mV, respectively. Traces for the holding potential of -80mV are shown with normalized control amplitude. I_{BR} evoked from -80mV was markedly larger than that obtained from -60mV, whereas the G protein-mediated inhibition showed similar effect at two holding potentials. The voltage-dependence of I_{BR} is interesting, in view of the recent reports that the phosphorylation of the membrane is facilitated by membrane hyperpolarization.

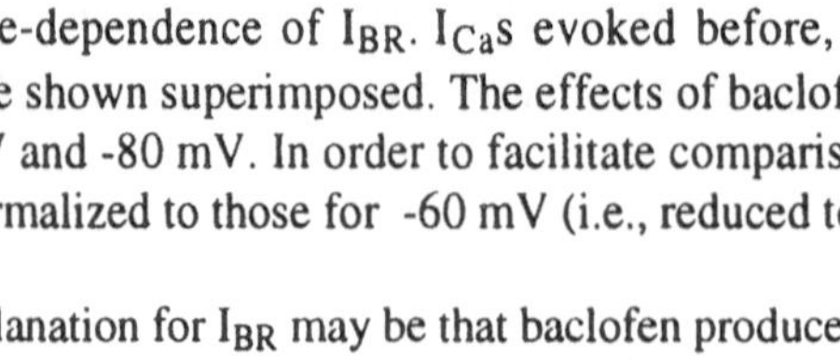

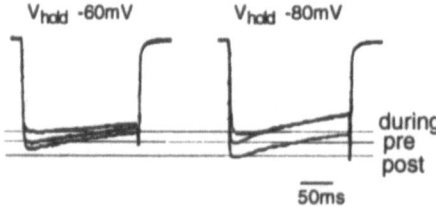

Figure 4. Voltage-dependence of I_{BR}. I_{Ca}s evoked before, during and after application of baclofen (50 μM) are shown superimposed. The effects of baclofen were examined under holding potentials of -60 mV and -80 mV. In order to facilitate comparison, the peak amplitudes of traces for -80 mV were normalized to those for -60 mV (i.e., reduced to 63%).

One possible explanation for I_{BR} may be that baclofen produces some metabolic alteration which could result in enhancement of I_{Ca}, and this metabolic change continues for a extended period of time even after removal of baclofen, and therefore, an enhancement of I_{Ca} become evident following the termination of the G protein-mediated inhibition of I_{Ca}. Another explanation is that I_{Ca} is usually inhibited by a tonic inhibition through G protein-mediated mechanism. An application of baclofen, which further increases this G protein-mediated inhibition, could result in temporary "desensitization" of this mechanism, thus inducing a rebound augmentation of I_{Ca}. In order to examine these possibilities, we tested whether or not I_{BR} was an "off-response" subsequent to a removal of the drug. Fig. 5 shows the recordings with very low concentration of baclofen. As shown before, this concentration was minimal for the activation of G protein-mediated inhibition. In this case, I_{Ca} during an application of baclofen was rather slightly augmented as compared with the control I_{Ca}. This indicates that the facilitation of I_{Ca} occurred already during an application of the drug. Actually, when I_{Ca}s were examined at the start of perfusion and during 1min perfusion, the latter was consistently larger than the former.

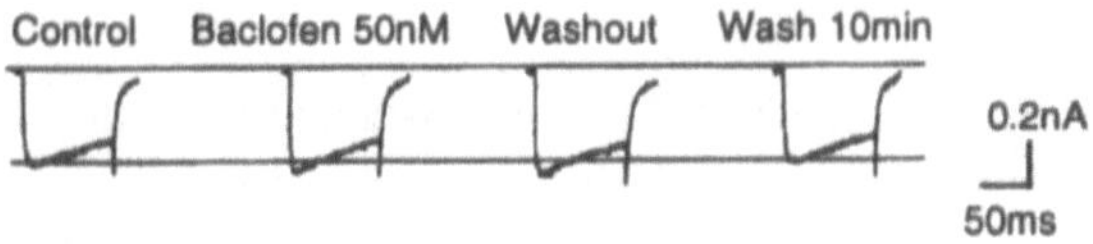

Figure 5. A low concentration of baclofen (50 nM) slightly increases the peal amplitude of I_{Ca}.

The possibility remains that the increase in I_{Ca} during a perfusion of baclofen may be due to a desensitization of the $GABA_B$ receptors linked to G protein-mediated inhibition. Baclofen was applied for 10 min in Fig. 6. I_{Ca}s were evoked, before application of the drug (a), immediately after introduction of the drug (b), just before washout of the drug (c), and immediately after washout of the drug (d). The ratio, b/a, represents the degree of G protein-mediated inhibition. It

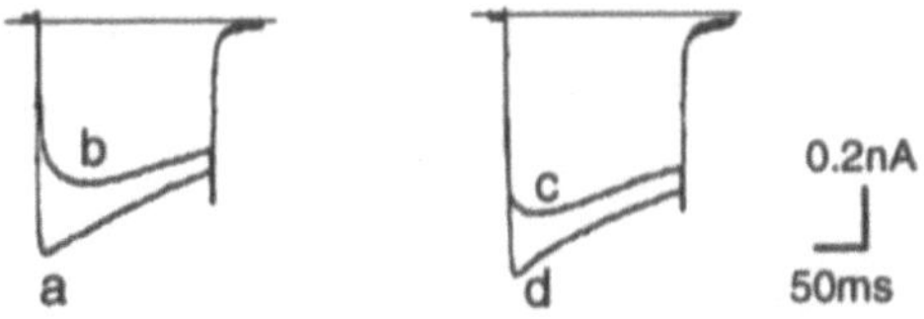

Figure 6. The inhibitory action of baclofen on I_{Ca} does not desensitize. Baclofen (50μM) was applied for 20 min. I_{Ca}s were evoked before application of the drug (a), immediately after introduction of the drug (b), before washout of the drug (c), and immediately after washout of the drug (d). Currents evoked at a-b, and c-d are shown superimposed, respectively.

should be noted that the recovery from G protein-mediated inhibition indicated by "c" and "d" was much the same as the initial inhibition shown by "a" and "b", indicating that the G protein-mediated inhibition does not desensitize. Thus, I_{BR} does not appear to be a "off-response" subsequent to removal of baclofen. These results suggest that the calcium channels responsible for I_{BR} may be distinct from those linked to G protein-mediated inhibition.

It is well known that the G protein-mediated inhibition is prevented by a prior large depolarization. Thus, the prepulse-induced facilitation of I_{Ca} is a good example for the release from

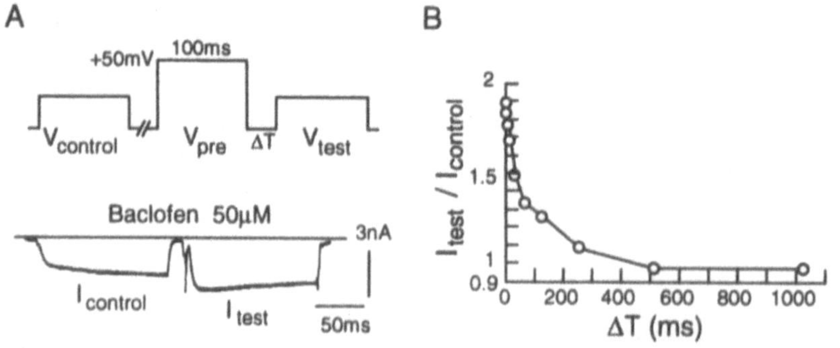

Figure 7. The time course of the prepulse-induced facilitation of I_{Ca} in the presence of baclofen. A, a pulse protocol and current responses. Two identical step depolarizations to -10 mV for 100 ms were applied 5 s before ($V_{control}$) and ΔT ms after (V_{test}) the large conditioning depolarizing prepulse to +50 mV. The current during the prepulse was not shown in the current trace due to a large sampling rate at this period. B, time dependence of the prepulse-induced facilitation of I_{Ca}. The peak amplitude of I_{test} was measured at 7 ms after the onset of the test pulse. The ratio, $I_{test}/I_{control}$, was plotted as a function of ΔT.

G protein-mediated inhibition. The time constant of this phenomenon was an order of ms (Fig. 7), and thus markedly different from the time course of I_{BR} whose time constant was an order of min.

In conclusion, the enhancement of calcium current following washout of baclofen was not due to desensitization of $GABA_B$ receptors or a liberation from tonic G protein-mediated inhibition. From its extremely prolonged time course, an involvement of some intracellular signal transduction system is strongly suggested.

References

1. Tatebayashi H, Ogata, N. Kinetic analysis of the $GABA_B$-mediated inhibition of the high-threshold Ca^{2+} current in cultured rat sensory neurones. J Physiol 1992; 447: 391-407.

2. Ogata N, Tatebayashi, H. A simple and multi-purpose "concentration-clamp" method for rapid superfusion. J Neurosci Methods 1991; 39: 175-183.

ed. by C. Tanaka & N.G. Bowery

STEREOSPECIFIC GABA$_B$ RECEPTOR ANTAGONISM

K. Frydenvang, B. Frølund, *U. Kristiansen and P. Krogsgaard-Larsen

PharmaBiotec Research Center, Departments of Medicinal Chemistry and *Biology, The Royal Danish School of Pharmacy, 2 Universitetsparken, DK-2100 Copenhagen, Denmark

Summary: (RS)-Phaclofen has been resolved chromatographically and the GABA$_B$ receptor affinity and antagonism shown to reside in the (R)-enantiomer. The absolute stereochemistry of (R)-phaclofen was established by an X-ray crystallographic analysis. The structural similarity of the agonist (R)-baclofen and the antagonist (R)-phaclofen suggests that these ligands interact with the GABA$_B$ receptors in a similar manner. Thus, it may be concluded that the opposite pharmacological effects of these compounds essentially results from the different spatial and protolytic properties of their acid groups. (R)-5-Amino-4-hydroxyvaleric acid [(R)-4-OH-DAVA] and (S)-2-OH-DAVA have been shown to antagonize GABA$_B$ receptor-mediated function in a stereoselective manner. Whereas none of the four stereoisomers of 5-amino-4-hydroxy-2-methylvaleric acid (2-Me-4-OH-DAVA) showed significant affinity for GABA$_A$ receptor sites or GABA uptake mechanisms in rat brain synaptic membranes, (2R,4R)-2-Me-4-OH-DAVA was shown to stereoselectively antagonize GABA$_B$ receptor-mediated function in the isolated guinea pig ileum. The structure of (2R,4R)-2-Me-4-OH-DAVA was established by an X-ray crystallographic analysis, and the solid state-conformation of (2R,4R)-2-Me-4-OH-DAVA was compared with the proposed receptor-active conformations of (R)-4-OH-DAVA and (S)-2-OH-DAVA.

Introduction

The inhibitory effects of the central amino acid neurotransmitter 4-aminobutyric acid (GABA) are mediated by at least two classes of receptors, GABA$_A$ and GABA$_B$ receptors, both of which appear to comprise a number of receptor subtypes (1-7). Whereas most

$GABA_A$ receptors seem to function as postsynaptic receptors, $GABA_B$ receptors probably are located predominantly on nerve terminals regulating the release of GABA (autoreceptors) or other neurotransmitters (heteroreceptors) (6,8,9). The (R)-enantiomer of 3-(4-chlorophenyl)-4-aminobutyric acid (baclofen) (Fig. 1) is a potent and selective $GABA_B$ agonist, whereas (S)-baclofen is inactive (10). (RS)-2-(4-Chlorophenyl)-3-aminopropane-1-phosphonic acid (phaclofen) is a $GABA_B$ antagonist (11), which, like (R)-baclofen, interacts selectively with cortical $GABA_B$ receptors, showing very low affinity for $GABA_B$ receptors in the spinal cord (9).

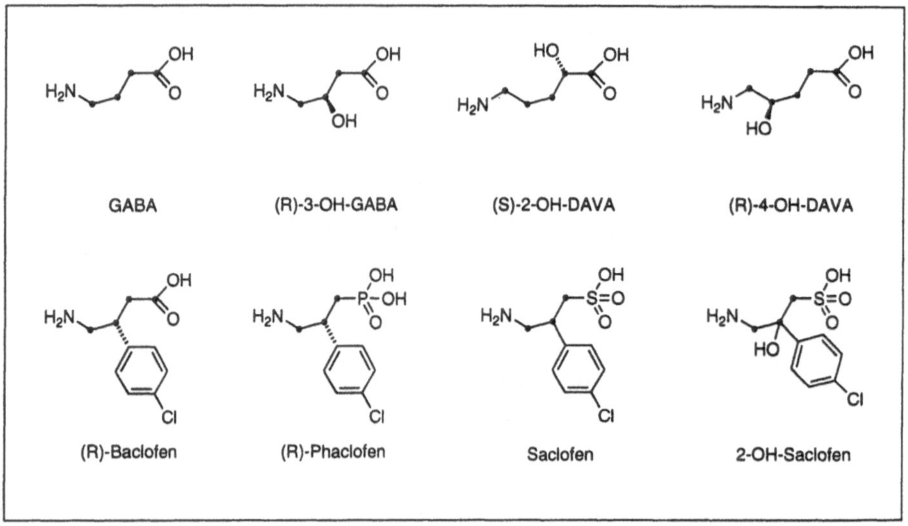

Figure 1. Structures of GABA, the $GABA_B$ agonists (R)-3-OH-GABA and (R)-baclofen, and the $GABA_B$ antagonists (R)-phaclofen, saclofen and 2-OH-saclofen. The structures of two hydroxylated analogues of 5-aminovaleric acid (DAVA) are shown.

In light of the structural similarity of the $GABA_B$ agonist, (R)-baclofen, and the $GABA_B$ antagonist, phaclofen, we assume that these compounds interact with the recognition site of the receptor in a similar manner. In order to study the validity of this assumption, we have now resolved phaclofen, established the absolute stereochemistry of its enantiomers, and studied the effects of these compounds for $GABA_B$ receptor sites in rat brain.

During the past few years, a number of $GABA_B$ antagonists, bioisosterically derived from baclofen, have been developed. Thus, phaclofen (11), or rather (R)-phaclofen (12), saclofen

(13) and, in particular, 2-hydroxysaclofen (2-OH-saclofen) (14) (Fig. 1) antagonize $GABA_B$ receptor-mediated functions with different potencies. More recently, a series of phosphinic acids (Fig. 2) has been shown to be very effective $GABA_B$ antagonists, and some of the more potent compounds within this series of antagonists contain a hydroxy group in a position equivalent to the hydroxylated position in 2-OH-saclofen (15).

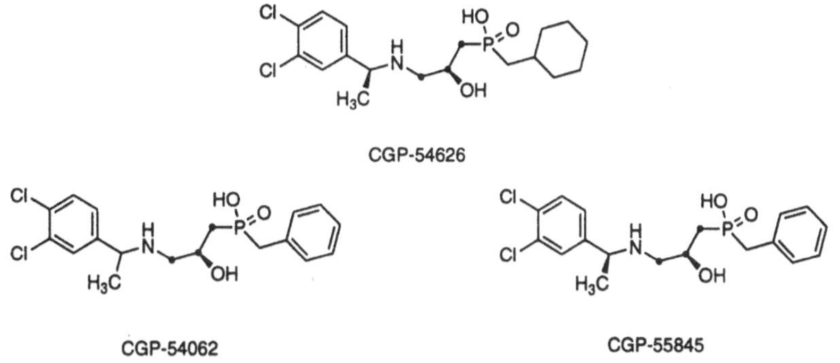

Figure 2. Structures of the new $GABA_B$ antagonists CGP 54626, CGP 54062, and CGP 55845, all of which contain a hydroxy group with (S)-configuration at C-2 of the amino acid backbone.

Whereas the (R)-form of 4-amino-3-hydroxybutyric acid [(R)-3-OH-GABA] is a $GABA_B$ agonist (16), we have recently shown that (S)-5-amino-2-hydroxyvaleric acid [(S)-2-OH-DAVA] as well as (R)-4-OH-DAVA are selective $GABA_B$ antagonists (17).

On the basis of these structure-activity relationships, we envisage that the $GABA_B$ receptors contain distinct binding sites for the hydroxy and 4-chlorophenyl groups in addition to charged groups complementary to the positively and negatively charged groups of the $GABA_B$ receptor ligands described. In continuation of this indirect elucidation of the topography of the $GABA_B$ recognition site(s), the four stereoisomers of 5-amino-4-hydroxy-2-methylvaleric acid (2-Me-4-OH-DAVA) (Fig. 3) have been synthesized (18). We here summarize the results of *in vitro* pharmacological studies of these compounds and an X-ray crystallographic analysis of the most potent isomer, the eutomer (19).

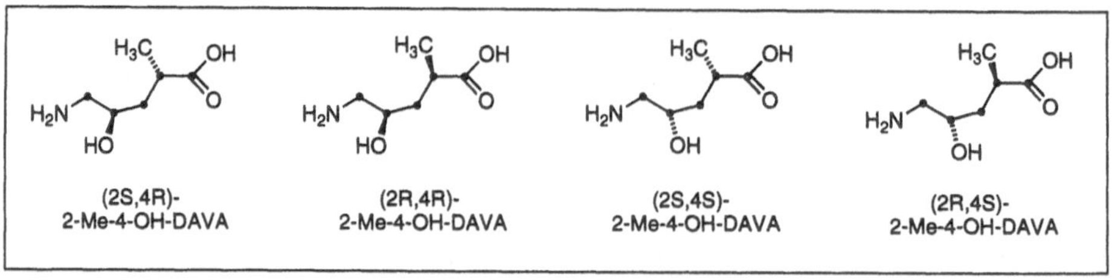

Figure 3. Structures of the four stereoisomers of 5-amino-4-hydroxy-2-methylvaleric acid (2-Me-4-OH-DAVA).

Structure-Activity Studies on (R)-Baclofen and (R)-Phaclofen

Substitution of a phosphonic acid group for the carboxyl group of the $GABA_B$ agonist baclofen produces phaclofen, a $GABA_B$ antagonist (11). Only the (R)-enantiomer of baclofen shows affinity for and agonist effect at the $GABA_B$ receptor (10). The present studies have disclosed that the $GABA_B$ receptor affinity and antagonist effect of phaclofen also resides in the (R)-enantiomer, suggesting that these ligands bind to the $GABA_B$ recognition site in a similar fashion. As shown in Figure 4, (R)-baclofen (20) and (R)-phaclofen (19) adopt similar conformations in the crystalline state. It may be suggested that these low-energy conformations to some extent reflect the conformations adopted by these compounds during their binding to the $GABA_B$ receptor.

The molecular mechanisms underlying the activation of $GABA_B$ receptors by (R)-baclofen and the antagonist binding of (R)-phaclofen are, however, still unknown. This paper describes the X-ray crystallographic analysis of zwitterionic (R)-phaclofen, but whether this form or the fully ionized species of the molecule (Fig. 5) actually is recognized by the $GABA_B$ receptor sites remains to be established.

The hydroxylated analogues of 5-aminovaleric acid (DAVA), (S)-2-OH-DAVA and (R)-4-OH-DAVA (Fig. 1), which are derived from the $GABA_B$ agonist, (R)-3-OH-GABA, are $GABA_B$ antagonists (17). Molecular modeling studies have shown that these two antagonists can adopt low-energy conformations with almost overlapping functional groups. These conformations are assumed to reflect the shapes of the two molecules when recognized by the

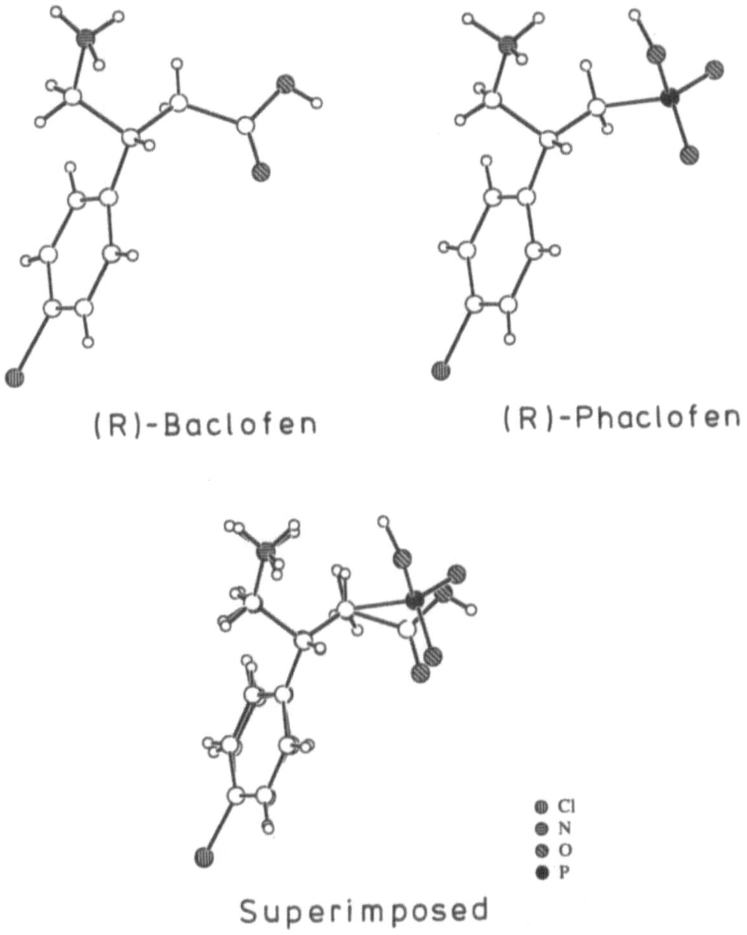

Figure 4. Structures of (R)-baclofen and (R)-phaclofen as determined by X-ray analyses, and a superposition of the two structures.

$GABA_B$ receptors (17). It is tempting to speculate that (R)-phaclofen interacts with $GABA_B$ receptor sites in the zwitterionic form, and that the nonionized phosphonic acid hydroxy group and the hydroxy groups of (S)-2-OH-DAVA and (R)-4-OH-DAVA bind to the same site of the antagonist conformation of the $GABA_B$ receptor (Fig. 5). These aspects are at present under investigation.

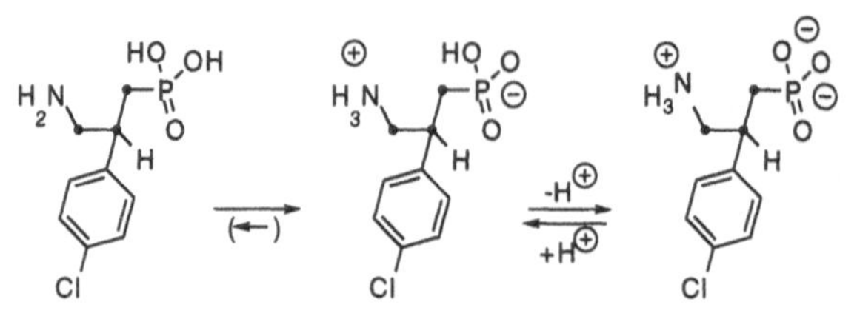

Figure 5. An illustration of the unionized, the zwitterionic (the central structure), and the fully ionized forms of phaclofen. At physiological pH, phaclofen will exist as an approximately equal mixture of the two ionized forms. It is hypothesized that phaclofen interacts with the $GABA_B$ receptor in the zwitterionic form and that the hydroxy group of this form of the molecule binds to the same site of the receptor as that recognized by the hydroxy groups of the hydroxylated DAVA analogues (see Figs. 6 and 7).

Structure-Activity Studies on the Stereoisomers of 5-Amino-4-hydroxy-2-methylvaleric Acid (2-Me-4-OH-DAVA)

The stereochemical orientations of the hydroxy and 4-Cl-phenyl groups of the $GABA_B$ agonists, (R)-3-OH-GABA and (R)-baclofen, respectively, are opposite (Fig. 1), suggesting that these groups interact with different substructures of the $GABA_B$ recognition site (17,21). The presence of a receptor substructure capable of binding hydroxy groups is further supported by the observation that 2-OH-saclofen is a more potent $GABA_B$ antagonist than saclofen (Fig. 1) (14). The $GABA_B$ antagonists, (R)-4-OH-DAVA and (S)-2-OH-DAVA, can adopt low-energy superimposable conformations (17), and we imagine that interaction of the hydroxy groups with this receptor substructure stabilizes the receptor binding of these antagonists.

Whereas (2S,4R)-2-Me-4-OH-DAVA is a weak $GABA_B$ antagonist, (2R,4R)-2-Me-4-OH-DAVA is relatively potent (Fig. 6), more potent than (R)-4-OH-DAVA. The enantiomers of these compounds are very weak or inactive. Thus, (R)-configuration of the hydroxy group at C-4 appears to be a necessary, but obviously not sufficient, condition for effective $GABA_B$

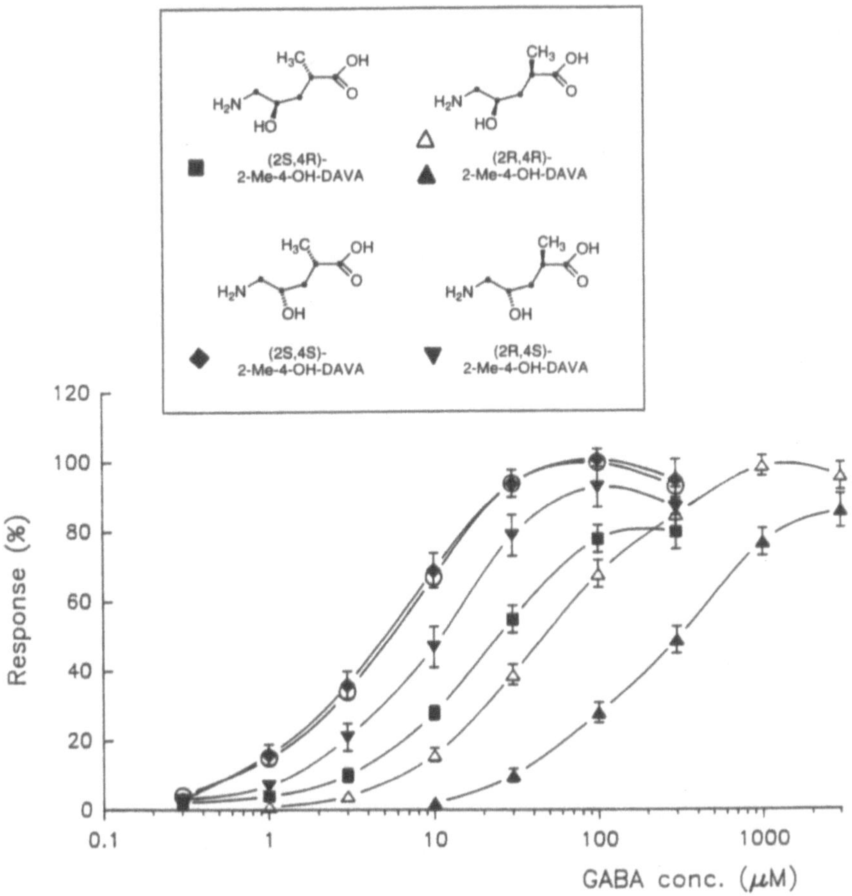

Figure 6. Concentration - response curves for GABA alone (○) and in the presence of (2S,4S)-2-Me-4-OH-DAVA (1 m*M*) (◆), (2R,4S)-2-Me-4-OH-DAVA (1 m*M*) (▼), (2S,4R)-2-Me-4-OH-DAVA (1 m*M*) (■), and (2R,4R)-2-Me-4-OH-DAVA (0.3 m*M*) (Δ), or (1 m*M*) (▲) using a guinea pig ileum preparation. Antagonists were added 5 - 7 min before the applications of GABA. The experiments were performed in the presence of the $GABA_A$ antagonist SR 95531 at a concentration of 10 μ*M*. Each experiment was performed on at least eight tissues from four animals (for details, see ref. 19).

antagonism of this class of hydroxy amino acids.

The stereochemical orientation of the methyl group of (2S,4R)-2-Me-4-OH-DAVA is identical with that of the hydroxy group of the $GABA_B$ antagonist, (S)-2-OH-DAVA. Thus,

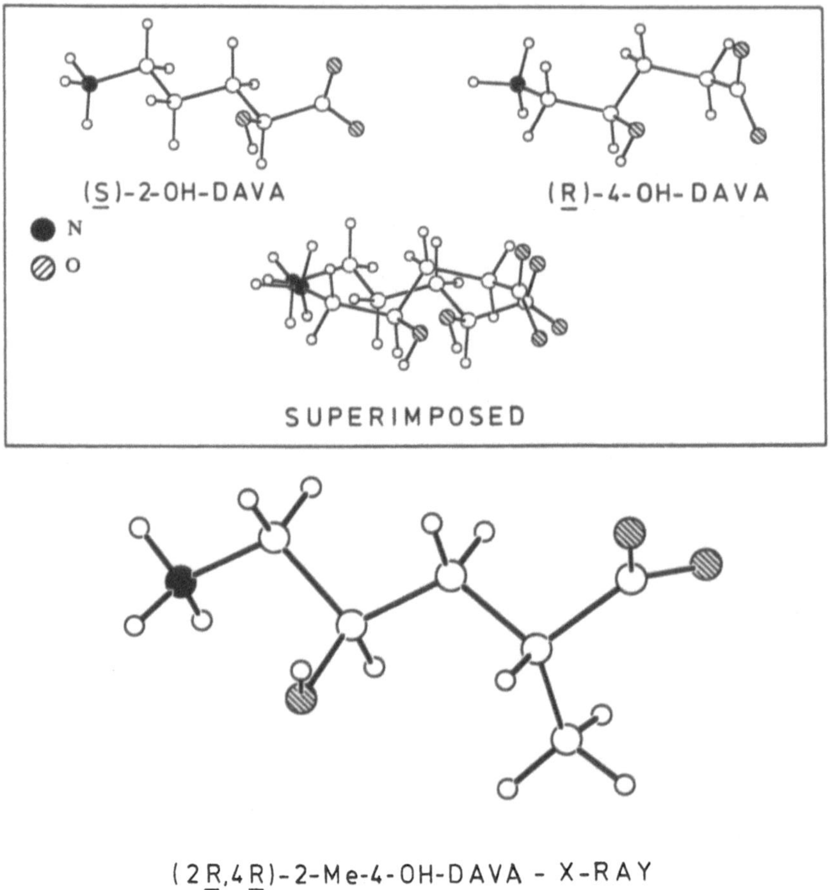

Figure 7. *Top*: Energy-minimized superimposable conformations of the $GABA_B$ antagonists, (S)-2-OH-DAVA (upper left) and (R)-4-OH-DAVA (upper right) and a four-point fit of these conformations (re-modulated from ref. 17) (SYBYL (22)). *Bottom*: The conformation of the $GABA_B$ antagonist, (2R,4R)-2-Me-4-OH-DAVA, in crystals as determined by an X-ray crystallographic analysis (for details see ref. 19).

unfavourable interaction of this methyl group with the proposed substructure of the $GABA_B$ receptor site capable of binding hydroxy groups may explain the low $GABA_B$ antagonist potency of (2S,4R)-2-Me-4-OH-DAVA. Along this line of reasoning, lack of interaction of the methyl group of (2R,4R)-2-Me-4-OH-DAVA with this particular receptor substructure may explain the much more potent $GABA_B$ antagonist effect of this compound. The superimposed conformations of (R)-4-OH-DAVA and (S)-2-OH-DAVA show a low degree of overlap of the α-carbon atoms (Fig. 7), suggesting that these parts of the molecules interact with areas of the $GABA_B$ recognition sites showing some steric tolerance. We imagine that the methyl group of the eutomer, (2R,4R)-2-Me-4-OH-DAVA may be accomodated by this part of the receptor.

It must be emphasized that all of these hydroxy amino acids possess a high degree of molecular flexibility, making predictions of receptor-active conformations difficult. Thus, the conformation of (2R,4R)-2-Me-4-OH-DAVA observed in crystals (Fig. 7) is primarily determined by intermolecular contacts and may not necessarily reflect the conformation recognized by the $GABA_B$ receptor. The conformation of (2R,4R)-2-Me-4-OH-DAVA in crystals has been compared with the conformations of (S)-2-OH-DAVA and (R)-4-OH-DAVA obtained by optimized flexible four-point fits (Fig. 7). Although this low-energy conformation of (2R,4R)-2-Me-4-OH-DAVA shows some similarity with that of (R)-4-OH-DAVA in the optimized fit, there are differences. Based on the small amounts of energy required for the transformation of both of these compounds (17), it is assumed that conversion of the depicted conformation of (2R,4R)-2-Me-4-OH-DAVA into a conformation similar to that shown for (R)-4-OH-DAVA only will require small amounts of energy. These aspects will be the subject of detailed computational studies in the future.

(2R,4R)-2-Me-4-OH-DAVA may be a useful pharmacological tool for studies of $GABA_B$ receptors, or receptor subtypes, and will be exploited by us as a lead in the design of new $GABA_B$ antagonists.

Acknowledgements

This work was supported by grants from the Danish Technical Research Council, the Alfred Benzon Foundation, the Lundbeck Foundation and the Danish State Biotechnology Programme. The secretarial assistance of Mrs. Anne Nordly is gratefully acknowledged.

References

1. Bowery NG, editor. Actions and Interactions of GABA and Benzodiazepine Receptor Subtypes. New York: Raven Press, 1984.

2. Olsen RW, Venter JC, editors. Benzodiazepine/GABA Receptors and Chloride Channels: Structural and Functional Properties. New York: Alan R. Liss, 1986.

3. Bowery NG, Nistico G, editors. GABA: Basic Research and Clinical Applications. Rome: Pythagora Press, 1989.

4. Biggio G, Costa E, editors. GABA and Benzodiazepine Receptor Subtypes. New York: Raven Press, 1990.

5. Drew CA, Johnston GAR. Bicuculline- and baclofen-insensitive γ-aminobutyric acid binding to rat cerebellar membranes. J Neurochem 1992; 58:1087-1092.

6. Bonanno G, Raiteri M. Multiple $GABA_B$ receptors. Trends Pharmacol Sci 1993; 14:259-261.

7. Krogsgaard-Larsen P, Frølund B, Jørgensen FS, Schousboe A. $GABA_A$ receptor agonists, partial agonists and antagonists. Design and therapeutic prospects. J Med Chem 1994; 37:2489-2505.

8. Pende M, Lanza M, Bonanno G, Raiteri M. Release of endogenous glutamic and aspartic acids from cerebrocortex synaptosomes and its modulation through activation of a γ-aminobutyric $acid_B$ ($GABA_B$) receptor subtype. Brain Res 1993; 604:325-330.

9. Bonanno G, Raiteri M. γ-Aminobutyric acid (GABA) autoreceptors in rat cerebral cortex and spinal cord represent pharmacologically distinct subtypes of the $GABA_B$ receptor. J Pharmacol Exp Ther 1993; 265:765-770.

10. Bowery NG, Hill DR, Hudson AL, Doble A, Middlemiss DN, Shaw J, Turnbull M. (-)-Baclofen decreases neurotransmitter release in the mammalian CNS by an action at a novel GABA receptor. Nature 1980; 283:92-94.

11. Kerr DIB, Ong J, Prager RH, Gynther BD, Curtis DR. Phaclofen: A peripheral and central baclofen antagonist. Brain Res 1987; 405:150-154.

12. Frydenvang K, Hansen JJ, Krogsgaard-Larsen P, Mitrovic A, Tran H, Drew CA, Johnston GAR. $GABA_B$ antagonists: resolution, absolute stereochemistry and pharmacology of (*R*)- and (*S*)-phaclofen. Chirality 1994; 6:583-589.

13. Kerr DIB, Ong J, Johnston GAR, Abbenante J, Prager RH. Antagonism at $GABA_B$ receptors by saclofen and related sulfonic acid analogues of baclofen and GABA. Neurosci

Lett 1989; 107:239-244.

14. Kerr DIB, Ong J, Johnston GAR, Abbenante J, Prager RH. 2-Hydroxy-saclofen: an improved antagonist at central and peripheral $GABA_B$ receptors. Neurosci Lett 1988; 92:92-96.

15. Bittiger H, Froestl W, Mickel SJ, Olpe H-R. $GABA_B$ receptor antagonists: from synthesis to therapeutic applications. Trends Pharmacol Sci 1993; 14:391-394.

16. Kristiansen U, Fjalland B. Ligand structural specificity of $GABA_A$ receptors in the guinea-pig ileum. Pharmacol Toxicol 1991; 68:332-339.

17. Kristiansen U, Hedegaard A, Herdeis C, Lund TM, Nielsen B, Hansen JJ, Falch E, Hjeds H, Krogsgaard-Larsen P. Hydroxylated analogues of 5-aminovaleric acid (DAVA) as $GABA_B$ receptor antagonists: stereostructure-activity relationships. J Neurochem 1992; 58:1150-1159.

18. Herdeis C, Lütsch K. Enantiopure DAVA-derivatives - Part III. Synthesis of all 4 stereoisomers of 2-methyl-4-hydroxy-5-aminopentanoic acid (2-Me-4-OH-DAVA). Tetrahedron: Asymmetry 1993; 4:121-131.

19. Frydenvang K, Kristiansen U, Frølund B, Herdeis C, Krogsgaard-Larsen P. $GABA_B$ antagonists: enantiopharmacology of 5-amino-4-hydroxy-2-methylvaleric acid and X-ray structure of the eutomer. Chirality 1995; 7 (in press).

20. Chang C-H, Yang DSC, Yoo CS, Wang B-C, Pletcher J, Sax M, Terrence CF. Structure and absolute configuration of (*R*)-baclofen monohydrochloride. Acta Cryst 1982; B38:2065-2067.

21. Falch E, Hedegaard A, Nielsen L, Jensen BR, Hjeds H, Krogsgaard-Larsen P. Comparative stereostructure-activity studies on $GABA_A$ and $GABA_B$ receptor sites and GABA uptake using rat brain membrane preparations. J Neurochem 1986; 47:898-903.

22. SYBYL, Version 6.1. Tripos Associates, Inc., St. Louis, Missouri 63144, USA, 1994.

GABA: Receptors, Transporters and Metabolism
ed. by C. Tanaka & N.G. Bowery

A NOVEL GABA RESPONSE IDENTIFIED OPTICALLY IN THE NUCLEUS OF THE *TRACTUS SOLITARIUS* OF THE CHICK EMBRYO

Y. Momose-Sato, K. Sato, T. Sakai, A. Hirota and K. Kamino

Department of Physiology, Tokyo Medical and Dental University School of Medicine, Bunkyo-ku, Tokyo 113, Japan

Summary: Using a multiple-site optical recording technique employing a fast voltage-sensitive dye, we found a novel type of GABA response (possibly $GABA_D$), which is insensitive to $GABA_A$ and $GABA_B$ antagonists, but is stimulated by either $GABA_A$ or $GABA_B$ agonists. This evidence was obtained in the early embryonic brainstem slice preparation.

Introduction

In various areas of the vertebrate central nervous system, it is known that release of γ-aminobutyric acid (GABA) provides inhibitory modulation of excitatory postsynaptic potentials. GABA acts postsynaptically at $GABA_A$ receptors, which are ligand-gated Cl^- channels (1), and at $GABA_B$ receptors, which activate K^+ channels by coupling to GTP-binding regulatory proteins (2). The $GABA_A$ receptors are blocked by the competitive antagonist bicuculline, and the non-competitive blocker picrotoxin (3). The $GABA_B$ receptors are specifically blocked by 2-hydroxysaclofen (4). In addition, bicuculline and baclofen-insensitive $GABA_C$ receptors have been described (5). Recently, possible novel classes of GABA receptors have also been reported (6-8), although it is suggested that they are similar to the $GABA_C$ receptors. We report here a novel type of GABA response, which was found for vagal glutaminergic excitatory postsynaptic potentials in the early embryonic chick brainstem, using an optical recording method.

Materials and Methods

The method that we used has been described in detail elsewhere (9, 10). The slice preparation (about 1000 μm thick) was made by transverse sectioning of 7- to 8-day old embryonic chick brainstems at the level of the vagus nerve. The preparation was stained with a voltage-sensitive dye (NK2761) (11). Positive square current pulses (8 μA/5.0 ms) were applied to the vagus nerve fibers by a micro-suction electrode.

Results and Discussion

Figure 1 shows examples of optical recordings of neural activity from brainstem slice preparations dissected from 8-day old chick embryos, in response to vagus nerve stimulation. These signals consist of two components: a fast spike-like signal and the following long-duration slow signal. As described previously (9, 12), the slow signals correspond to excitatory postsynaptic potentials which are glutamate mediated, and arise within the nucleus of the tractus solitarius. The slow signals are first detected unambiguously from 7-day old embryonic brainstem preparations in normal Ringer's solution (10, 12). The fast spike-like signals themselves consist of two components: one is the antidromic action potential in the motoneurons, and the other is the action potential in the nerve terminals of sensory neurons (9).

When GABA was applied to the bathing solution, the evoked slow signals were reversibly reduced in amplitude (Fig. 1: middle). This suppressive effect of GABA on the slow component of the optical signal was dose dependent, and the slow signals were completely suppressed by application of 50 μM GABA in the bathing solution. There were no regional differences in this effect of GABA.

We then examined the effects of $GABA_A$ and $GABA_B$ antagonists on the GABA-induced inhibition of the slow optical signal. The results obtained are shown in Fig. 1 (bottom). In the presence of picrotoxin (a non-competitive antagonist of $GABA_A$ receptors: 100 μM), the slow signals appeared to be slightly recovered, but, for the most part, the slow signal was insensitive to picrotoxin. Similar results were obtained for dieldrin (a non-competitive antagonist of $GABA_A$ receptors: 10 μM), bicuculline (a competitive $GABA_A$ antagonist: 200 μM), and SR95531 (a competitive $GABA_A$ antagonist: 100 μM). These results strongly suggest that most of the GABA-inhibitory action on the slow component of the optical signals observed in the present experiments is distinct from that of $GABA_A$ receptors, and that, if there is a component due to $GABA_A$ receptors, it is extremely small.

The GABA-inhibitory action was almost unaffected by 2-hydroxysaclofen (200 μM), suggesting that this action of GABA is also not mediated by $GABA_B$ receptors (Fig. 1: bottom). Similar results were obtained in the experiment with CGP35348 (100 μM), and phaclofen (500 μM). Furthermore, the GABA-induced reduction was not blocked in the presence of picrotoxin (200 μM) and 2-hydroxysaclofen (400 μM) (13). Thus we conclude that most of the GABA-inhibitory action is not mediated by $GABA_A$ or $GABA_B$ receptors. Since $GABA_C$ receptors are insensitive to

$GABA_A$ competitive antagonists (e.g. bicuculline) and to $GABA_B$ antagonists (e.g. 2-hydroxysaclofen), but they are sensitive to non-competitive antagonists of $GABA_A$ receptors (e.g. picrotoxin, dieldrin), the present results are inconsistent with the pharmacological nature of $GABA_C$ receptors (5-8).

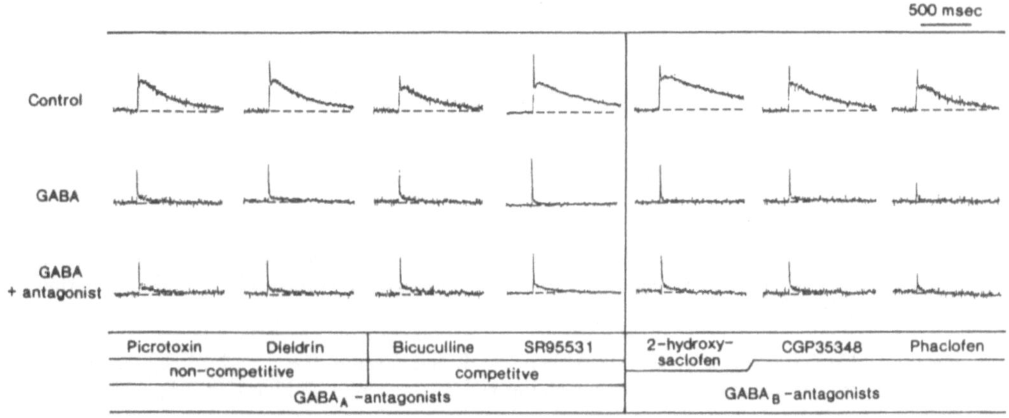

Figure 1. Effects of GABA and GABA antagonists on glutaminergic excitatory postsynaptic potentials (EPSPs) evoked by vagal stimulation. The slow signals (related to glutaminergic EPSPs) identified in the control solution (upper traces) were eliminated in the presence of GABA (50 μM) in the bathing solution (second traces). The GABA-dependent reduction in the amplitude of the slow optical signals was not reversed by picrotoxin (100 μM), dieldrin (10 μM), bicuculline (200 μM), SR95531 (100 μM), 2-hydroxysaclofen (200 μM), CGP35348 (100 μM), or phaclofen (500 μM) (third traces). Data were obtained from 8-day old embryonic brainstem slice preparations.

We also examined the effects of agonists of $GABA_A$ receptors (muscimol: 10 μM) and $GABA_B$ receptors (baclofen: 10 μM) on the slow optical signal. Both muscimol and baclofen reduced partly the amplitude of the slow optical signal, in the presence of picrotoxin (200 μM) and 2-hydroxysaclofen (400 μM) (13). When muscimol and baclofen were applied together, the slow signals were completely suppressed in the same manner as with GABA (13).

The results obtained in the present experiments possibly suggest that a novel class of GABA receptors (possibly $GABA_D$) is generated in the early embryonic chick brainstem. On the other hand, the possibility of an early prototype of $GABA_A$ and $GABA_B$ receptors has not been ruled out.

Acknowledgements

Uehara Memorial Life Science Foundation and Mochida Memorial Medical and Pharmaceutical Foundation.

References

1. Sivilotti L and Nistri A. GABA receptor mechanisms in the central nervous system. Prog Neurobiol 1991; 36: 35-92.

2. Dutar P and Nicoll RA. A physiological role for $GABA_B$ receptors in the CNS. Nature 1988; 332: 156-158.

3. Barker JL, McBurney RN and Mathers DA. Convulsant-induced depression of amino-acid responses in cultured mouse spinal neurones studied under voltage clamp. Br J Pharmacol 1983; 806: 19-629.

4. Kerr DI, Ong J, Johnston GAR, Abbenante J and Prager RH. 2-hydroxy-saclofen: an improved antagonist at central and peripheral $GABA_B$ receptors. Neurosci Lett 1988; 92: 92-96.

5. Drew CA, Johnston GAR and Weatherby RP. Bicuculline-insensitive GABA receptors: studies on the binding of (-)-baclofen to rat cerebellar membranes. Neurosci Lett 1984; 52: 317-321.

6. Feigenspan A, Wassle H and Bormann J. Pharmacology of GABA receptor Cl^- channels in rat retinal bipolar cells. Nature 1993; 361: 159-162.

7. Lukasiewicz PD, Maple BR and Werblin FS. A novel GABA receptor on bipolar cell terminals in the tiger salamander retina. J Neurosci 1994; 14: 1202-1212.

8. Qian H and Dowling JE. Novel GABA responses from rod-driven retinal horizontal cells. Nature 1991; 361: 159-162.

9. Komuro H, Sakai T, Momose-Sato Y, Hirota A and Kamino K. Optical detection of postsynaptic potentials evoked by vagal stimulation in the early embryonic chick brain stem slice. J Physiol (Lond) 1991; 442: 631-648.

10. Momose-Sato Y, Sakai T, Hirota A, Sato K and Kamino K. Optical mapping of early embryonic expression of Mg^{2+}-/APV-sensitive components of vagal glutaminergic EPSPs in the chick brainstem. J Neurosci 1994; 14: 7572-7584.

11. Kamino K, Hirota A and Fujii S. Localization of pacemaking activity in early embryonic heart monitored using voltage-sensitive dye. Nature 1981; 290: 595-597.

12. Momose-Sato Y, Sakai T, Komuro H, Hirota A and Kamino K. Optical mapping of the early development of the response pattern to vagal stimulation in embryonic chick brain stem. J Physiol (Lond) 1991; 442: 649-668.

13. Momose-Sato Y, Sato K, Sakai T, Hirota A and Kamino K. A novel γ-aminobutyric acid response in the embryonic brainstem as revealed by voltage-sensitive dye recording. Neurosci Lett 1995; 191: 193-196.

GABA: Receptors, Transporters and Metabolism
ed. by C. Tanaka & N.G. Bowery

A NOVEL BENZODIAZEPINE PARTIAL INVERSE AGONIST, S-8510, AS A COGNITIVE ENHANCER

Kazuo Kawasaki, Tsuyoshi Kihara, Shunji Murata, Katsumi Koike,
Masato Ikeda, Masami Eigyo & Susumu Takada

Shionogi Research Laboratories, Shionogi & Co., Ltd., Osaka 561, Japan.

Summary: Pharmacological actions of a novel benzodiazepine partial inverse agonist, S-8510 were examined. S-8510 ameriolated memory impairment induced by cholinergic deficit in rats in addition to augmentation of the LTP in hippocampal slice preparation. Furthermore, S-8510 increased in the extracellular level of acetylcholine in the hippocampus. S-8510 selectively potentiated pentylenetetrazol-induced convulsion without affecting minimum electrical shock- or strychnine-induced convulsion in mice. These results suggest that S-8510 is a therapeutic drug for senile dementia, including Alzheimer's disease with little risk for inducing anxiety or convulsion.

Introduction

Benzodiazepines (BDZs) and its inverse agonists are known to exert their pharmacological actions via action on the macromolecule in $GABA_A$ receptor complex. BDZs show clinical effects such as anti-epileptic, sedating, anxiolytic and sleep-inducing actions (1). On the contrary, BDZ inverse agonists show various pharmacological actions opposite to those of conventional BDZs depending on their intrinsic activities. BDZ inverse agonists with high intrinsic activity such as methy-β-carboline-3-carboxylate (β-CCM) or methyl-6,7-dimethoxy-4-ethyl-β-carboline-3-carboxylate (DMCM) can cause convulsion by themselves (2,3), while ones with less intrinsic activity like ethyl-β-carboline-3-carboxylate (β-CCE) and FG7142 show proconvulsant activity only when administered with a subconvulsive dose of pentylenetetrazole (PTZ) (4,5). Furthermore, FG7142 has been reported to produce anxiety in human (6). Contrary to these adverse actions, BDZ inverse agonists can enhance not only the long-term potentiation (LTP) in the hippocampus (7) but the memory and learning tasks *in vivo* (8) as well as can ameriolate memory impairment produced by several drugs or surgical procedures (4, 9, 10, 11).

Here we discovered a novel BDZ inverse agonist, S-8510 (2-(3-isoxazolyl)-3,6,7,9-tetrahydroimidazo [4,5-d]pyrano[4,3-b] pyridine monophosphate monohydrate) with a low intrinsic activity. In this study, we mainly focused our studies on the hippocampus which is considered as one of the major centers relating with memory function and epilepsy, to examine pharmacological profile of S-8510, and found that S-8510 can ameriolate memory impairment without producing apparent adverse effects like non-selective proconvulsant and anxiogenic one.

Materials and Methods

1. Receptor binding studies.

Conventional receptor binding studies were used for examining biochemical characters of S-8510 *in vitro*. Procedure for these studies was the same as that described elsewhere (3, 12) in detail.

2. Proconvulsant actions of S-8510 on PTZ-, strychnine- or electrical shock-induced convulsion in mice.

To determine pharmacological characters of S-8510 as a proconvulsant, we examined proconvulsive characters of S-8510 in combination with a subconvulsive dose of convulsant (PTZ or strychnine) or minimal electro-convulsive shock (MES) in mice. Each convulsant was administrated immediately and MES was applied 5 min after intravenous injection of each inverse agonist. Each group consisted of 8 - 16 mice. ED_{50} values for producing lethal convulsion in 50% of mice were calculated by the probit method.

3. LTP study in rat hippocampal slices.

Effects of S-8510 on the LTP and synaptic transmission of the Schaffer collateral/commissural fibers-CA1 synapses were examined in rat hippocamapal slice preparations. Most of experimental procedures except for a concentration of extracellular K^+ (3.5 mM) and tetanic condition (33 Hz, 8 trains) were the same as those described elsewhere (7, 13).

4. Anti-amnesic actions of inverse agonists in rats.

Anti-amnesic actions of inverse agonists were examined in the water maze paradigm of Morris type (15). Rats were given 4 trials on each day for consecutive 4 days (acquisition period). On the 5th day, each animal had the retention test for 60 sec. Amnesia was produced by intraperitoneally injected scopolamine at a dose of 0.5 mg/kg. Scopolamine and each inverse agonist were given 15 min and 30 min before the retention trial, respectively.

5. Measurement of extracellular level of acetylcholine by microdialysis in rat hippocampus.

Extracellular level of acetylcholine in the hippocampus was measured by conventional microdialysis method described elsewhere (16) in detail.

6. Anxiogenic effects of inverse agonists in rats.

Anxiogenic actions of each inverse agonist were examined in the punished water licking method of Vogel's type (14). Each inverse agonist was administered orally 30 min before the test.

7. Drugs.

S-8510 and CGS8216 were dissolved in physiological saline and 0.1 N NaOH respectively.

FG7142, DMCM, β-CCE, β-CCM, diazepam and flumazenil were dissolved in hydrogenated castor oil, polyethyleneglycol-200 and physiological saline.

Results and Discussion

1. Receptor binding studies.

S-8510 had relatively high affinity to BDZ receptors as shown in Table 1. The γ-aminobutyric acid (GABA) ratio is defined as the ratio of Ki values measured with or without GABA (3, 12). The GABA ratio for each inverse agonist is shown in Table 1. The GABA ratio for S-8510 was nearly equal to that for CGS8216 and was larger than that for other inverse agonists used here, indicating that S-8510 and CGS8216 have smaller intrinsic activity than other inverse agonists. This result is in agreement with our previous observations based on results from *in vivo* crossed extensor reflex test (4). The rank of order of intrinsic activity as an inverse agonist is suggested to be as follows; DMCM ≈ FG7142 ≈ β-CCM ≈ β-CCE > CGS8216 ≈ S-8510.

Table 1. Pharmacological characterization of S-8510. Inhibition of [^{3}H]flunitrazepam binding and GABA-induced shifts of Ki value, and proconvulsant effects of S-8510 and other inverse agonists.

			ED_{50} (mg/kg, i.v.)[c]						
		GABA[b]	Dose of PTZ (mg/kg, s.c.)					Other convulsants	
	Ki(nM)[a]	ratio	0	52	62.5	75	90	Strychnine[d]	MES[e]
S-8510	**3.53**	**0.96**				**>50**	**2.87**	**>50**	**>50**
CGS8216	**0.44**	**0.93**				**>25**	**0.20**		
β-CCE	**1.28**	**0.53**		**>30**	**20**	**1.00**	**0.46**		
FG7142	**144**	**0.50**		**>40**	**18.66**	**7.34**	**4.19**		
β-CCM	**2.89**	**0.53**	**0.24[f]**	**4.52**	**1.22**	**0.49**	**0.29**		
DMCM	**9.65**	**0.50**	**1.93[f]**	**0.92**	**0.59**	**0.34**	**0.23**		

[a] Displacing potential to [^{3}H]flunitrazepam binding in membrane fraction of rat cerebral cortex. [b] GABA ratio is defined as the ratio of two Ki values, measured with or without GABA on the affinity of ligands for [^{3}H]flunitrazepam in membrane fraction of rat cerebral cortex. [c] Proconvulsive activity in mice (n = 8-16). [d] Strychnine: 0.25 mg/kg, s.c. [e] MES = 23 mA, 0.9 msec width, 0.2 sec duration. [f] ED_{50}s show occurrence of clonic convulsion.

2. Selective proconvulsant actions of S-8510 on PTZ-induced convulsion.

Proconvulsant actions of BDZ inverse agonists were examined in combination with a subconvulsive dose of PTZ (52, 62.5, 75 or 90 mg/kg, s.c.). Table 1 summarizes results of proconvulsive actions of BDZ inverse agonists. DMCM and β-CCM with the highest intrinsic activity (3,4) among BDZ inverse agonist used here caused convulsion without PTZ. However,

FG7142, β-CCE, CGS8216 or S-8510 failed to produce convulsion by itself. These inverse agonists could produce lethal convulsion only when administered with each subconvulsive dose of PTZ depending on their intrinsic activities as an inverse agonist (4). FG7142 and β-CCE produced convulsion in combination with PTZ at higher doses than 62.5 mg/kg, however, CGS8216 or S-8510 could cause convulsion only in combination with more than 90 mg/kg of PTZ. From these results, the rank of intrinsic activity as a BDZ inverse agonist of six compounds is concluded as follows; DMCM > β-CCM > FG7142 ≅ β-CCE > CGS8216 ≅ S-8510. This is nearly equal to our results from the binding studies.

Even at a higher dose (50 mg/kg) of S-8510 failed to produce lethal convulsion in combination with strychnine (0.25 mg/kg, s.c.) or MES.

3. Enhancement of LTP by S-8510.

Effects of S-8510 on the LTP were examined in the rat hippocampal slices. S-8510 did not affect the field potential induced by the stimulation of input fibers to CA1 pyramidal cells up to 10^{-5} M. S-8510 augmented the field excitatory postsynaptic potential and population spike at a dose of 10^{-4} M (data not shown). When S-8510 was applied at a concentration of 10^{-7} M, the LTP observed after the tetanic stimulation of input fibers was augmented and this augmentation was antagonized by BDZ receptor antagonist, flumazenil as shown in Fig. 1. Flumazenil itself failed to influence the LTP as well as the evoked potential prior to the tetanic stimulation.

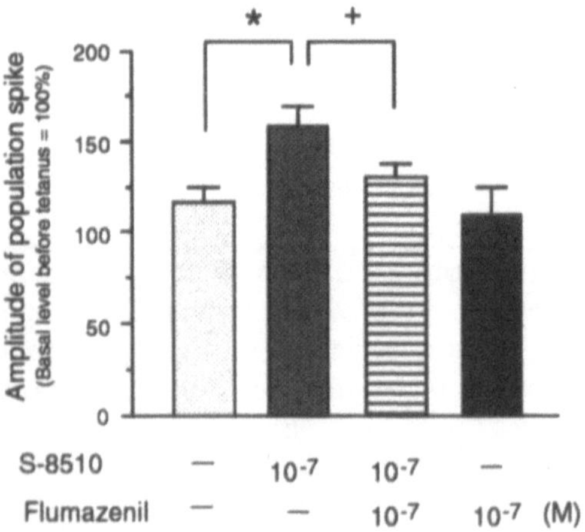

Fig. 1. Augmentation of LTP by S-8510, and antagonism of S-8510 by flumazenil. S-8510 was applied for 40 min since 10 min before the tetanus. Flumazenil was applied for 50 min since 20 min before the tetanus. Data (mean ± S.E.M.) are expressed as % of amplitude of population spike, recorded between 10 min and 30 min after the tetanus. * $P < 0.05$ (Dunnett's t-test), + $P < 0.05$ (Student's t-test). n = 6-10.

4. Anti-amnesic actions of S-8510.

Fig. 2 shows the anti-amnesic actions of inverse agonists in the Morris water maze paradigm. Scopolamine decreased time spend in the area around the platform. S-8510 increased it, indicating that S-8510 ameliorated amnesia produced by cholinergic deficiency, however, FG7142 failed to ameliorate scopolamine-induced amnesia in this model (data not shown).

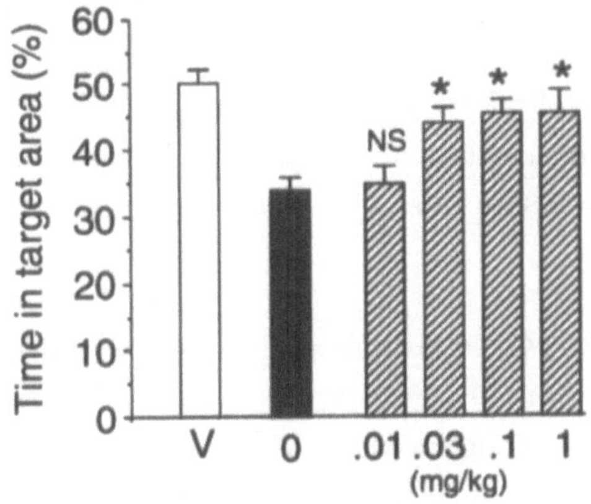

Fig. 2 Ameriolating effects of S-8510 on memory impairment induced by scopolamine in the water maze paradigm of rats. S-8510 (p.o.) was administered 30 min and scopolamine (0.5 mg/kg, i.p.) was treated 15 min before the retention test. Data (mean ± S.E.M.) are expressed as % of total time spend in a circular area centered on the platform position when the goal was removed. V: vehicle without scopolamine. NS: no significant. * $P < 0.05$, ** $P < 0.01$ significantly different from vehicle + scopolamine group (Dunnett's t-test). n = 8-15.

5. Increase in the extracellular level of acetylcholine by S-8510.

Fig. 3 shows the increase in the extracellular level of acetylcholine by S-8510. S-8510 at a dose of 1 or 10 mg/kg increased in the level of acetylcholine with a time-course shown in Fig. 3A.

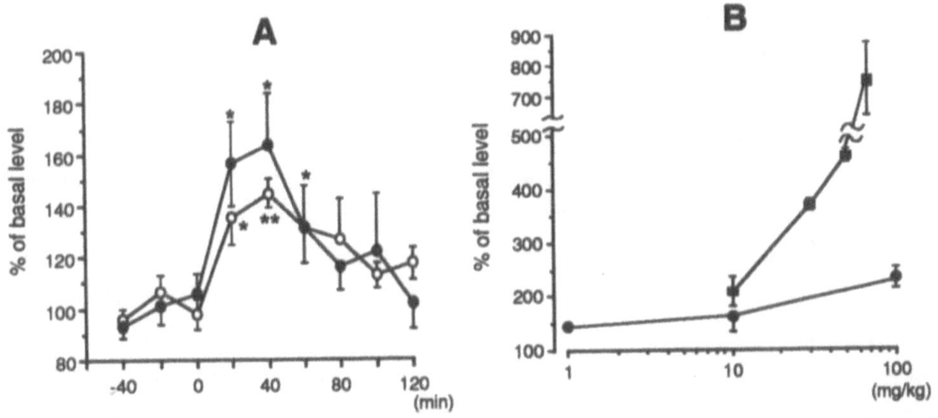

Fig. 3 Increase in extracellular level of acetylcholine in the hippocampus induced by S-8510 (A), and the difference in the slope of dose-response relationship between S-8510 and PTZ (B). Data (mean ± S.E.M.) are expressed as % of the basal level in 3 fractions before drug administration. A: The animals were given 1 mg/kg (O, n = 8) or 10 mg/kg (●, n = 5) of S-8510 p.o. at time 0. * $P < 0.05$, ** $P < 0.01$ compared with before drug administration (Fisher's PLSD test). B: Data are indicated as the values at 40 min after administration of the drug. S-8510 (●, n = 4-8) or PTZ (■, s.c., n = 3-4).

PTZ also increased in the level of acetylcholine straight up to a subconvulsive dose (70 mg/kg), however, the slope of dose-response relationship for PTZ was much steeper than that for S-8510 (Fig. 3B). In addition to decreasing effect of S-8510 on the $GABA_A$ receptor function in the hippocampus, potentiating action on the cholinergic system in the septum via inhibition of $GABA_A$ system can explain the anti-amnesic action of S-8510 observed in *in vivo* experiment.

6. Anxiogenic actions of S-8510.

Anxiogenic actions of S-8510, CGS8216 and FG7142 were examined in the punished water licking paradigm in rats. FG7142 at doses more than 10 mg/kg decreased the number of responses during the punished period, indicating the anxiogenic action of FG7142, however, S-8510 and CGS8216 failed to affect on this behavioral paradigm up to a dose of 30 mg/kg (data not shown).

From the present results in combination with our previous reports (4), S-8510 is characterized as a partial BDZ inverse agonist with a low intrinsic activity similar to CGS8216, which is reported as a vigilance-enhancing compound without producing any sign of anxiety and convulsion in human. We have confirmed that S-8510 failed to produce any sign of anxiety in the rat Vogel's model as well as CGS8216 did not. In this model FG7142 which has been reported as an anxiogenics in human (6) was characterized as an anxiogenics. Furthermore, proconvulsant action of S-8510 was specific for PTZ and S-8510 failed to produce any sign of convulsion by itself. These results suggest that S-8510 has little risk to produce anxiety and convulsion in human.

S-8510 can augment the LTP in the Schaffer collateral/commissural fiber-CA1 system in *in vitro* as well as ameriolate the memory impairment induced by muscarinic antagonist, scopolamine in *in vivo* study. Furthermore, S-8510 increased the extracellular level of acetylcholine in the hippocampus which is possibly related to the spatial memory function (17). S-8510 selectivity potentiated PTZ-induced convulsion. PTZ is reported to have beneficial effects on cognitive disorders in geriatric subjects at lower does than convulsive ones (18). The dose-response relationships of PTZ and S-8510 for increasing in the level of acetylcholine are quite different. The slope of dose-response curve for S-8510 is much gentler than that for PTZ. Moreover, ceiling effect is observed for S-8510 at 10 mg/kg, however, PTZ straight increased the level of acetylcholine up to a subconvulsive dose as shown in Fig. 3. This result indicates that S-8510 is a much safer medicine than PTZ.

From these results, S-8510 is suggested to be a therapeutic drug for senile dementia including Alzheimer's disease with little risk for adverse actions.

References

1. Müller WE. New trends in benzodiazepine research. Drugs of today 1988; 24: 649-663.

2. Oakley NR, Jones BJ. Differential pharmacological effects of β-carboline-3-carboxylic acid esters. Neuropharmacol 1982; 21: 587-589.

3. Braestrup C, Schmiechen R, Neef G, Nielsen M, Petersen EN. Interaction of convulsive ligands with benzodiazepine receptors. Science 1982; 216: 1241-1243.

4. Matsushita A, Kawasaki K, Matsubara K, Eigyo M, Shindo H, Takada S. Activation of brain function by S-135, a benzodiazepine receptor inverse agonist. Prog Neuro-Psychopharmacol & Biol Psychiat 1988; 12: 951-966.

5. Little HJ, Nutt DJ, Taylor SC. Acute and chronic effects of the benzodiazepine receptor ligand FG 7142: proconvulsant properties and kindling. Br J Pharmacol 1984; 83: 951-958.

6. Dorow R, Horowski R, Paschelke G, Amin M, Braestrup C. Severe anxiety induced by FG 7142, a β-carboline ligand for benzodiazepine receptors. Lancet 1983; 11: 98-99.

7. Yasui M, Kawasaki K. Benzodiazepine inverse agonists augment long-term potentiation in CA1 and CA3 of guinea pig hippocampal slices. Neuropharmacol 1993; 32: 127-131.

8. Venault P, Chapouthier G, Carvalho LP, Simiand J, Morre M, Dodd RH, Rossier J. Benzodiazepine impairs and β-carboline enhances performance in learning and memory tasks. Nature 1986; 321: 864-866.

9. Galliani G, Nencioni A, Maggioni A, Miller P. Reversal of scopolamine-induced amnesia by the benzodiazepine receptor inverse agonists RU 34332 and FG 7142 in the mouse. Med Sci Res 1991; 19: 441-442.

10. Lorez HP, Martin JR, Keller HH, Cumin R. Effect of aniracetam and the benzodiazepine receptor partial inverse agonist Ro 15-3505 on cerebral glucose utilization and cognitive function after lesioning of cholinergic forebrain nuclei in the rat. Drug Develop Res 1988; 14: 359-362.

11. Sarter M, Steckler T. Spontaneous exploration of a 6-arm radical tunnel maze by basal forebrain lesioned rats: effects of the benzodiazepine receptor antagonist β-carboline ZK 93 426. Psychopharmacol 1989; 98: 193-202.

12. Wastek GJ, Speth RC, Reisine TD, Yamamura HI. The effect of γ-aminobutyric acid on ^{3}H-flunitrazepam binding in rat brain. Eur J Pharmacol 1978; 50: 445-447.

13. Yasui M, Kawasaki K. Vulnerability of CA1 neurons in SHRSP hippocampal slices to ischemia, and its protection by Ca^{2+} channel blockers. Brain Res 1994; 642: 146-152.

14. Vogel JR, Beer B, Clody DE. A simple and reliable conflict procedure for testing anti-anxiety agents. Psychopharmacologia 1971; 21: 1-7.

15. McNaughton N, Morris RGM. Chlordiazepoxide, an anxiolytic benzodiazepine, impairs place navigation in rats. Behav Brain Res 1987; 24: 39-46.

16. Kihara T, Ikeda M, Miyazaki H, Matsushita A. Influence of potassium concentration in microdialysis perfusate on basal and stimulated striatal dopamine release: effect of ceruletide, a cholecystokinin-related peptide. J Neurochem 1993; 61: 1859-1864.

17. Tarricone BJ, Keim SR, Simon JR, Low WC. Intrahippocampal transplants of septal cholinergic neurons: high-affinity choline uptake and spatial memory function. Brain Res 1991; 548: 55-62.

18. Leckman J, Ananth JV, Ban TA, Lehmann HE. Pentylenetetrazol in the treatment of geriatric patients with disturbed memory function. J Clin Pharmacol 1971; 11: 301-303.

SPATIAL BENZODIAZEPINE RECEPTOR IMAGING IN THE HUMAN BRAIN

S Pauli, H Hall, C Halldin and G Sedvall

Karolinska Institutet, Department of Clinical Neuroscience, Psychiatry Section, Karolinska Hospital, S-171 76 Stockholm, Sweden

Summary High affinity ligands have been used to examine neuroreceptor distribution in the human brain by in vivo positron emission tomography (PET) and in vitro autoradiography. So far the visualization of the receptor distribution has been limited to two dimensional presentations of the anatomic distribution of the receptors. The development of high resolution PET has increased the number of planes to about 50. From a cryosectioned human postmortem brain about 1500 of 100 micron thick sections can be obtained for autoradiographic studies. 3D graphical techniques have been used to display and analyze the results. We have now used these techniques and data from PET and autoradiographic experiments with the high affinity benzodiazepine receptor ligand flumazenil, labeled with [^{11}C] or [^{3}H] to visualize benzodiazepine receptors in human brain structures. From such 3D-images, receptor containing structures can be correctly presented with regard to volume, form and anatomical localization. Discrimination and seeding techniques was used to define volumes and quantities. The total amount of BZ receptors in the human brain examined with cerebellum excluded was calculated to 15.5 nMoles. This technical development represents a new dimension in clinical neuroscience and can be used for quantitative studies in neuropsychiatric disorders and in relation to drug treatment.

Key words

$GABA_A$ receptor, benzodiazepine receptor, positron emission tomography, autoradiography, human brain, [^{11}C]flumazenil, [^{3}H]flumazenil, 3D-imaging.

Introduction

Benzodiazepine (BZ) receptors represent important components of the $GABA_A$ receptor/chloride channel complex, regulated by GABA, the most abundant inhibitory transmitter in the brain. The quantitative distribution of the BZ receptor in the human brain may be used as an index of the potential for inhibitory transmission in different regions. Such a dimension of a global transmittory function in defined brain regions in a human individual has not previously been grasped or measured by scientific methodology.

During the past two decades the development of selective radioligands with high affinity for neuroreceptors has allowed the characterization and visualization of the anatomical receptor distribution in the animal and human brain during in vitro and in vivo conditions. By using [^{11}C]

labeled radioligands and positron emission tomography (PET), the binding can be directly recorded in the living brain of healthy subjects and patients with neuropsychiatric disorders (1). By using tritium labeled radioligands and in vitro autoradiography two-dimensional receptor distributions in whole hemisphere cryosections from the human postmortem brain can also be described (2). To gain insight into the role of GABA transmission in neuropsychiatric disorders the detailed exploration of GABA receptor mechanisms in the human brain is required. The BZ receptors are vital components in this analysis.

Previously we described the use of PET and the BZ antagonist [^{11}C]flumazenil to examine BZ receptor binding in the living human brain (11). In vitro we used tritium labeled flumazenil and [^{3}H]Ro15-4513 in order to examine benzodiazepine receptor binding in the human postmortem brain (2).

In previous PET and autoradiographic studies on BZ receptors their distribution has been presented as a series of two dimensional planes. The development of high resolution PET technology (HRPET) has significantly increased the number of planes, presently about 50 (9). To extract all the information from such a set of 2D images into meaningful quantitative information concerning neuroreceptor populations in defined volumes and structures, new approaches are called for. This is irrespective of whether the acquisition is made in 2D or 3D mode.

During the past two, three decades various techniques for 2D and 3D computer graphics have been developed (3-5). Among those, volume rendering has a number of advantages (6, 7).

The aim of the present study was to apply 3D computer graphic technique to visualize BZ receptor binding within the whole human brain volume as reconstructed from the compound sets of 2D planes obtained from PET as well as autoradiographic experiments.

Materials and Methods

The PET experiment was performed on a 29-year-old healthy man. To obtain the same positioning of the brain in MRI as in PET investigations a helmet was individually prepared to fixate the head in the cameras (8).

The PET experiment was performed using a high resolution PET camera (Siemens Exact HR47). The radial resolution in the plane (FWHM) is 4.3 mm (9). The radioligand [^{11}C]flumazenil was synthesized as previously described (10). Details of the PET experiments are described in Pauli and Sedvall 1995.

The autoradiographic studies were made using a whole human postmortem brain hemisphere cryosectioned in the horizontal plane (2). The hemisphere was obtained according to the ethical rules of the Karolinska Institute from a 55 year old woman with no evidence for brain disorder [^{3}H]flumazenil was incubated with the 100 micron thick cryosections as described by Hall et al. In the present communication data from 24 sections taken through the occipital pole of the brain at 1 mm intervals are presented.

Image processing

The 47 PET images were transferred from the PET scanner to a Silicon Graphics computer (Onyx, RealityEngine2) to be stored on the hard disc (Fig. 1). The 24 human postmortem brain sections processed by autoradiography were recorded using a high resolution video-camera (Kontron, Progress 3012). These autoradiographic images were also transferred to the Silicon Graphics computer as shown in Fig. 1. Each section was scanned using 1532 x 640 pixels. In order to obtain cubic volume elements, voxels, interpolation was applied. For the PET data 1 mm^3 voxels were calculated giving 10 million voxels. For the autoradiograms 0.1 mm^3 voxels were used giving 20 million voxels. The program VoxelView (Vital Images, Iowa, USA) was applied to visualize these voxels by using volume rendering (12, 7). This technique takes all voxel information in account when constructing the final images.

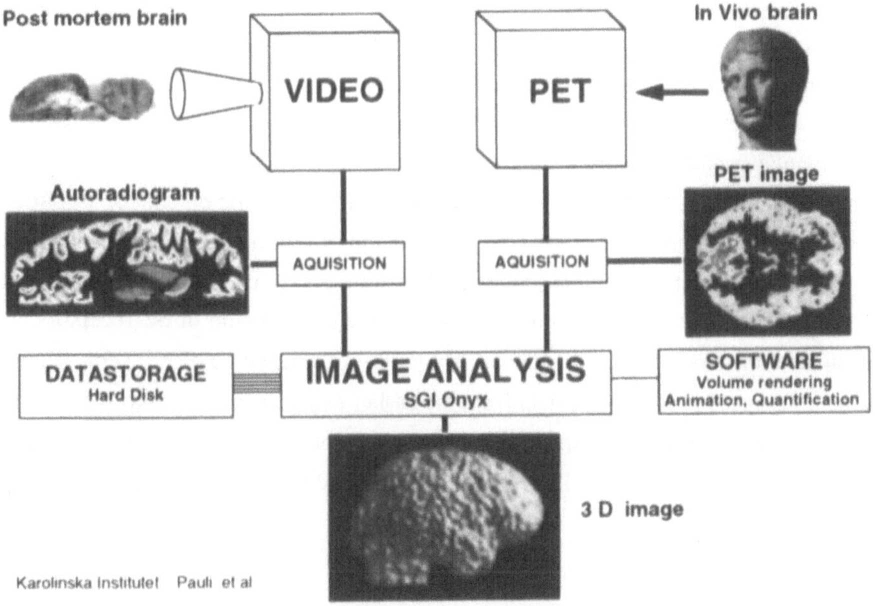

Fig 1. Graphical representation of the techniques used for PET and videorecording of radioligand binding data using a positron camera for studying in vivo [^{11}C]flumazenil binding and autoradiography to examine [^{3}H]flumazenil binding in the postmortem human brain.

Results

PET experiment with [^{11}C]flumazenil

A typical example of a two-dimensional PET image from a plane through the brain is shown in Fig. 2. It is apparent that the radioactivity distribution is localized to the benzodiazepine receptor rich neocortical regions. Lower densities of benzodiazepine receptor binding are seen in the basal ganglia.

In Fig. 3 a three-dimensional reconstruction of the exterior and interior surfaces of the BZ populations of the brain of the healthy volunteer participating in the PET experiment is shown. It is obvious how details of the exterior and interior surfaces of the rich BZ receptor populations in these brain regions can be visualized by this technique.

Autoradiographic studies with [^{3}H]flumazenil

In Fig 4 a typical autoradiogram obtained from a 100 micron thick cryosection of the human brain is presented. Similarly to PET, the distribution of benzodiazepine receptors as recorded in vitro is predominantly to neocortical areas of the brain.

Fig 5 shows a three-dimensional reconstruction of the BZ receptor distribution in the occipital pole of the postmortem brain. The topology of BZ receptor populations are spatially presented in the convolutions of the neocortex in the occipital region.

Quantification of spatial receptor populations

Using the technique of volume rendering and data from the specific binding of [^{11}C]flumazenil in the PET experiment quantification of the total amount of BZ receptors could be obtained in the brain of the male subject participating in the PET experiment. Binding data from previous studies giving values for K_d were used in order to calculate the total amount of BZ receptors (11). For this calculation the binding data from the present PET experiment with high specific activity of the radio tracer was used in relation to data from the earlier experiments with two PET experiments, one with high and one with low specific radioactivity. The total amount of BZ receptors in the human brain examined with cerebellum excluded was calculated to be 15.5 nMoles (7).

Discussion

In the present communication radioligand binding to BZ receptors in the human healthy brain were examined using both in vivo PET and in vitro autoradiographic methods. Computer graphic technique was utilized to visualize the vast amount of digitized 2D image information obtained from the series of images produced by PET and autoradiography. Until recently most efforts to visualize 3D information utilized conventional geometric graphic technique.

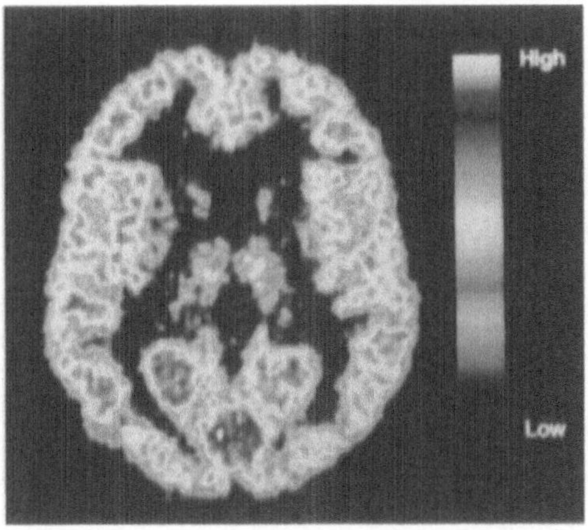

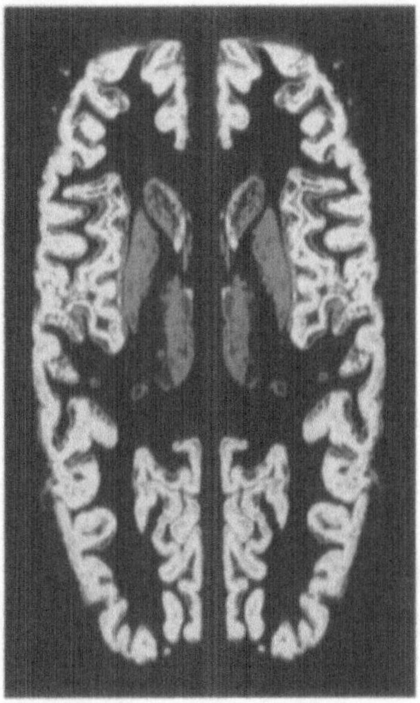

Fig 2. (Top) PET scan section showing the distribution of radioactivity in a healthy human man after intravenous injection of [^{11}C]flumazenil.

Fig 4. (Bottom) Autoradiogram of a human postmortem brain cryosection incubated with [^{3}H]flumazenil.

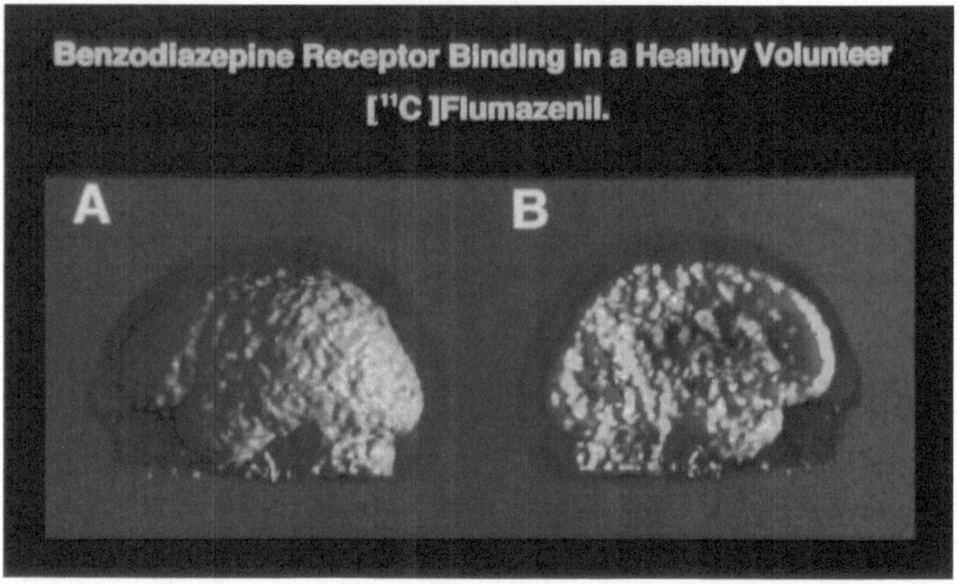

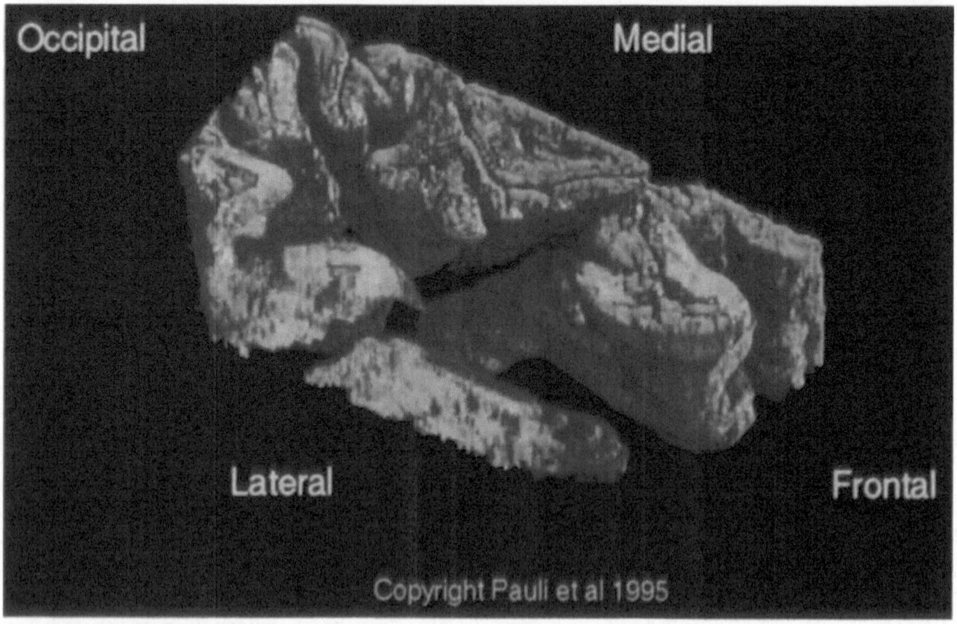

Fig 3. (Top) Three-dimensional visualization of [^{11}C]flumazenil binding data in a human healthy volunteer. In A the exterior surface of the brain is visualized. In B the interior aspect of the hemisphere showing the insular cortex is shown. The PET image is merged with the MR image of the scull.

Fig 5. (Bottom) Three-dimensional reconstruction of [^{3}H]flumazenil receptor binding in a series of autoradiograms of consecutive cryosections of the human brain.

This involves the organization of the data to be presented into a model composed of geometric primitives such as points, lines or polygons to build up surfaces that describe the volume. In the present communication we have used volume rendering as a more efficient and practical technique to record, present and quantify the ligand binding data from of voxels obtained from the PET and autoradiographic experiments (12, 13). The illustrations represent the first application of this technique to radioligand binding data.

These methodological advantages should represent a rational approach to the recording of the spatial distribution of neuroreceptor populations in vast areas of the human brain. This technique allows the visualization of receptor containing structures in a correct form with regard to localization in the anatomical three-dimensional structure. Discrimination and seeding techniques can be used with the approach presented to define volumes and quantities. The use of voxel (volume elements) instead of polygons for volume rendering allows a smoother presentation of volumes and radioactivity within the volume. Structures with high receptor density are observed as objects that can be rotated and examined from different angles giving a strong visual perception of their 3D extension using computer graphic methodology. This technique also allowed the rapid quantification of total BZ receptor amounts in the whole human cerebrum. Such a measure from the whole brain or its subregions may turn out to be a useful individual characteristic for future comparative studies of patients with neuropsychiatric disorders or vulnerabilities.

The techniques outlined in this communication give the possibility to reconstruct spatial representation of specific biochemical systems and of detailed anatomical extensions and quantities within the human brain. This development should represent a new dimension in clinical neuroscience and can be used for quantitative studies in neuropsychiatric disorders and in relation to drug treatment.

Acknowledgements

Supported by grants from the Swedish Medical Research Council (03560 and 09114), the National Institute of Mental Health, USA (NIMH, Grant No MH 41205 and 44814) and the Karolinska Institute.

References

1. Sedvall G, Brené S, Farde L, Hall H, Halldin C, Hurd Y, Karlsson P, Pauli S: Utilization of radioligands in schizophrenia research. Clin Neurosci 1995; 3:112-121.

2. Hall H, Litton J-E, Halldin C, Kopp J, Sedvall G: Studies on the binding of [^{3}H]flumazenil and [^{3}H]sarmazenil in post mortem human brain. Human Psychopharmacol 1992; 7:367-377.

3. Toga A W: Three-dimensional neuroimaging. Raven Press, New York, 1990.

4. Russ J C: The image processing handbook 2nd edition, 1995; 629-658.

5. Watt A: 3D computer graphics 2nd edition, 1991; 313-319.

6. Pauli S, Farde L, Halldin C, Sedvall G: Spatial neuroreceptor imaging in the living human brain. Neuropsychopharmacology 1994; 10:29S.

7. Pauli S, Sedvall G: Three-dimensional visualization and quantification of the benzodiazepine receptor population in a living human brain. Eur Arch Psychiatr Clin Neurosci In press

8. Bergström M, Boethius J, Eriksson L, Greitz T, Ribbe T, Widén L. Head fixation device for reproducible position alignment in transmission CT and positron emission tomography. J Comput Assist Tomogr 1992; 5:136-141.

9. Weinhard K., Dahlbom M., Eriksson, L., M C Bruckbauer T: The ECAT EXACT HR: Performance of a New High Resolution Positron Scanner. J Comput Assist Tomogr 1994; 18:110-118.

10. Halldin C, Stone-Elander S, Thorell J-O, Persson A, Sedvall G: ^{11}C-labelling of Ro 15-1788 in two different positions, and also ^{11}C-labelling of its main metabolite Ro 15-3890, for PET studies of benzodiazepine receptors. Appl Radiat Isot 1988; 39:993-997

11. Persson A, Pauli S, Halldin C, Stone-Elander S, Farde L, Sjögren I, Sedvall G: Saturation analysis of specific ^{11}C Ro 15-1788 binding to the human neocortex using positron emission tomography. Hum Psychopharmacol 1989; 4:21-31

12. Drebin R A, Carpenter L, Hanrahan P: Volume rendering. Comput Graph 1988; 22:4:65-74

13. Argiro V J: Seeing in volume. Pixel 1990; 1:3:35-39

GABA: Receptors, Transporters and Metabolism
ed. by C. Tanaka & N.G. Bowery

DOES THE GABA$_A$ RECEPTOR HAVE A ROLE TO PLAY IN ANGELMAN SYNDROME?

DeLorey, T.M, Minassian, B.*, Tyndale, R., and Olsen, R.W.

Department of Molecular and Medical Pharmacology, UCLA, Los Angeles and *California Comprehensive Epilepsy Program, West Los Angeles VA Medical Center, Los Angeles, California, U.S.A.

Summary

Angelman syndrome (AS) is a genetic disorder which produces severe mental retardation with concurrent epileptic seizures. AS typically results from a large deletion on maternal chromosome 15q11-13. The mutant mouse strain p^{cp} (pink-eyed cleft palate) has a deletion in murine chromosome 7, syntenic to the human chromosomal region on 15 associated with AS, and may prove to be a valuable asset in addressing some of the pharmacological and behavioral consequences associated with AS. Changes in the density and properties of GABA$_A$ receptors in these mice is consistent with a loss of a GABA$_A$ gene cluster (subunits α_5, β_3, γ_3) on murine chromosome 7. A large loss of inhibitory receptors could conceivably contribute to altered brain function possibly leading to seizure activity.

Angelman syndrome

The neurodevelopmental disorder Angelman syndrome (AS) is characterized by arrest of motor development, absence of language, happy demeanor, inappropriate laughter, and epilepsy in childhood and adolescence with a incidence of 1:20,000 [1,2]. The seizure activity in AS is highly variable, as is the extent of the deletion (Fig. 1). While physical development progresses relatively normally, mental and motor development do not. Brain maturation seems to arrest at about a one year old level. The adult patient's behavior and demeanor are that of a child not having reached age 2 years. The patient continuously smiles, has frequent easily elicited bursts of laughter, has no language beyond a few words, and is generally pleasant and happy. This happy state is one of the characteristic features of the syndrome. The other characteristic feature is the apparent ataxia that these patients have. The gait of the AS patient is wide-based with forward stooping at the pelvis. The arms are flexed at the elbows.

All limb movements are jerky so much so that the patients have been likened to "puppets". When one observes the growth of normal children, one may see that a newly walking toddler resembles a "happy puppet". While the normal 2 year old child starts walking more "normally" and acquires language, the AS child remains a 1 year old in mental and motor function, with a relatively normally maturing body. Patients in their late 60's still walk and smile like toddlers. This conceptualization of AS should not mislead one to think that a sudden arrest occurs at one year of age, rather there is a gradual slowing of brain development to a virtual standstill at what is the approximate equivalent of a one year old. Developmental delay can be noticed as early as 6 months of age and walking is usually acquired after 3 years of age. Feeding difficulties are commonly observed in early infancy. Decreased sleep, microcephaly, wide-spaced teeth and ophthamalogical problems such as strabismus are other common features. Magnetic Resonance Imaging is generally normal. Several reviews of the clinical features of AS are available [3-5].

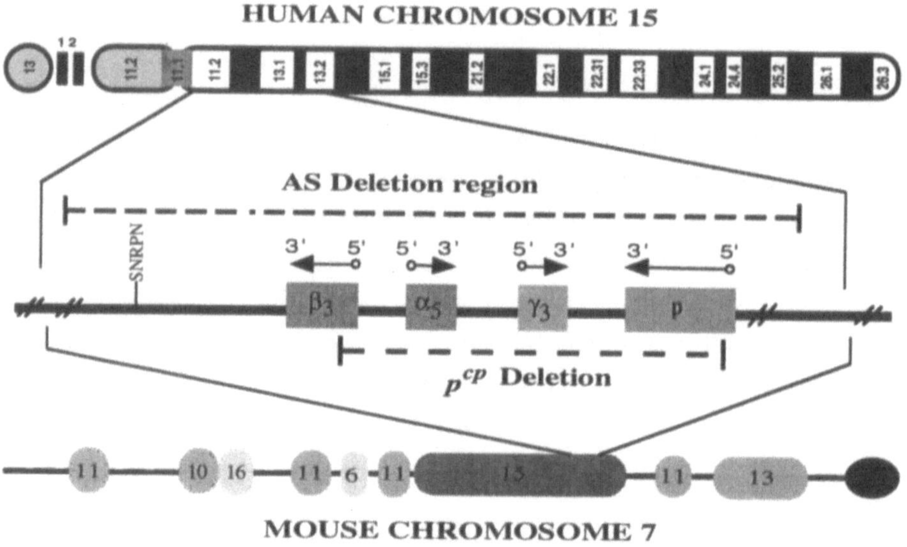

Figure 1. Arrangement of the $GABA_A$ receptor cluster (β_3, α_5, γ_3) on human chromosome 15 and mouse chromosome 7. The majority of AS probands have a large deletion in 15q11-13 (the upper extent of the deletion, which is in the range of 5-8 megabases, is indicated on the diagram). The deletion in the mouse extends from the first intron of the p gene to the third intron of the β_3 subunit gene (~ 3 Megabases). Numbers on the mouse chromosome indicate human chromosomal synteny. SNRPN: Small nuclear ribonucleoprotein polypeptide N.

Seizure disorders are observed in the majority of AS probands. The following discussion is based on a review of the literature [6-8], and on our experience with patients with AS who received 48 hours of video monitored electroencephalography at our institution. Absence attacks commonly appear in the second year of life, persist till midadolescence and tend to abate later in life. If untreated, the seizures are often continuous in the form of absence status epilepticus. Other types of seizures such as myoclonic, tonic-clonic, drop attacks and others have been described. The background EEG shows frequent diffuse though frontally predominant high amplitude (200 μV) slow waves of 2 to 3 Hz in frequency. The absence attacks are generally accompanied with continuous diffuse even higher amplitude (~ 500 μV) slow waves of 2.5 to 3.5 Hz frequency mixed with spikes. Sometimes, short runs of single spike and slow waves are seen during the attack. The above EEG findings are seen with greater frequency during sleep. Most patients are treated with the antiepileptic drugs clonazepam or valproic acid with comparable degrees of improvement. Most continue to have some seizures and some continue to have many seizures despite different combinations of antiepileptic drugs.

In the majority of AS cases the deletion is large, with the loss of many genes which may contribute to the full AS phenotype. Within this region is a cluster of $GABA_A$ receptor subunit genes (α_5, β_3, γ_3) which would be eliminated on one allele in the majority (~75 %) of AS cases [9]. $GABA_A$ receptors, which mediate the majority of inhibitory synapses in brain, are ligand-gated ion channels each composed of five related but different polypeptide subunits from a gene family of currently 17 members [10]. In the rodent brain the subunit subtypes show regional and developmental expression control [11,12]. Considering the importance of GABA in nervous system function, mutations leading to the loss of genes encoding subunits of the $GABA_A$ receptor family could lead to deleterious effects on the central nervous system (CNS). Interestingly, the α_5 and β_3 subunit mRNAs belong to a group of prenatal $GABA_A$ receptor subunits, characterized by predominant expression in fetal and neonatal rodent brain and dramatically reduced expression in many areas of adult brain [12]. $GABA_A$ receptors have long been associated with seizure phenomena and a large loss of inhibitory receptors could conceivably contribute to seizure activity [13]

Mouse mutant

To study whether $GABA_A$ receptor expression is reduced in AS, human postmortem tissue from AS probands would be required. Unfortunately postmortem tissue is rare, therefore, finding a suitable animal model is desirable. The mouse pink-eyed cleft palate (p^{cp}) mutation was discovered in the progeny of a neutron-irradiated male. The mutation results from a deletion on murine chromosome 7, syntenic to the region on human chromosome 15 associated with AS (Fig. 1). It is characterized as having hypopigmentation associated with cleft-palate, neurological disorders and runting [14,15]. Most homozygote mutants die soon after birth most likely due to feeding problems associated with the cleft-palate [16]. A few survive to maturity (~3-5 %), presumably with an unaffected palate and are fertile. Those that do survive display tremor and jerky gait [15].

RT-PCR revealed the absence of PCR product for the β_3, α_5, and γ_3 subunits of the $GABA_A$ receptor in the p^{cp}/p^{cp} mutant mouse (-,-) compared to the control mouse p^d/p^d (pink-eyed, dark) (+,+) (Fig. 2). Upon further evaluation it was determined that the deletion in p^{cp} mice extends from the third intron of the β_3 gene to the first intron of the *p* gene, thus completely eliminating the α_5 and γ_3 subunit genes, the promoter and start codons for the β_3 gene and the entire coding region of the *p* gene [17]. There appears to be no compensation for the loss of these three $GABA_A$ receptor subunits by an increase in the other receptor subunit gene products (data not shown). The working hypothesis is that a deficit of α_5, β_3, and γ_3 subunits results in an altered $GABA_A$ physiology and brain development. This may contribute to seizures observed in the mutant mice (p^{cp}/p^{cp}) and, perhaps, be responsible for the epilepsy observed in Angelman syndrome.

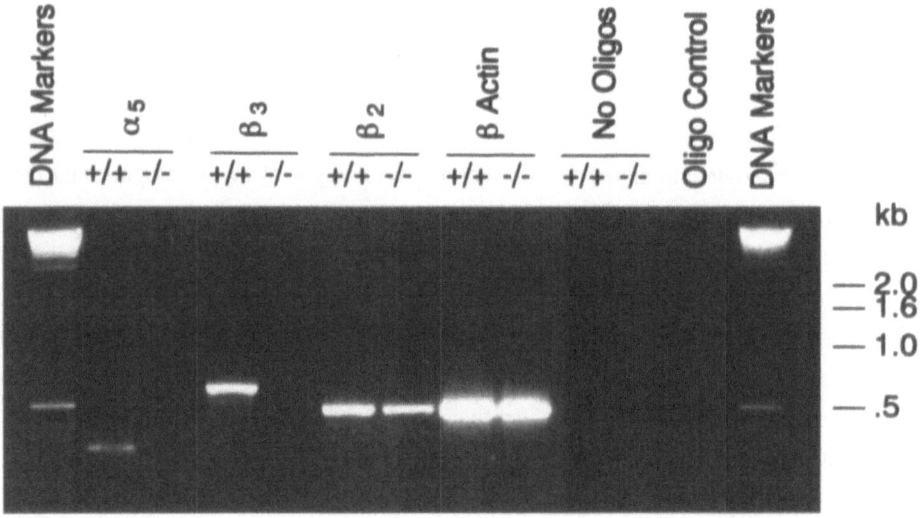

Figure 2. Analysis of transcripts of β_3, α_5, and β_2 in P0 (day of birth) homozygous mutant p^{cp}/p^{cp} (-,-) and control p^d/p^d (+,+) mice. PCR products from the control p^d/p^d (+,+) mouse and the mutant (p^{cp}/p^{cp}) (-,-) mouse amplified using $GABA_A$ α_5, β_3, β_2, and β-actin primers. Controls include template amplification without oligomer primers and oligomer primer amplification from blank cDNA synthesis. A 1Kb DNA ladders is run in both outside lanes. Methods: Total cellular RNA was extracted from p^d/p^d control (+,+) and mutant (p^{cp}/p^{cp}) mice (-,-) by the acid phenol method. First-strand cDNA synthesis was performed using random hexamers as primers and murine leukemia virus reverse transcriptase. For PCR amplification, the samples were brought to final concentrations of 10 mM Tris (pH 8.3), 50 mM KCl, 4% dimethylsulfoxide, 2.5 mM $MgCl_2$, 500 μM nucleotide triphosphate, 50 pmol of both the forward and reverse primers, 30 ng cDNA template and 2.5 units of Taq I polymerase, in a total volume of 100 μl. Each amplification cycle consisted of denaturing at 94°C for 45 sec, annealing at 55°C for 60 sec, and extending at 72°C for 60 sec. PCR was carried out for 40 cycles. The PCR products were visualized by electrophoresis in 1.2% agarose gels containing 0.5 μg/ml ethidium bromide.

Data derived from film autoradiograms of sagittal mouse brain sections labeled with either the GABA analog [^{3}H]muscimol or the benzodiazepine ligand [^{3}H]Ro15-1788 revealed a reduction in binding in all regions tested in the P0 mutant mouse (p^{cp}/p^{cp}) as compared to the P0 control mouse (p^d/p^d) (Fig 3 A and B). [^{3}H]Muscimol binding in mutant mice, as compared to control mice, was reduced by 10-45 % depending on the brain region tested. Reduction of [^{3}H]Ro15-1788 binding was approximately 50-75 % for P0 mice (fig. 3) and close to 80 % in E-18 mice in the brain regions tested (Data not shown). Of the regions tested, the cerebellum has been reported to have the lowest level of α_5, β_3, γ_3 mRNA message expressed. This is reflected in the less pronounced reduction in binding of both [^{3}H]muscimol and [^{3}H]Ro15-1788 in the autoradiograms (Fig. 2A and B).

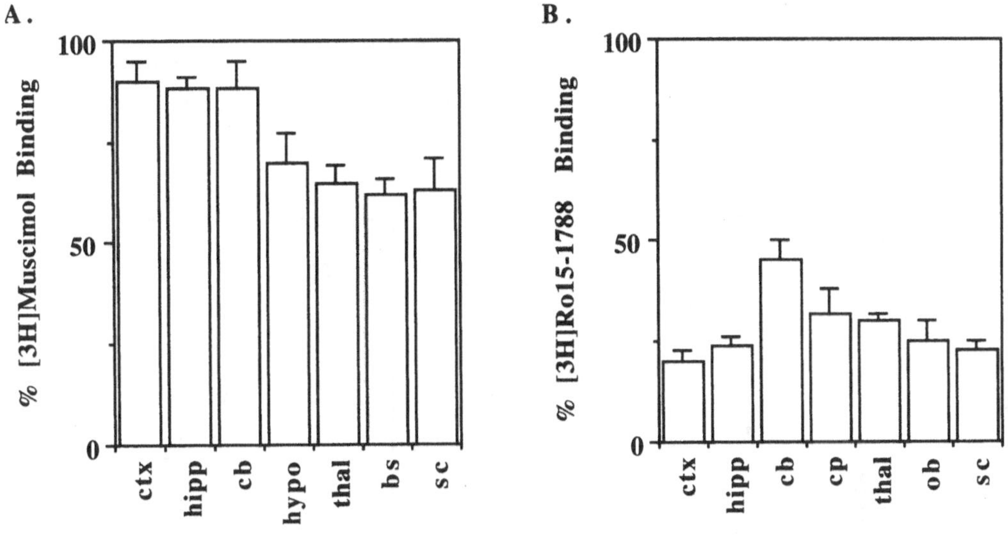

Figure 3. Binding of **A.** [^{3}H]muscimol and **B.** [^{3}H]Ro15-1788 in thin (12 μm) sagittal brain sections of P0 mice, as determined by autoradiography. Graphical representation of the specific binding of [^{3}H]muscimol and [^{3}H]Ro15-1788 to various brain regions of mutant mice (p^{cp}/p^{cp}) relative to control (p^d/p^d) Legend: ctx, cerebral cortex; hipp, hippocampus; cb, cerebellum; cp, caudate-putamen; thal, thalamus; ob, olfactory bulb; sc, superior colliculus. Methods are described elsewhere [17,18].

Differential displacement of [^{3}H]muscimol by the GABA analog THIP in the thalamus reveals a shift to the right in the mutant mouse as compared to the control mouse (fig. 4A). The lower potency for

THIP in the mutant mouse as compared to the control mouse is consistent with autoradiography studies showing brain regional variation in THIP affinity [19] and THIP has a higher affinity for proteins containing the β_3 subunit polypeptide rather than the β_2 [20].

Pritchett and Seeburg [21] have demonstrated that the co-expression of the α_5 with β and γ_2-subunits, in cultured cells, produces receptors with essentially no affinity for the benzodiazepine ligand zolpidem as compared to receptors containing α_1, α_2, or α_3 subunits. In addition, the γ_3 receptor in combination with other subunits has been found to convey an insensitivity to zolpidem [22]. Using this information we believe we are able to observe, in a more specific manner, the consequences of the loss of α_5 subunit by measuring the inhibition of [^{3}H]Ro15-1788 by zolpidem. Differential displacement of [^{3}H]Ro15-1788 by the benzodiazepine zolpidem in cerebral cortex of sagittal brain slices from control (p^d/p^d) mice and mutant (p^{cp}/p^{cp}) mice produces two different dose response curves (Fig 4 B). The dose response curve for the mutant mouse lacking the zolpidem-insensitive α_5 and γ_3 subunits, is fully inhibited by the presence of high concentrations of zolpidem (1-10 μM), unlike the control mouse which shows a plateau in inhibition at about 50 % binding even at 10 μM zolpidem.

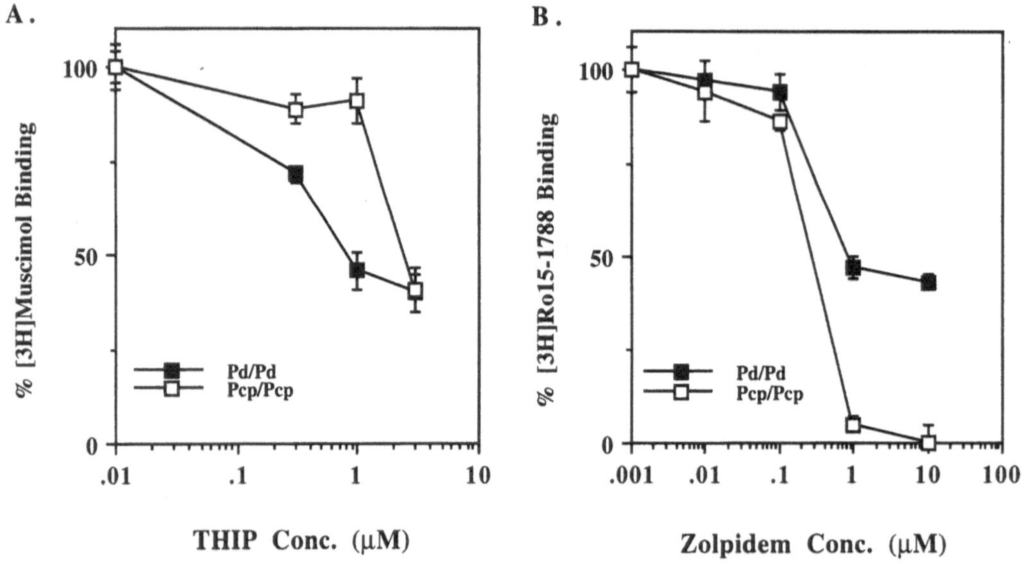

Figure 4. A. Differential concentration-dependent displacement by the GABA analog THIP of [^{3}H]muscimol binding measured in thalamus, showing that THIP inhibits [^{3}H]muscimol binding much less potently in the mutant mouse brain slice than in the control. **B.** Differential displacement by the benzodiazepine site ligand zolpidem of [^{3}H]Ro15-1788 in control (p^d/p^d) and mutant (p^{cp}/p^{cp}) mouse thalamus, showing complete inhibition of [^{3}H]Ro15-1788 binding by zolpidem (1-10 μM) in the mutant mouse brain slice as compared to the control. All samples are at P0. Similar results were noted for the other six brain regions measured in fig 3.

In summary, the mouse mutant (p^{cp}/p^{cp}) contains a deletion in chromosome 7 that includes a cluster of three GABA$_A$ receptor subunit genes ($\alpha_5/\beta_3/\gamma_3$). Brain slices from homozygous (p^{cp}/p^{cp}) mice show reduced GABA$_A$ receptor binding in all regions tested, relative to control (p^d/p^d), for the GABA analog muscimol and the benzodiazepine ligand Ro15-1788, suggesting a loss of GABA$_A$ receptors. Comparison of dose-response curves for both zolpidem and THIP in the mutant mouse (p^{cp}/p^{cp}) relative to the control (p^d/p^d) indicates GABAergic pharmacology is altered. Evaluating these data in conjunction with data collected on AS probands should provide us with a clearer understanding of whether the GABA$_A$ receptor plays a role in phenotypic characteristics, such as epilepsy, associated with Angelman syndrome.

Acknowledgment:

We would like to thank Dr. Michel Philippart on whose impressions the above clinical understanding of AS is based, and Dr. Antonio Delgado-Escueta for his support. Supported by NIH grants NS21908 and NS22071 to Richard W. Olsen.

References

1. Clayton-Smith J, Pembrey ME. Angelman syndrome. J Med Genet 1992; 29:412-415.

2. Angelman H. "Puppet" children: a report on three cases. Dev Med Child Neurol 1965; 7:681-688.

3. Clayton-Smith J. Clinical research on Angelman syndrome in the United Kingdom: Observations on 82 affected individuals. Am J Med Genet 1993; 46:12-15.

4. Williams CA, Frias JL. The Angelman ("Happy Puppet") syndrome. Am J Med Genet 1982; 11:453-460.

5. Zori RT, Hendrickson J, Woolven S, Whidden EM, Gray B, Williams CA. Angelman syndrome: Clinical profile. J Child Neurol 1992; 7:270-280.

6. Sugimoto T, Yasuhara A, Ohta T, Nishida N, Saitoh S, Hamabe J, Niikawa N. Angelman syndrome in three siblings: Characteristic epileptic seizures and EEG abnormalities. Epilepsia 1992; 33:1078-1082.

7. Matsumoto A, Kumagai T, Miura K, Miyazaki S, Hayakawa C, Yamanaka T. Epilepsy in Angelman syndrome associated with chromosome 15q deletion. Epilepsia 1992; 33:1083-1090.

8. Boyd SG, Harden A, Patton MA. The EEG in early diagnosis of the Angelman (Happy Puppet) syndrome. Eur J Pediatr 1988; 147:508-513.

9. Buxton JL, Chan CJ, Gilbert H, Clayton-Smith J, Burn J, Pembrey M, Malcolm S. Angelman syndrome associated with a maternal 15q11-13 deletion of less than 200 kb. Hum Mol Genet 1994; 3:1409-1413.

10. DeLorey TM, Olsen RW. γ-Aminobutyric acid$_A$ receptor structure and function. J Biol Chem 1992; 267:16747-16750.

11. Laurie DJ, Seeburg PH, Wisden W. The distribution of 13 GABA$_A$ receptor subunit encoding mRNAs in the rat brain. II. Olfactory bulb and cerebellum. J Neurosci 1992; 12:1063-1076.

12. Laurie DJ, Wisden W, Seeburg PH. The distribution of 13 GABA$_A$ receptor subunit mRNAs in the rat brain III. Embryonic and postnatal development. J Neurosci 1992; 12:4151-4172.

13. Olsen RW, Wamsley JK, Lee RJ, Lomax P. Benzodiazepine / barbiturate / GABA receptor-chloride ionophore complex in a genetic model for generalized epilepsy. In: Delgado-Escueta AV, Ward AA, Woodbury DM, Porter RJ, eds. Advances in Neurology 44. Basic Mechanisms of the Epilepsies: Molecular and Cellular Approaches. New York: Raven Press, 1986:365-378.

14. Silvers WK. A model for Mammalian gene action. In: The Coat Colors of Mice. New York: Springer, 1979:83-108.

15. Lyon MF, King TR, Gondo Y, Gardner JM, Nakatsu Y, Eicher EM, Brilliant MH. Genetic and molecular analysis of recessive alleles at the pink-eyed dilution (p) locus of the mouse. Proc Natl Acad Sci 1992; 89:6968-6972.

16. Phillips RJS. Mouse News Lett 1977; 56:38-38.

17. Nakatsu Y, Tyndale R, DeLorey TM, Durham -Pierre,D., Gardner JM, McDanel HJ, Nguyen Q, Wagstaff J, Lalande M, Sikela JM, Olsen RW, Tobin AJ, Brilliant MH. A cluster of three GABA$_A$ receptor subunit genes is deleted in a neurological mutant of the mouse p locus. Nature 1993; 364:448-450.

18. Olsen RW, McCabe RT, Wamsley JK. GABA$_A$ receptor subtypes: Autoradiographic comparison of GABA, benzodiazepine, and convulsant binding sites in the rat central nervous system. J Chem Neuroanat 1990; 3:59-76.

19. Bureau M, Olsen RW. Multiple distinct subunits of the gamma-aminobutyric acid-A receptor protein show different ligand binding affinities. Mol Pharm 1990; 37:497-502.

20. Ebert B, Wafford KA, Whiting PJ, Krogsgaard-Larsen P, Kemp JA. Molecular pharmacology of γ-aminobutyric acid type A receptor agonists and partial agonists in oocytes injected with different α, β, and γ receptor subunit combinations. Mol Pharm 1994; 46:957-963.

21. Pritchett DB, Seeburg PH. Gamma-aminobutyric acidA receptor alpha 5-subunit creates novel type II benzodiazepine receptor pharmacology. J Neurochem 1990; 54:1802-1804.

22. Herb A, Wisden W, Luddens H, Puia G, Vicinci S, Seeburg PH. The third γ subunit of the γ-aminobutyric acid type A receptor family. Proc Natl Acad Sci 1992; 89:1433-1437.

GABA: Receptors, Transporters and Metabolism
ed. by C. Tanaka & N.G. Bowery

MODULATORS OF BOTH GABAA RECEPTOR AND CYTOSOLIC Ca^{2+} LEVEL RELATED TO A NOVEL TREATMENT FOR EPILEPSY

Y. Watanabe and T. Shibuya

Department of Pharmacology, Tokyo Medical College,
6-1-1 Shinjuku, Shinjuku-Ku, Tokyo 160, Japan

Summary: In order to offer a key to the development of novel anti-epileptic agents, the synthesized GABA-peptides, Piv-Leu-GABA (PLG) and Ser-contained PLG, were examined. In in vitro studies, these peptides blocked a high K- and a 25 μM glutamate-increased cytosolic Ca^{2+} levels in a dose-dependent manner. Moreover, PLG blocked ^{3}H-muscimol binding site, although the inhibitory potency of other peptides was much less. In in vivo studies, these peptides reduced the pentylentetrazole and BAYK 8644 induced seizure in a dose-dependent manner. Furthermore, the dose-dependent nature of the reduction was much more apparent after a injection of PLG.

Introduction

The major biological phenomena are regulated by inhibitory-excitatory mechanism, e.g., γ-amino acidergic (GABA) and glutamatergic (Glu) neurons, in the brain. Furthermore, considering the significant role of cytosolic Ca^{2+} on the regulation of neuronal signal transduction, the intracellular Ca^{2+} ($[Ca^{2+}]i$)levels might be another important factor in maintaining a homeostasis of neuronal function and control of the neurologic condition.

For instance, the neuronal dysfunctions including the severe seizure induced by a neurotoxin is implicated in excessive amounts of cytosolic Ca^{2+} levels and extracellular Glu contents (1,2). These extraordinary hyperactivities can be attenuated by treatment of Ca^{2+}channel antagonists and/or GABA agonist (1,3). These results lead us to consider the possible application of the double pharmacological effects acting on intracellular Ca^{2+} modulation and GABA neuronal activity in the development of new classes of anti-epileptic agents.

In light of the above information, we focused on the pharmacological function of the GABA-peptide which was reported by Galzigna et al (4). This peptide is an N-protected GABA-dipeptide, i.e., N-pivaloyl-leucyl-GABA (PLG), and has an anticonvulsant action. Lacking, however, is an understanding of the pharmacological mechanisms of this peptide on convulsions, since it might be difficult to explain its anticonvulsant action by the GABAA agonistic action alone. Moreover, we also considered the modulation of brain L- and D-serine (Ser) on the N-methyl-D-aspartate (NMDA)-induced seizure susceptibility due to the inhibition of glycine binding to NMDA receptors(5). Therefore we newly synthesized PLGs containing L- and D-Ser, i.e., N-pivaloyl-L-Ser-leucyl-GABA (PSLG) and N-pivaloyl-D-Ser-leucyl-GABA (PDSLG), whose structures are shown in Fig.1.

To achieve our main purpose, first we examined the pharmacological properties of these GABA-peptides on the affinity to each receptor, i.e., GABAA, GABAB and benzodiazepines (BZP), and the neuronal Ca^{2+} channel activities by means of a radio receptor assay (RRA) and the fura-2 method, respectively. Next, the antagonistic effects of these peptides on several drugs-induced abnormal behaviours, including the convulsions, were tested. In the final analysis we try to suggest that the double pharmacological action might be important in developing a new generation of antiepileptics.

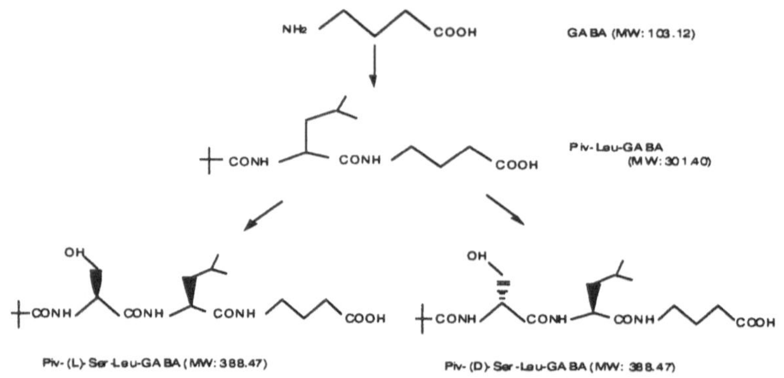

Figure. 1 Newly Synthesized N-protected GABA peptide

Materials and Methods

<u>RRA for $GABA_A$, $GABA_B$ and benzodiazepine receptors</u>: The rapid filtration methods described previously (6) were used for the binding study. After the crude synaptosome fraction of rat cortices were prepared, the protein concentrations were adjusted to 0.2-0.5 mg protein/each sample tube. The binding of [^{3}H] muscimol (for $GABA_A$ receptor; 10nM) and [^{3}H] diazepam (for BZP receptor; 5nM) were carried out at 0 °C for 10min and 30 min, respectively. [^{3}H] Baclofen (for $GABA_B$ receptor; 30mM) binding experiments were carried out at 25 °C for 10 min. Non-specific binding was defined by parallel incubations in the presence of 100 μM of muscimol, 30 μM of diazepam and 100 μM baclofen, and the ten different concentrations between 10^{-9} and 10^{-4}M of PLG, PSLG, PDSLG and GABA were incubated with each radio-ligand.

<u>Measurement of $[Ca^{2+}]i$ levels</u>: The determination of $[Ca^{2+}]i$ in cultured rat cerebellar granule cells was acheived by using the previously described fura-2 method by means of the c-image system (C-Image, Conpix Inc., PA, U.S.A.) (7). The cultured cerebellar granule cells were prepared as described previously (8). Briefly, ten cerebella from 8 day old Wistar rats were dissected and the meninges were carefully dissected away in the sterile media. Cells were isolated by mild trypsinization followed by trituration in a DNAase solution containing a trypsin inhibitor. Finally, the cells were suspended in modified Basal Medium Eagle (Kyokuto Pharm Ind Co Ltd, Tokyo, Japan) containing 10%(V/V) fetal calf serum (Bockneck, Canada) at a concentration of 1.25×10^6 cells/ml. Within 16 hrs after suspension of the cells, the prepared cells were treated with a final concentration of 10 μM cytosine arabinoside (Ara-C; Sigma, MO, U.S.A.). The cells were then incubated with 5 μM fura-2 at 37 °C for 45 min. The resting $[Ca^{2+}]i$ levels in the cultured cells was 100±8.6 nM. The superfusion of 50 mM KCl and 25 μM Glu in Mg^{2+} omitted Krebs'HEPES buffered solution for 90-sec maximally enhanced the $[Ca^{2+}]i$ levels to about 800 nM and 400 nM, respectively. These ligands were chosen as the stimulating agents in this experiment. The dose-dependent inhibition of GABA-peptides on both $[Ca^{2+}]i$ levels increased by stimulants were examined.

<u>Measurement of dopamine release from rat striatal slice</u>: The rat brain slice was prepared by using the method of Bull et al (8). Briefly, throughout the brain dissection the striatum was liberally bathed in ice-cold artificial cerebrospinal fluid. The s;ice was placed on a 35 mm cultured disk, and held in place with nylon mesh. A carbon fibre microelecrode (diameter; 35 μm) was located in the caudate putamen. And the dopamine (DA) release was detected and analyzed by a fast cyclic voltammetry (Millar Voltammeters).

<u>Intra-caudate putamen injections</u>: Rats were anaesthetized with sodium pentobarbitone (40 mg/kg; intraperitoneally) and positioned in a stereotaxic frame (1). Stainless steel guide cannulae (20G, Ontramedic luer stub adapter, Becton Dickinson and Co, NJ, U.S.A.) were

implanted chronically into either the caudate putamen (A: 1.70, L: 2.40, V: 5.00). The microinjection into both brain regions (2.7 μl/min) were made on free moving animals using a micro perfusion pump (IP-2, BRC, Nagoya, Japan) 7-9 days after recovery from the implantation procedure.
Measurement of locomotor activity increased by MAP, PCP and BAYK 8644: The rat spontaneous locomotor activity increased by methamphetamine (MAP; 1 mg/kg., i.p.), phencyclidine (PCP; 10 mg/kg, i.p.) and BAYK 8644 (100 nmol/caudate putamen) were recorded by Automex II animal activity monitor (Columbus Instruments Inc., OH, U.S.A.). The total count of locomotor activity was automatically monitored at 15-min intervals for 120 min after MAP, PCP and BAYK 8644 injection. Before the microinjections of GABA or the synthesized peptides were applied, animals were adapted in a transparent polycarbonate cage (LxWxH: 36x23x32 cm) to be used for the measurement of locomotor activity. Exactly five min after the microinjection, stimulants were injected into each rat and the locomotor activity was recorded. Each experiment was conducted using more than eight rats.
Measurement of convulsions and electroencephalography: The intensity of convulsions was determined by the appearences of clonic and/or tonic convulsions and fatalities induced by four different kinds of stimulants. Furthermore, the frontal cortical electroencephalography was measured by a telemetry system and analyzed by a Brain Wave system.
Statistical analysis: All results obtained from the in vitro and in vivo examinations were analyzed with the Dunnett's multiple comparison and Newman Keulus tests. Statistical significance was set at $P<0.05$.

Results and Discussion

Affinities of GABA-peptides to $GABA_A$, $GABA_B$ and BZP receptors: The binding affinities of GABA-peptides to three deferent kinds of receptors in the brain were summarized in Fig.2. In case of GABA-peptides to $GABA_A$ receptors, GABA and PLG significantly inhibited the ^{3}H-muscimol binding site in a dose-dependent manner, although other newly synthesized GABA-peptides showed a weak inhibition. The 50% inhibition concentration (IC_{50}) of GABA and each peptide to $GABA_A$ receptors were displayed in Fig.2. In $GABA_B$ and BZP binding experiments, the lower concentration of 100 μM of three examined GABA-peptides did not show any inhibition.

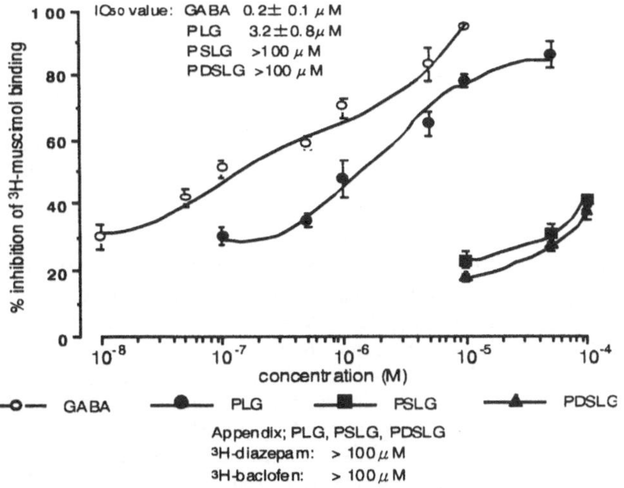

Figure. 2 Affinity of GABA peptides to $GABA_A$ receptor

These results suggest that PLG has a much higher affinity to GABAA receptor than to PSLG and PDSLG, although the GABA-peptides have no affinity to GABAB and BZP receptors.
Ca^{2+} channel antagonistic effects of GABA-peptides: The inhibitory effects of GABA-peptides on cytosolic Ca^{2+} levels increased by 50 mM KCl and 25 μM Glu were examined in the cultured neuronal cells. These results were summarized in Fig.3. The examined GABA-peptides inhibited both $[Ca^{2+}]i$ levels, which were increased by stimulants, in a dose-dependent manner. The inhibitory effects of PLG to both stimulants were much stronger than those of PSLG and PDSLG.

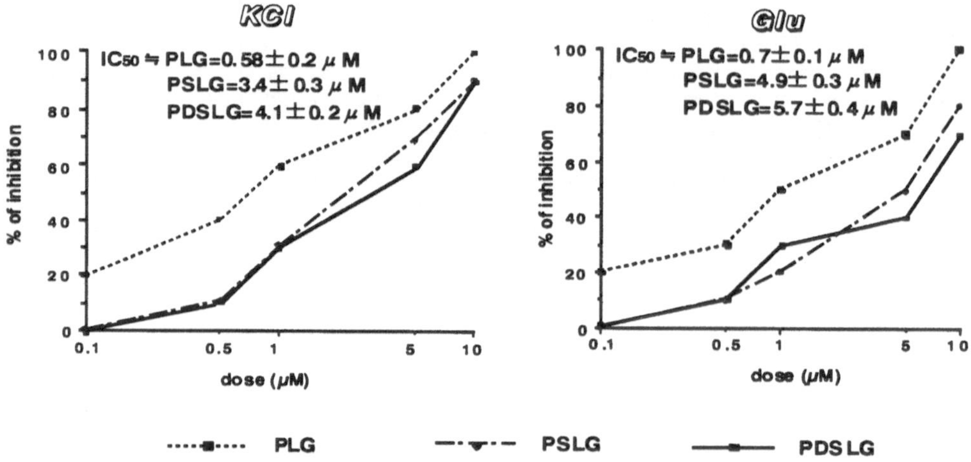

Figure. 3 Inhibition curve of GABA-peptides on KCl or Glu increased cytosolic Ca^{2+} levels

These results indicate that PLG has very potent inhibitory effects on the post-synaptic neuronal Ca^{2+} channel than PSLG and PDSLG, since it has been demonstrated that the quantitative results of $[Ca^{2+}]i$ alteration using the cultured neuronal cell can be cited in discussing the changes of post-synaptic neuronal Ca^{2+} channel activity.
Inhibition of GABA-peptides on DA release from rat striatal slice: To determine the inhibitory effects of GABA-peptides on pre-synaptic neuronal Ca^{2+} channel activity, we examined the effects of GABA-peptides on high-K evoked DA release. As shown in Fig.4, PSLG and PDSLG significantly inhibited the DA release, although the administration of PLG has no effects on this release. In these experiments, these peptides required the higher dosages than those examined in other experiments, since these peptides might adhese to the applied tubes.

These results can be explained in that PSLG and PDSLG, but not PLG, block the pre-synaptic Ca^{2+} channel activity.
Inhibition of GABA-peptides on CNS stimulants increased SLMA: In examining the results of the above three in vitro experiments, we can summarize the specificity of GABA-peptides on GABA receptor subtype and pre-/post-synaptic neuronal Ca^{2+} channel activity. Moreover, the pharmacological properties of these peptides were examined by the behavioral tests as shown in Fig.5. The significant inhibition of three peptides were measured in all of behavioral experiments. And in cases of BAYK 8644 or PCP, but not in MAP, induced hyperactivity, the inhibitory effects of PSLG and PDSLG were much stronger than that of PLG. Moreover the inhibitory effect of abnormal behavior induced by PDSLG on PCP was significantly more potent compared to those seen after administration of PSLG or PLG.

In the behavioral tests, the stronger effects of PSLG and PDSLG than PLG might be due to their inhibition of DA release.

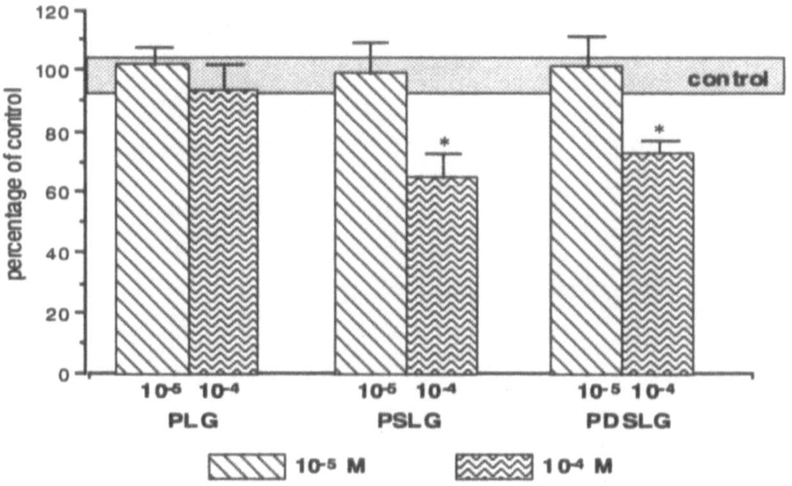

Figure. 4 Inhibitory effects of GABA-peptides on high K evoked DA release from striatal slice

Antagonistic effects of GABA-peptides on stimulants-induced convulsions: In order to evaluate the antagonistic actions of GABA-peptides, their inhibitory effects on PTZ, PQ, and BAYK 8644 induced seizures were tested by using the behavioral markers (appearance of clonic and/or tonic convulsions and mortality) and EEG. PTZ is a well-known CNS stimulant which blocks the $GABA/Cl^-$ channel complex, and PQ, a popular herbicide, has been demonstrated as a novel neurotoxin (1), and the i.c.v. treatment of PQ enhanced the Ca^{2+} influx and increased the extracellular Glu amounts (9). In Table.1, the antagonistic effects of GABA-peptides (three dosages; 20, 40 and 80 mg/kg, i.p.) were summarized. In PTZ-induced convulsions, a dosage of 45 mg/kg, i.p. of PTZ appeared to be only a clonic convulsion, and a dosage of 75 mg/kg, i.p. of PTZ induced both clonic and tonic convulsions following death. The dosages of larger than 40 mg/kg of PLG were completely diminished in a 45 mg/kg of PTZ-induced convulsion, although a 80 mg/kg i.p. injection of PSLG and PDSLG attenuated this convulsion. However, convulsions and death induced by 75 mg/kg of PTZ were blocked by a treatment of 80 mg/kg PLG and PSLG, but not PDSLG. In a dosage of 100 nmol, i.c.v. of PQ-induced convulsions, a severe clonic convulsion and a tonic convulsion appeared, and these treated rats were dead within 3 days after a PQ injection. Three GABA-peptides attenuated these convulsions and staved off death in a dose-dependent manner. In particular, a 80 mg/kg PLG completely diminished all signs of convulsions and death induced by PQ. In a larger dosage of BAYK, a clonic convulsion can be seen, but a tonic convulsion and a PQ-induced death appeared in a few animals. Therefore the marker of clonic convulsion was regarded as an important index in this experiment. The administration of each GABA-peptides attenuated the severe markers caused by PQ treatment in a dose-dependent manner. In addition, we examined the inhibitory effects of GABA-peptides on a glutamine synthetase inhibitor (methionine sulfoximine: MSO), which induced abnormal behaviours. The abnormal behaviours, including the clonic and/or tonic convulsions, appeared no later than 4 hrs after the intraperitoneal injection of a dosage of 150 mg/kg MSO, and these behaviours were sustained for more than 8 hrs. In this experiment, the administration of MSO was done at 7:00 p.m., and beahviour and EEG measured until 6:00 a.m., since the circadian rhythm of spontaneous locomotor activity was considered (8). As results of EEG, in the MSO treated group, the appearance of delta waves increased significantly after a 4th hr of injection, and these increases were ongoing in excess of 8 hrs as seen in Fig.6.

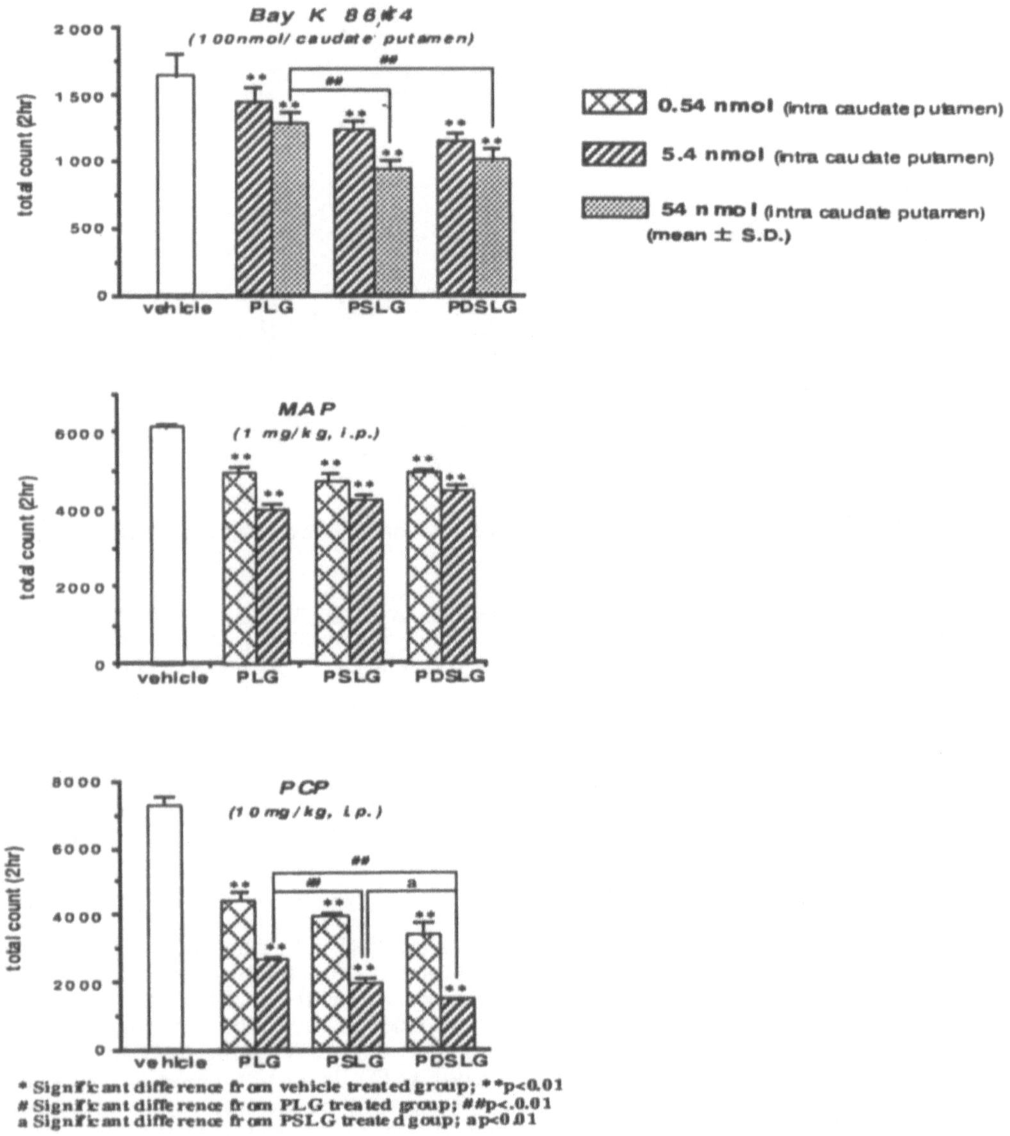

Figure. 5 Inhibition of GABA-peptides on CNS stimulants increased SLMA

Table 1. Anti-convulsive actions of GABA-peptieds

1) pentylenetetrazole (i.p. administration)

		PLG (mg/ kg)			PSLG (mg/ kg)			PDSLG (mg/ kg)		
	45 mg/kg	20	40	80	20	40	80	20	40	80
clonic convulsion	+	+	-	-	+	+	-	+	+	-
tonic convulsion	-	-	-	-	-	-	-	-	-	-
dead	-	-	-	-	-	-	-	-	-	-

		PLG (mg/ kg)			PSLG (mg/ kg)					
	75 mg/kg	20	40	80	20	40	80	20	40	80
clonic convulsion	+++	++	+	-	+++	++	-	+++	+++	+
tonic convulsion	+++	+++	++	+	+++	+++	+	+++	+++	++
dead	+	+	-	-	+	+	-	+	+	-

2) paraquat (i.c.v. administration)

		PLG (mg/ kg)			PSLG (mg/ kg)			PDSLG (mg/ kg)		
	100 mg/kg	20	40	80	20	40	80	20	40	80
clonic convulsion	+++	+++	++	-	+++	+++	++	+++	+++	++
tonic convulsion	++	+	+	-	++	++	+	++	++	+
dead	+	+	-	-	+	+	-	+	+	-

3) Bay K 8644 (i.c.v. administration)

		PLG (mg/ kg)			PSLG (mg/ kg)			PDSLG (mg/ kg)		
	200 mg/kg	20	40	80	20	40	80	20	40	80
clonic convulsion	+++	++	+	-	+++	++	-	+++	++	-
tonic convulsion	+	+	-	-	+	-	-	+	-	-
dead	+	-	-	-	+	-	-	+	-	-

Interestingly, the appearance of theta waves was decreasing, but not significantly. In contrast, such significant changes induced by MSO did not arise in the saline-treated rat (control). These phenomena of EEG might be a specific effect which was seen in the treatment of a larger dosage of MSO, since the appearance of delta waves did not increase after administration of the other three stimulants. In the test for the antiepileptic actions of GABA-peptides, the administration of a 80 mg/kg PLG weakly inhibited not only the clonic and/or tonic convulsions but also the increase of delta waves induced by MSO, but not significantly. Surprisingly, the administration of a 80 mg/kg PDSLG potentiated the toxicities of MSO, and this treatment severely heightened the mortality rate.

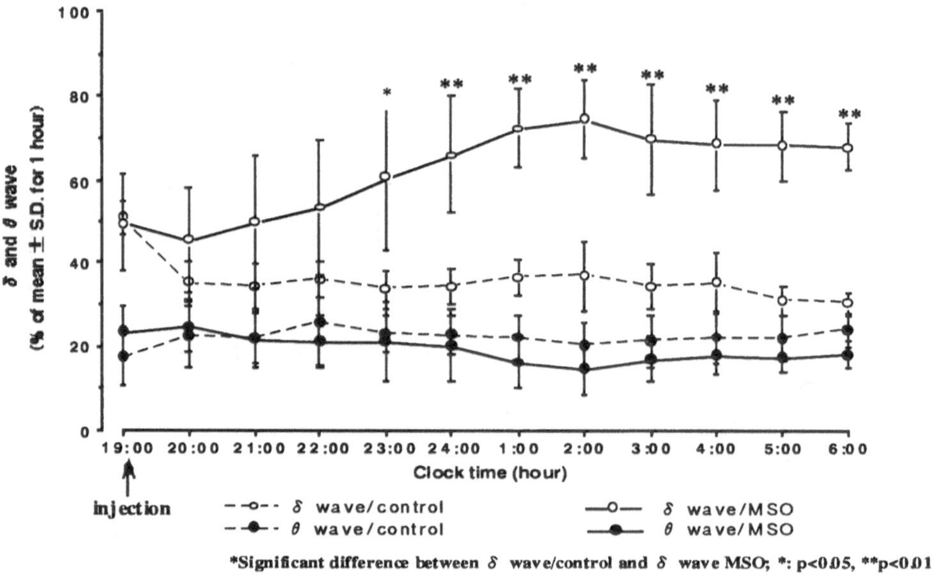

Figure. 6 Compared to changes of δ and θ wave induced by with or without metionine sulfoximine

In experiments for the determination of the antiepileptic actions of GABA-peptides, we chose four different kinds of stimulants. The administration of PLG showed the significant and dose-dependent inhibition on the PTZ, BAYK 8644 and PQ induced convulsions, and its potency was much higher than those seen in the administration of PSLG and PDSLG. However, MSO-induced convulsions were not attenuated by these GABA-peptides significantly, in contrast the administration of PDSLG potentiated MSO-induced toxicity. These results suggest that the GABA-peptides may not be useful for the epilepsy caused by the metabolic disorder syndrome, since the MSO-induced convulsions might be due to the increase of cytosolic ammonium levels and the excessive amounts of Glu levels in not only glia cells but also synaptic cleft.

In the comparison of the antiepileptic effects of three synthesized peptides, it is evaluated that the administration of PLG, which might have a GABAA agonistic action and a post-synaptic neuronal Ca^{2+} channel antagonistic action, can attenuate the convulsions induced by several types of CNS stimulants.

Conclusion

The administration of newly synthesized Ser contained N-protected GABA-peptides into the caudate putamen did show the remarkable inhibitory effects on the MAP, PCP and BAYK 8644 induced-hyperactivity, and these peptides reduced not only the ambulation but also the stereotypy (eg., rearing and so on). The inhibitory potency of PSLG and PDSLG on drug-induced convulsions appeared to have been less potent than those seen in the treatment of PLG. The results obtained from the behavioral studies and binding assay could suggest that the enzyme degradation of GABA was much faster than that of the synthesized GABA-peptides which were resistant to the degrading enzyme. In addition, a pharmacological drug with a double effect, such as PLG, might be a useful key for developing a new class of antiepileptic drugs.

Acknowledgments

We thank Ms. Seiko Yajima for her secretarial work. Partial financial supports from The Science Research Promotion Fund of Japan Private School Promotion Foundation and from the Japan Human Health Sciences Foundation are gratefully acknowledged.

References

1. Yoshimura Y, Watanabe Y, and Shibuya T. Inhibitory effects of calcium channel antagonists on motor dysfunction induced by intracerebroventricular administration of paraquat. Pharmacol & Toxicol 1993; 72: 229-235
2. Vriend J, Alexiuk NA, Green-Johnson J, Ryan E. Determination of amino acids and monoamine neurotransmitters in caudate nucleus of seizure-prone BALB/c mice. Journal of Neurochemistry 1993; 60(4): 1300-1307.
3. Shuaib A, Ijiaz S, Hasan S, Kalra, J. Gamma-vinyl GABA prevents hippocampal and substantia nigra reticulata damage in repetitive transient forebrain ischemia. Brain Research 1992; 590: 13-17.
4. Galzigna L, Bianchi M, Bertazzon A, Barthez A, Quadro G, Coletti-Previero MA. An N-protected γ-aminobutyric acid dipeptide with anticonvulsant action. Journal of Neurochemistry 1984; 42: 1762-1766.
5. Singh L, Oles RJ, Tricklebank MD. Modulation of seizure susceptibility in the mouse by the strychnine-insensitive glycine recognition site of the NMDA receptor/ion channel complex. British Journal of Pharmacology 1990; 99: 285-288.
6. Watanbe Y, Shibuya T, Salafsky B, Hill HF. Prenatal and postnatal exposure to diazepam; Effects on opioid receptor binding in rat brain cortex. Eur J Pharmacol 1983; 96: 141-144.
7. Zhang XQ, Watanabe Y, Ohnishi M, Baba T, Shibuya T. Comparative studies on the inhibitory effects of calcium antagonists on cytosolic Ca^{2+} levels increased by high-potassium or glutamate in cultured rat cerebellar granule cells. Japan J Pharmacol 1993; 62: 411-414.
8. Bull DR, Palij P, Sheehan MJ, Millar J, Stamford JA at al. Application of fast cyclic voltammetry to measurement of electrically evoked dopamine overflow from brain slices in vitro. J Neurosci Methods 1990; 32: 37-44.
9. Kiso K, Watanabe Y, Shibuya T. Central mechanism of a novel neurotoxin, paraquat, and its relation-ship to increased amounts of excitatory amino acids. Neurosciences 1994; 20: 169-179.
10. Watanabe Y, Nishimura T, Shibuya T, Spain JW. Comparison of motor activity circadian rhythm for SHR and WKY during moderate audiogenic stress. Proc West Pharmacol Soc 1988; 31: 273-276.

GABA: Receptors, Transporters and Metabolism
ed. by C. Tanaka & N.G. Bowery

MUSCIMOL INDUCES STATE-DEPENDENT LEARNING WHEREAS BACLOFEN IMPAIRS PLACE LEARNING IN MORRIS WATER MAZE TASK IN RATS

Yutaka Nakagawa*,**, Yoshinori Ishibashi*, Toshio Yoshii*, Eijiro Tagashira* and Tsuneo Iwasaki**

*Tsukuba Research Laboratories, Experimental Biomedical Research Inc. (Jisseiken), 8-5-1, Chuo, Ami-machi, Inashiki-gun, Ibaraki 300-03, Japan
**Institute of Psychology, University of Tsukuba, Tsukuba, Ibaraki 305, Japan

Summary: Effects of muscimol and baclofen on place learning in the Morris water maze task were examined. Rats were given 4 training trials per day for 4 days. Performance was impaired on day 4 in the rats treated with muscimol on day 1 to 3 and saline on day 4, suggesting that muscimol induces state-dependent learning (SDL). Baclofen injected on day 1 to 4 dose-dependently increased latency, indicating that baclofen induces the deficit of place learning. Thus it is concluded that GABA-A and GABA-B receptors play different roles in learning and memory: activation of GABA-A receptors may induce SDL, whereas GABA-B receptor stimulation may impair learning and memory.

Introduction

Animals trained on a task under some drugs often show a failure of learning performance when they are tested in the absence of the drugs. The failure, however, is ameliorated when the drugs are re-introduced. These phenomena are termed state-dependent learning (1). In state-dependent learning (SDL), it is hypothesized that acquisition of a task under a drug may require the same or similar drug state for recall, because the drug serves as a relevant internal cue (2).

γ-Aminobutyric acid (GABA) is the main inhibitory neurotransmitter in the brain. GABA acts on two pharmacologically distinct receptor subtypes, GABA-A and GABA-B (3, 4). GABA-A receptors are coupled with benzodiazepine receptors and Cl^- channels, and mediate neuronal inhibition through Cl^- channels (5). On the other hand, GABA-B receptors, which also have inhibitory property, are associated with adenylate cyclase activity and G protein system (6). The activation of GABA-B receptors decreases the amplitude of Ca^{2+} currents and increases K^+ conductance (3, 4).

We previously reported that muscimol (GABA-A agonist) induced SDL and that baclofen (GABA-B agonist) impaired the learning and memory in the passive avoidance task in rats (7).

These findings suggest that GABA-A and GABA-B receptors may influence learning and memory in a different manner.

In this article, we reviewed our recent studies (8, 9), where we assessed the effects of muscimol and baclofen on place learning in the Morris water maze task in rats.

Materials and Methods

A circular pool (150 cm in diameter, 35 cm high) was filled 24 cm deep with clear water (21±1 °C). A transparent platform (13 cm in diameter) was submerged 1 cm below the water surface at a fixed location.

Male Wistar rats (Charles River Japan) were given 4 training trials each day for 4 consecutive days. For each training trial, the rats were allowed to swim until they reached the platform. Latency to reach the platform was recorded up to 60 sec.

Results and Discussion

As shown in Figure 1, there were no significant changes in the latency to reach the platform when muscimol was injected on day 1 to 4. However, in the rats treated with muscimol on day 1 to 3 and treated with saline on day 4, the latency increased on day 4. These results suggest that muscimol may not impair learning processes but induce SDL in the Morris water maze task. Taking into the consideration that muscimol also induced SDL in the passive avoidance task (7), it is unlikely that muscimol-induced SDL is task-dependent.

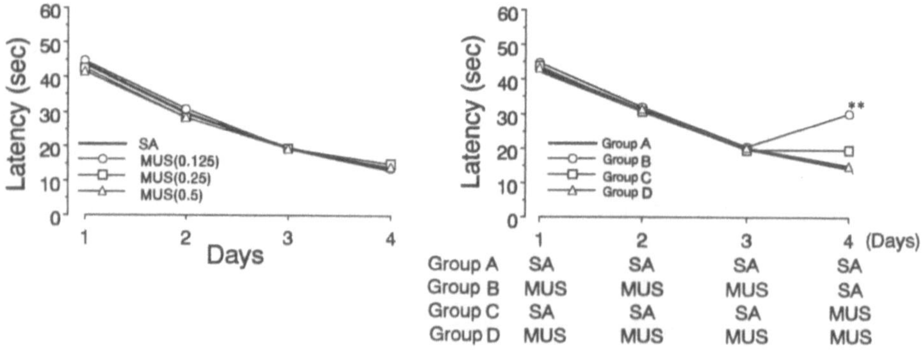

Figure 1. Effects of muscimol on mean latency to reach the platform in the Morris water maze. (Left panel) Muscimol (MUS) was injected i.p. 30 min before the training trials. Ten rats were used in each group. Numbers in parentheses: drug dose in mg/kg. SA, saline. (Right panel) Muscimol (MUS, 0.5 mg/kg) was injected i.p. 30 min before the training trials. ** $p<0.01$ versus Group A (Dunnett test). (From reference No. 8)

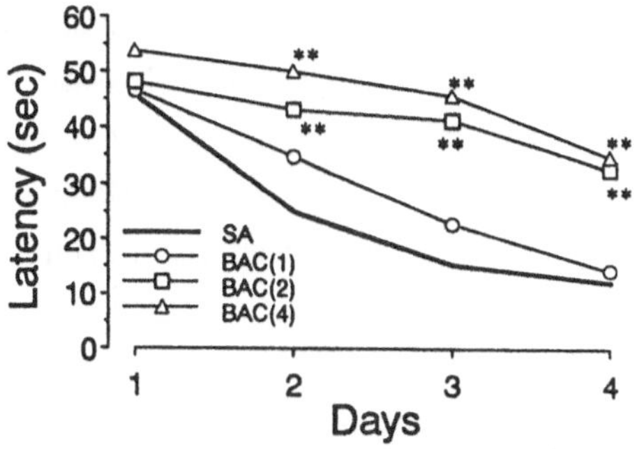

Figure 2. Effects of baclofen on mean latency to reach the platform in the Morris water maze. Baclofen (BAC) was injected i.p. 30 min before the training trials. Numbers in parentheses: drug dose in mg/kg. ** $p<0.01$ versus saline control (Dunnett test). See Figure 1 for further information. (From reference No. 9)

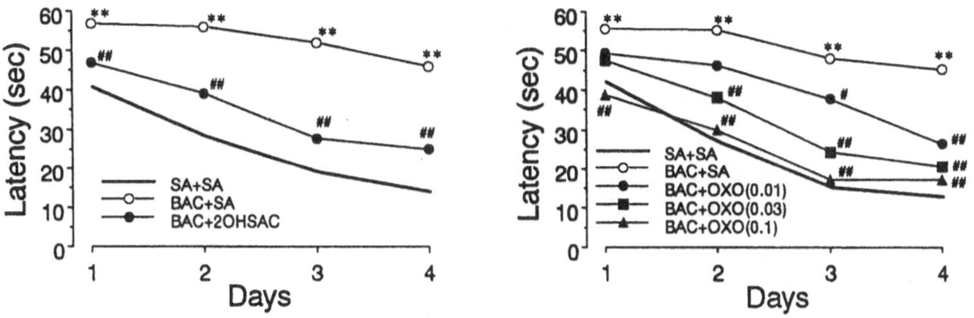

Figure 3. Effects of 2-hydroxysaclofen and oxotremorine on the baclofen-induced deficit of place learning in the Morris water maze. Baclofen (BAC, 4 mg/kg) was injected i.p. 30 min before the training trials. (Left panel) 2-Hydroxysaclofen (2OHSAC, 25 μg/2.5 μl/rat) was injected i.c.v. 35 and 5 min before the training trials. (Right panel) Oxotremorine (OXO) was injected i.p. 15 min before the training trials. Numbers in parentheses: drug dose in mg/kg. ** $p<0.01$ versus SA+SA; # $p<0.05$, ## $p<0.01$ versus BAC+SA (Dunnett test). See Figure 1 for further information. (From reference No. 9)

On the other hand, baclofen increased the latency in a dose-dependent manner (Figure 2). These findings indicate that baclofen may impair learning and memory in agreement with our previous data (7). 2-Hydroxysaclofen antagonized the deficit of place learning induced by baclofen (Figure 3). Thus, it is confirmed that the baclofen-induced deficit of place learning is mediated by GABA-B receptors.

As shown in Figure 3, oxotremorine attenuated the deficit of place learning in the baclofen-treated rats. Castellano and McGaugh (10) reported that oxotremorine attenuated the baclofen-impaired passive avoidance learning. These findings suggest that cholinergic systems may interact with GABA-B receptor systems in the brain. It is possible that GABA-B receptor systems may modulate learning and memory through cholinergic systems.

Oxotremorine at a dose of 0.1 mg/kg failed to improve the motor incoordination induced by baclofen (4 mg/kg) in the rotarod test, but improved the baclofen-induced deficit of place learning (9). Thus, it is concluded that ameliorative effects of oxotremorine on the deficit of learning and memory induced by baclofen may not be due to the improvement of motor incoordination.

In summary, these findings (7, 8, 9) provide the direct evidence that GABA-A and GABA-B receptors play different roles in learning and memory: activation of GABA-A receptors may induce SDL whereas GABA-B receptor stimulation may impair learning and memory.

References

1. Overton DA. State-dependent or "dissociated" learning produced with pentobarbital. J Comp Physiol Psychol 1964; 57: 3-12.

2. Bliss DK. Theoretical explanations of drug-dissociated behaviors. Fed Proc 1974; 33: 1787-1796.

3. Bowery N. GABA-B receptors and their significance in mammalian pharmacology. Trends Pharmacol Sci 1989; 10: 401-407.

4. Matsumoto RR. GABA receptors: are cellular differences reflected in function? Brain Res Rev 1989; 14: 203-225.

5. Olsen RW. GABA-benzodiazepine-barbiturate receptor interactions. J Neurochem 1981; 37: 1-13.

6. Hill DR, Bowery NG, Hudson AL. Inhibition of GABA-B receptor binding by guanyl nucleotides. J Neurochem 1984; 42: 652-657.

7. Nakagawa Y, Iwasaki T, Ishima T, Kimura K. Interaction between benzodiazepine and GABA-A receptors in state-dependent learning. Life Sci 1993; 52: 1935-1945.

8. Nakagawa Y, Ishibashi Y, Yoshii T, Tagashira E. Muscimol induces state-dependent learning in Morris water maze task in rats. Brain Res 1995; 681: 126-130.

9. Nakagawa Y, Ishibashi Y, Yoshii T, Tagashira E. Involvement of cholinergic systems in the deficit of place learning in Morris water maze task induced by baclofen in rats. Brain Res 1995; 683: 209-214.

10. Castellano C, McGaugh JL. Oxotremorine attenuates retrograde amnesia induced by post-training administration of the GABAergic agonists muscimol and baclofen. Behav Neural Biol 1991; 56: 25-31.

$GABA_B$ RECEPTOR ANTAGONISTS: POTENTIAL THERAPEUTIC APPLICATIONS

H.Bittiger, W.Froestl, C.Gentsch, J.Jaekel, S.J.Mickel, C.Mondadori, H.R.Olpe and M.Schmutz

Research and Development Department of the Pharmaceutical Division, CIBA Ltd.
CH 4002 Basel, Switzerland

Summary: GABA acts as a modulator at postsynaptic $GABA_B$ receptors where it mediates late inhibitory postsynaptic potentials and at presynaptic $GABA_B$ receptors where it inhibits the release of several neurotransmitters and neuropeptides. Thus $GABA_B$ receptor antagonists differing in their preferential interactions with potential subtypes of receptors are expected to produce different profiles of CNS effects. Therefore a spectrum of therapeutic applications may be expected. In mice, rats and Rhesus monkeys the $GABA_B$ receptor antagonist CGP 36742 improved cognitive functions in a number of relevant tests. Positive effects on cognitive functions were found with some, but not all of the $GABA_B$ receptor antagonists examined. However, in animal models of absence epilepsy all those $GABA_B$ receptor antagonists were active which antagonized the effects of baclofen in vivo. With selected compounds potential anxiolytic and antidepressant effects were also observed.

Introduction

$GABA_B$ receptor antagonists act as modulators of neurotransmission: Suppression of the late inhibitory postsynaptic potentials partially disinhibits postsynaptic target neurons and may increase their responsiveness (1). Antagonism of the inhibitory action of GABA on the release of several neurotransmitters (e.g. glutamate, GABA, noradrenaline, serotonin) and neuropeptides (e.g. substance P, CCK, somatostatin) may enhance neurotransmission by augmenting neurotransmitter/neuropeptide release (2). Depending on the preferential interactions of $GABA_B$ receptor antagonists for potential subtypes of receptors, different activity profiles of compounds may be expected. Therefore agents showing a variety of potential therapeutic applications may originate from this class of compounds. As there are no methods yet developed allowing the reliable characterization of compounds according to their receptor subtype preference and the $GABA_B$ receptor has not yet been cloned, a number of selected compounds were tested in suitable animal behavioural models to find potential therapeutic applications.

Characterization of $GABA_B$ receptor antagonists according to their receptor interactions in vitro and in vivo

Compounds were selected differing in their chemical structures, their affinities to the $GABA_B$ receptors as measured in vitro by bindings assays in which [^{3}H]CGP 27492 (APPA, 3-aminopropyl phosphinic acid) was used as radioligand (3), in their potency to antagonize the effects of the $GABA_B$ receptor agonist baclofen in behavioural (e.g. induction of sedation and muscle relaxation) and electrophysiological models (reduction of cell firing by baclofen in microiontophoretic experiments) in vivo (See Table 1) and in the induction of gross behaviour in

Table 1. Interactions of $GABA_B$ receptor antagonists with $GABA_B$ receptors

Compound	Receptor binding IC_{50} μM	Baclofen antag. ID_{50} mg/kg	
		i.p.	p.o.
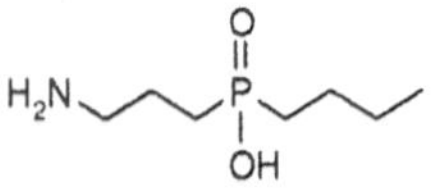 CGP 36742	32	50	100-200
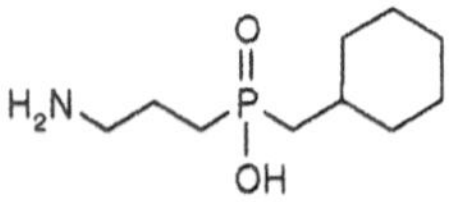 CGP 46381	5	10	100
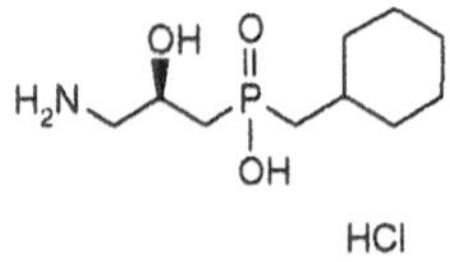 CGP 51176A	6.3	8	100-150
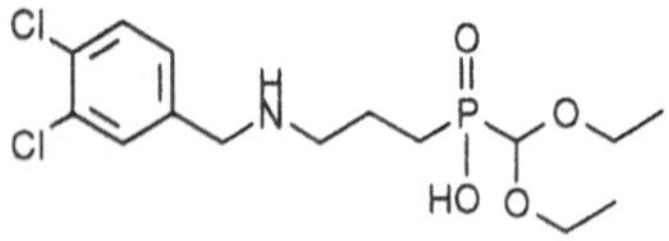 CGP 52432	0.07	100-200	not tested
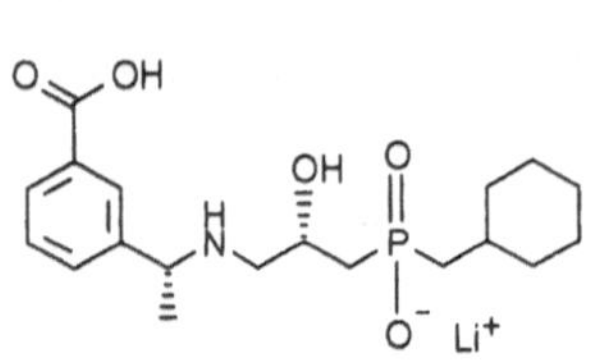 CGP 56999A	0.002	0.3	< 30

animals. CGP 36742 is a molecule with a structure closely related to GABA. Its affinity to the $GABA_B$ receptor with an IC_{50} of 32 μM in receptor binding is quite low. Nevertheless its activity as a $GABA_B$ antagonist in vivo with an ID_{50} of 50mg/kg i.p. and about 100-200 mg/kg p.o. is acceptable. It is a compound which is very well tolerated and does not induce overt behavioural changes of experimental animals even at doses sufficient to occupy most of the $GABA_B$ receptors and only induces convulsions at high intraperitoneal doses (> 800mg/kg). CGP 46381 is about 5 times more potent in vitro and in vivo (i.p.). It shows a trend towards sedative effects at higher doses and causes convulsions only at very high doses (starting at 1600mg/kg i.p.). CGP 51176A is similar in structure and affinity, but shows a trend towards stimulatory behavioural effects. In spite of an intense chemical effort, the IC_{50}'s of this class of compounds at the $GABA_B$ receptor could not be decreased below the micromolar range. This limit was overcome with new types of compounds which were substituted at the amino group (4). As an example CGP 52432 shows an IC_{50} of 70 nM. Its potency in vivo was, however, not proportionally improved and it was unexpectedly weak. This was true for all compounds of this series showing IC_{50} values in the low nanomolar range and very weak in vivo activity i.e. most probably they only poorly crossed the blood-brain barrier. Unexpectedly carboxy derivatization of the aromatic rings making the compounds more water soluble resulted in compounds showing strong activity in vivo. CGP 56999 had an IC_{50} of 2 nM and an ID_{50} of 0.3 mg/kg i.p.. The compound was convulsive at a dose of 3 mg/kg i.p.

Characterization of compounds according to potential therapeutic applications

$GABA_B$ receptor antagonists augmenting the responsiveness of target neurons due to suppression of the late ipsp and inducing increased release of glutamate and other neurotransmitters are expected to amplify signal transmission in the brain. This may lead to improved cognitive processes. As the different forms of epilepsies are considered to be strongly linked to the GABA system, the compounds were examined in models for different epilepsies. In view of the modulatory role of $GABA_B$ receptor antagonists, tests established to detect anxiolytic and antidepressant activities were also performed.

Effects on cognitive functions

As a primary test the one way passive avoidance, step down test in mice was performed. The antagonists were tested at 0.3, 3, and 30 mg/kg p.o. The results are given in Table 2. CGP 36742 showed activity at all doses and at 100 mg/kg. CGP 46381, CGP 51176A and CGP 52432 (at

100 mg/kg i.p.) were inactive and CGP 56999A was as active as CGP 36742. However, since the duration of action of CGP 56999A was less than 5 hours and since the compound was convulsive, it was not further characterized. CGP 36742, however, was profiled in a number of paradigms and in different species (5). In the one trial step down test it was active not only after one hour pretreatment, but also after 5 hours. Posttrial treatment similarly improved cognitive functions. The compound was also active in the electroshock induced amnesia in mice at the same doses as in the passive avoidance test. CGP 36742 also enhanced the spatial learning performance of mice in the eight arm radial maze paradigm at doses of 10 mg/kg and 100 mg/kg,

Table 2. Activity of $GABA_B$ receptor antagonists in the step down passive avoidance test

Compound	dose (mg/kg p.o.)	effect
CGP 36742	0.3, 3, 30, 100	++
CGP 46381	0.3, 3, 30	0
CGP 51176A	0.3, 3, 30	0
CGP 52432	(100 mg/kg i.p.)	0
CGP 56999A	0.3, 3, 30	++

but not 1 mg/kg i.p. (6). In aged rats CGP 36742 improved the learning performance in a multiple trial one-way active avoidance test: the animals had to jump from compartment A to compartment B to avoid a footshock in compartment A. Positive effects in social learning were also observed. Social interactions between two experimental animals are influenced by their degree of familiarity i.e. the better they know each other, the less time they spend on mutual scrutiny at each meeting. CGP 36742 facilitated recollection of an already encountered partner during retest at a dose of 3 mg/kg p.o., but not of 0.3 mg/kg p.o. Rhesus monkeys showed impressive improvements in the conditional colour-spatial test involving a series of tasks of increasing complexity . On an experimental table in front of Rhesus monkeys three beakers with different colours were placed upside down. The colour corresponded to the position of the beakers i.e. red to left, yellow to middle and blue to right. If the monkey selects the right colour in the correct position he will find a reward in form of a food pellet hidden under the beaker. In a cross-over blind study the monkeys were trained for two weeks (20 animals). Impressive improvements in the performance were found in the animals at the dose of 0.5 mg/kg p.o. of CGP 36742.

In conclusion CGP 36742 exerts positive effects on cognitive functions in three different species and diverse paradigms.

Effects in models of epilepsy

No protective effects of $GABA_B$ receptor antagonists were observed in models for grand mal epilepsy for example electroshock induced seizures. This was, however, different in a model for absence epilepsy. A rat strain has been bred by C.Marescaux et al. at the University of Strasbourg (GAERS, Genetic Absence Epilepsy Rats from Strasbourg). These animals show typical spike and wave discharges (SWD) in the EEG similar to those in human absence epilepsy EEG and also the typical absence behaviour. Most of the compounds enforcing GABA function like GABA uptake inhibitors, GABA-T inhibitors, the $GABA_B$ receptor agonist baclofen increase the duration of spike and wave discharges (7). The $GABA_B$ receptor antagonist CGP 35348 not only antagonized the effects of baclofen, but in addition dose dependently suppressed SWD's. All $GABA_B$ receptor antagonists presented in this paper suppressed SWD's at the same doses as they antagonized the effects of baclofen (Table 3).

Table 3. Activity of $GABA_B$ receptor antagonists in models of absence epilepsy (8,9)

Compound	Baclofen antag. ID_{50} mg/kg		Suppression of SWD in GAERS ID_{50} mg/kg	
	i.p	p.o.	i.p.	p.o.
CGP 36742	50	100-200	50	100
CGP 46381	10	100	12	100
CGP 51176A	8	100-150	12	100-150
CGP 52432	100-200	n.t.	100-200	n.t.
CGP 56999A	0.3	< 30	0.25	n.t.

Potential anxiolytic actions

CGP 52432 was described by M.Raiteri et al (10) as a compound interacting much more potently with presynaptic $GABA_B$ receptors at GABA nerve terminals as compared to those located at glutamate nerve terminals. In his synaptosome model the IC_{50} values for antagonism of the release inhibiting effects of baclofen were 80 nM and 9000 nM respectively. This suggested an increased release of GABA and could cause benzodiazepine-like effects in animals due to an amplification of GABA transmission. Therefore the compound was assayed in two tests established to detect potential anxiolytic actions: the antagonism of stress induced hyperthermia in mice (11) and the study of the typical behaviour of mice originating from the Balb/c strain. These animals do not enter an unfamiliar compartment due to their inbred neophobia and preference to remain in a familiar place (12). Benzodiazepines are effective in both types of tests. CGP 52432 had no effect at the lower doses of 10 and 50 mg/kg i.p. but antagonized fully the stress induced temperature increase at 200 mg/kg . In the exploration test with Balb/c mice, CGP 36742 showed no effect whereas CGP 52432 was active at 200 mg/kg i.p.; also CGP 46381 showed effects at the higher dose of 50 mg/kg i.p. Thus, CGP 52432 showed anxiolytic-like effects in two tests and CGP 46381 in one paradigm, showing that anxiolytic activity can in principle be obtained with $GABA_B$ receptor antagonists Due to its poor solubility and its weak in vivo activity CGP 52432 is not suited as a development compound.

Table 4. Models for anxiolytic action

Stress induced hyperthermia in mice (R Porsolt, 13)		
CGP 52432	10, 50 mg/kg i.p 200 mg/kg i.p.	no effect antagonism
Free exploration test with Balb/c mice (R.Misslin,12)		
CGP 52432	10, 50, 100 mg/kg i.p. 200 mg/kg i.p.	no effect reduction of neophobia
CGP 36742	10, 50, 100, 200 mg/kg	no effect
CGP 46381	3, 10 mg/kg i.p. 50 mg/kg i.p.	no effect reduction of neophobia

Potential antidepressant activity

Potential antidepressant activity may be predicted from the forced-swimming test(14) in which rats or mice are immersed in water. During a second exposure the animals give up swimming and show phases of floating immobility. The time of immobility is reduced following chronic administration of various antidepressants. CGP 36742 was weakly active in this test, CGP 46381 was inactive, whereas CGP 51176A showed a clear effect (Table 5). To further test the latter compound's potential antidepressant activity one of the most reliable tests to predict antidepressant-like activity, the chronic mild stress model was assessed (15). In this test classical antidepressants as well as newer, atypical compounds are active. Basically during chronic mild stress the sucrose intake of rats is attenuated and such a reduction can be reversed following chronic administration of antidepressants. In these studies imipramine (10 mg/kg p.o.) was active after 3-4 weeks of treatment. CGP 51176A after a 4 weeks treatment with 3 and 30, but not 0.3 mg/kg p.o. fully normalized sucrose consumption (16). In view of the negligible side effects of this compound, it may be a novel type of antidepressant.

Table 5. CGP 51176A - a potential antidepressant

Effects in rat swim test

CGP 36742	0.1, 10 mg/kg i.p.	0	30, 100 mg/kg i.p.	+
CGP 46381	0.1, 10, 100 mg/kg i.p.	0		
CGP 51176A			0.01, 0.1, 1 mg/kg i.p.	++

Effects in chronic mild stress model (M.Papp,Krakow, 16)

CGP 51176A	0.3 mg/kg p.o.	0	3, 30 mg/kg p.o.	++
Imipramine			10 mg/kg p.o.	++

Discussion

A number of possible therapeutic applications of GABA$_B$ receptor antagonists were explored. There is therapeutic potential as cognition enhancers, antiepileptics in absence epilepsy, possibly also as anxiolytics and antidepressants. However, since the classification of compounds according to GABA$_B$ receptor subtype specificity is at present not possible and since GABA$_B$ receptors could not yet be isolated and cloned, a correlation between receptor subtype interactions, pharmacological profile and potential therapeutic application of GABA$_B$ receptor antagonists cannot be established.

References

1. Olpe H-R, Steinmann MW, Ferrat T, Pozza M F, Greiner G, Brugger G, Froestl W, Mickel SJ, Bittiger H. The actions of orally active $GABA_B$ receptor antagonists on GABAergic transmission in vivo and in vitro. Eur J Pharmacol 1993;233:179-186

2. Bowery N G, $GABA_B$ receptor pharmacology. Annu Rev Pharmacol Toxicol.1993;33:109-147
3. Bittiger H, Reymann N, Hall R, Kane P. CGP 27492, a new potent and selective radioligand for $GABA_B$ receptors. Eur J Neurosci (Suppl) 1988:Abstr.16.10

4. Bittiger H, Froestl W, Mickel SJ, Olpe H-R. $GABA_B$ receptor antagonists: from synthesis to therapeutic applications. TiPS 1993;14:391-394

5. Mondadori C, Jaekel J, Preiswerk G. CGP 36742: The first orally active $GABA_B$ blocker improves the cognitive performance of mice, rats and Rhesus monkeys. Behavioural and Neural Biology 1993; 60:62-68

6. Carletti R, Libri V, Bowery NG. The $GABA_B$ antagonist CGP 36742 enhances spatial learning performance and antagonizes baclofen-induced amnesia in mice. Brit J Pharmacol 1993; 109 : Suppl.Abstr 74P

7. Marescaux C, Vergnes M, Bernasconi R. $GABA_B$ receptor antagonists: potential new antiabsence drugs. J Neural Transm. 1992; 35:179-188

8. Marescaux C, Liu Z, Bernasconi R, Vergnes M. $GABA_B$ receptors are involved in the occurence of absence seizures in rats. Pharmacology Communications 1992;2:57-62

9. Marescaux C, INSERM Unité 398, Neurobiologie et Neuropharmacologie des Epilepsies générales Strasbourg, France, personal communication

10. Lanza M, Fassio A, Gemignani A, Bonanno G, Raiteri M. CGP 52432: a novel and selective $GABA_B$ autoreceptor antagonist in rat cerebral cortex. Eur J Pharmacol 1993;237:191-195

11. Lecci A, Borsini F, Volterra G, Meli A . Pharmacological validation of a novel animal model of anticipatory anxiety in mice. Psychopharmacology 1990;101:255-261

12. Misslin R, University of Strasbourg, Laboratoire de Psychophysiologie, personal communication

13. Porsolt R, I.T.E.M.- Labo, Le Kremlin-Bicêtre, France, personal communication

14. Porsolt RD, Bertin A, Jalfre M. Immobility induced by forced-swimming in rats. Effects of agents which modify central catecholamine and serotonin activity. Eur J Pharmacol 1979; 57:201-210

15. Willner P, Towell A, Sampson D, Muscat R, Sophokleous S. Reduction of sucrose preference by chronic mild stress and its restoration by a tricyclic antidepressant. Psychopharmacology 1987; 93: 358-364

16. Papp, M, Polish Academy of Sciences, Institute of Pharmacology, Krakow, Poland, personal communication

ed. by C. Tanaka & N.G. Bowery

EXPECTED THERAPEUTIC USE OF GABA AGONISTS AND ANTAGONISTS FOR GUT MOTILITY DISORDERS

K. Taniyama[1], M. Kusunoki[2], S. Matsuyama[3], K. Yamashita[1] and M. Kaibara[1]

[1]Department of Pharmacology II, Nagasaki University School of Medicine, Nagasaki 852, [2]Department of Surgery, Hyogo College of Medicine, Nishinomiya 663 and [3]Department of Pharmacology, Kobe University School of Medicine, Kobe 650, Japan

The gut motility is controled by the autonomic nervous system, such as sympathetic and parasympathetic neurons, and the enteric nervous systems. Parasympathetic neurons are very important for regulation of gut motility. Acetylcholine (ACh) released from parasympathetic postganglionic cholinergic nerve terminals produces contractions of smooth muscles and stimulates gut motility. Most postganglionic sympathetic adrenergic nerves form a terminal network, mainly around the myenteric ganglia (1, 2). Noradrenaline (NA) released from adrenergic nerve terminals presynaptically inhibits the release of ACh from the pre- and postganglionic parasympathetic cholinergic nerve terminals, through α_2-adrenoceptors (3, 4). The enteric nervous system contains a variety of intrinsic neurons, and has structural and chemical features similar to those of the central nervous system (2). Several reviews have focussed on the possible role of GABA as a neurotransmitter in the peripheral nervous systems, particularly in the enteric nervous system (5-10), and the criteria required to establish GABA as a neurotransmitter have been fulfilled. These criteria are: 1) the occurrence of GABA and its decarboxylating enzyme, glutamic acid decarboxylase (GAD) within neurons; 2) the neuronal release of GABA; 3) the neuronal and glial uptake of GABA and its inactivation by GABA transaminase (GABA-T) as mechanisms by which GABA is removed from the synaptic cleft and conserved for reuse; and 4) electrical stimulus-induced depolarization of the neuronal membrane, which is mimicked by GABA and antagonized by blockers.

Actions of GABA on the myenteric neurons

GABA is shown to have dual effects on the membrane of myenteric neurons, determined using intracellular recording techniques (11, 12). The first is that GABA applied by iontophoresis led to a membrane deporalization of the after-hyperpolarizing (AH type 2) myenteric neurons which have no nicotinic fast excitatory postsynaptic potential (EPSP) but did have a prolonged Ca^{2+}-dependent after-hyperpolarization. This depolarization was associated with an increase of chloride conductance and prevented by bicuculline (11) and potentiated by benzodiazepines (13), indicating that the effect is typical of $GABA_A$ receptor-mediated action. The second finding is that GABA depolarizes myenteric neurons, even in the presence of bicuculline, which is Ca^{2+} sensitive and is mimicked by baclofen, thereby suggesting the involvement of $GABA_B$ receptors. It is considered that the depolarization may result from suppression of a constant inward calcium current occurring at a nonsomatic location and that GABA inhibits the release of ACh and the transmitter mediating slow EPSP. GABA has been shown to cause bicuculline-insensitive chloride-mediated current response in rod bipolar cells (14, 15). Since the response was not activated by baclofen, but activated selectively by cis-4-aminocrotonic acid, the receptor was designated $GABA_C$ receptor. As yet, the presence of receptor similar to $GABA_C$ receptor has not been reported in the myenteric neurons. GABA produced a transient increase in ACh release from the guinea pig ileum (16, 17) and the excitatory effect of GABA was mimcked by muscimol and antagonized by bicuculline, while GABA inhibited the stimulus-evoked release of ACh and the effect of GABA was mimicked by baclofen (Fig. 1) (16). Excitatory and inhibitory effects of GABA are mediated by stimulation of $GABA_A$ receptor and $GABA_B$ receptor, respectively. Since the $GABA_A$ receptor-mediated response was tetrodotoxin-sensitive (17), $GABA_A$ receptors may be located on the soma/dendritic regions of postganglionic cholinergic neurons. Baclofen inhibited the high K^+-evoked tetrodotoxin-insensitive release of ACh, therefore $GABA_B$ receptor may be located on nerve terminals of cholinergic neurons.

Actions of GABA on the gut motility

There are many reports on the effects of GABA on the motility of mammalian intestine (18-32) including human intestine (20). The ileal strips of guinea pig produced biphasic response to GABA, a transient contraction followed by relaxation (21-23). The nerve stimulated twitch contractions of isolated preparations from guinea pig ileum are inhibited by GABA and baclofen (22, 23). Guinea pig intestine other than ileum, such as duodenum, jejunum and colon produced relaxing response to GABA (24, 25). These excitatory and inhibitory responses are mediated by stimulation of $GABA_A$ receptor and $GABA_B$ receptor, respectively. In the colon of several species, such as guinea pig (26-28) and dog (29), the predominant effect of GABA was a bicuculline- and tetrodotoxin-sensitive relaxation, indicating that the response is mediated by stimulation of $GABA_A$ receptor. GABA neurons have been shown to participate in the regulation of peristaltic activity (30-32). Activation of $GABA_A$ receptor accelerates and inactivation of

$GABA_A$ receptor slows the rate of propulsion of fecal pellets in isolated segments of guinea pig colon (30, 31). GABA augments ascending contraction mediated by stimulation of ACh release and descending relaxation mediated by stimulation of vasoactive intestinal polypeptide (VIP) release, both GABA effects activaing $GABA_A$ receptor (32).

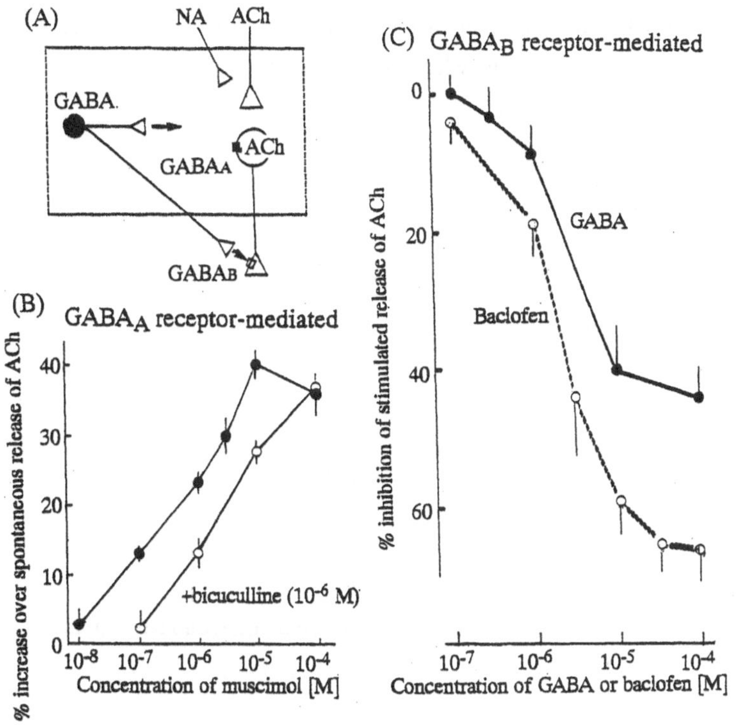

Figure 1. GABA receptors-mediated modulation of acetylcholine release from guinea pig intestine. (A) Possible localization of GABA receptors on cholinergic neurons in the intestine, (B) Concentration-response curves of muscimol in the absence and presence of bicuculline, (C) Concentration response curves of GABA and baclofen.

When GABA was administered intravenously, distal colon caused relaxation (Fig. 2), as noted by Giotti et al. (27). The effect of GABA was mimicked by baclofen, and antagonized by 2-OH-saclofen, a $GABA_B$ antagonist, but not by bicuculline, thereby indicating the relaxing responses are mediated by stimulation of $GABA_B$ receptor.

Possibility that GABA agonists and antagonists are used for treatment of digestive disorders related to motility

In vitro and *in vivo* studies have confirmed physiological role of GABAergic neurons in the gut motility. Both $GABA_A$ and $GABA_B$ receptors are located on the cholinergic neurons.

Activation of $GABA_A$ receptor stimulates ACh release and gut motility, and $GABA_B$ receptor activation inhibits ACh release and inhibits gut motility. The $GABA_A$ receptor-mediated response is transient, while the $GABA_B$ receptor-mediated response lasts longer. *In vitro* studies has shown that $GABA_A$ receptor is more predominant than $GABA_B$ receptor in the regulation of peristaltic activity (30-32), while *in vivo* studies demonstrated more predominant of $GABA_B$ receptor function than $GABA_A$ receptor in gut motility. At any rate, either hyperactivity or hypoactivity of GABAergic neurons may lead to gut motility disorders.

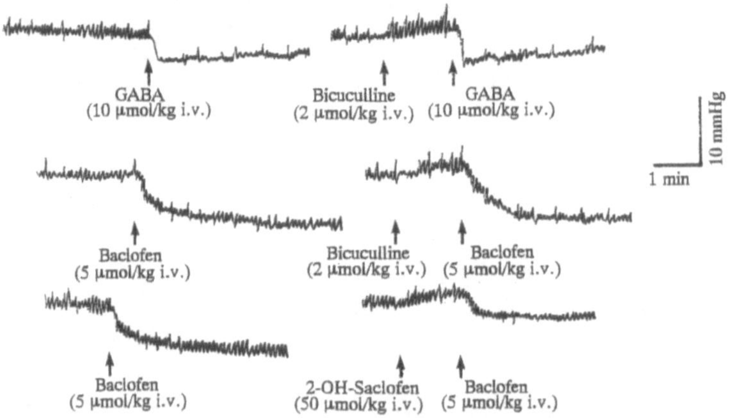

Figure 2. Effects of GABA and baclofen administered intravenously on the resting tone of guinea pig distal colon.

GABA mimetic has been tried for treatment of incontinence of alimentary tract (33). Valproate sodium, a GABA transaminase inhibitor was useful in the treatment of minor incontinence after ileoanal anastomosis. Frequent defecation and soiling and the resulting perianal skin complications occur in patients with adenomatosis coli and ulcerative colitis after receiving ileoanal anastomosis. Valproate sodium seems to ameliorate postoperative incontinence by increasing anal resting pressure through GABA receptors (34).

Dopaminergic receptor antagonists (metoclopramide, sulpiride and domperidone) (35), opioid receptor agonist (trimebutine) (36) and serotonin receptor antagonist/agonist (cisapride) (37) have been prescribed to treat digestive disorders related to motility. Dopaminergic receptor antagonists prevent the inhibitory effect of dopamine on ACh release. Cisapride antagonizes the inhibitory $5HT_1$ receptor and activates the excitatory $5HT_4$ receptor (38). All these drugs exert effects on gut motility through cholinergic neurons. Since GABA also controls gut motility by modulating cholinergic nerve activity, GABA agonists and antagonists may be useful for treatment of gut motility disorders.

Acknowlegement

This work was supported by grants from the Ministry of Education, Science and Culture, Japan.

References

1. Jacobowitz D. Histochemical studies of the autonomic innervation of the gut. J Pharmacol Exp Ther 1965; 149: 358-364.
2. Costa M, Gabella G. Adrenergic innervation of the alimentary cannal. Z Zellforsch 1971; 122: 357-377.
3. Furness JB, Costa M. Types of nerves in the enteric nervous system. Neuroscience 1980; 5: 1-20.
4. Alberts P, Stjarne L. Facilitation and muscarinic and α-adrenergic inhibition of the secretion of [^{3}H]-acetylcholine and [^{3}H]-noradrenaline from guinea-pig ileum myenteric nerve terminals. Acta Physiol Scand 1982; 116: 83-92.
5. Wikberg JES, Lefkowitz RJ. α2 Adrenergic receptors are located prejunctionally in the Auerbach's plexus of the guinea pig small intestine: Direct demonstration by radioligand binding. Life Sci 1982; 31: 2899-2905.
6. Erdo SL. Peripheral GABAergic mechanisms. Trends Pharmacol Sci 1985; 6: 205-208.
7. Tanaka C. γ-Aminobutyric acid in peripheral tissues. Life Sci 1985; 37: 2221-2235.
8. Jessen KR, Mirsky R, Hills M. GABA as an autonomic neurotransmitter: studies on intrinsic GABAergic neurons in the myenteric plexus of the gut. Trends Neurosci 1987; 6: 255-262.
9. Erdo SL, Wolff JR. γ-Aminobutyric acid outside the mammalian brain. J Neurochem 1990; 54: 363-372.
10. Tanaka C, Taniyama K. In: GABA outside the CNS. Erdo SL ed. Berlin: Springer, 1992: 3-17.
11. Cherubini E, North RA. Action of γ-aminobutyric acid on neurones of guinea pig myenteric plexus. Br J Pharmacol 1984; 82: 93-100.
12. Cherubini E, North RA. Inhibition of calcium spikes and transmitter release by γ-aminobutyric acid in the guinea pig myenteric plexus. Br J Pharmacol 1984; 82: 101-105
13. Cherubini E, North RA. Benzodiazepines both enhance γ-aminobutyrate responses and decrease calcium action potentials in guinea-pig myenteric neurones. Neuroscience 1985; 14: 309-315.
14. Feigenspan A, Wassle H, Bormann J. Pharmacology of GABA receptor Cl^- channels in rat retinal bipolar cells. Nature 1993; 361: 159-162.
15. Qian H, Dowling JE. Novel GABA responses from rod-driven retinal horizontal cells. Nature 1993; 361: 162-164.
16. Kleinrok A, Kilbinger H. γ-Aminobutyric acid and cholinergic transmission in the guinea pig ileum. Naunyn-Schmied Arch Pharmacol 1983; 322: 216-220.

17. Taniyama K, Kusunoki M, Saito N, Tanaka C. GABA evoked ACh release from isolated guinea pig ileum. Life Sci 1983; 32: 2349-2353.

18. Hobbiger F. Effect of γ-aminobutyric acid on the isolated mammalian ileum. J Physiol (Lond) 1958; 142: 147-164.

19. Inouye A, Fukuya M, Tsuchiya K, Tsujioka T. Studies on the effect of γ-aminobutyric acid on the isolated guinea pig ileum. Jpn J Physiol 1960; 10: 167-182.

20. Gentilini G, Franchi-Micheli S, Pantalone D, GABAB receptor-mediated mechanisms in human intestine in vitro. Eur J Pharmacol 1992; 217: 9-14.

21. Krantis A, Costa M, Furness JB, Orbach J. γ-Aminobutyric acid stimulates intrinsic inhibitory and excitatory nerves in the guinea pig intestine. Eur J Pharmacol 1980; 67: 461-468.

22. Kaplita PV, Waters DH, Triggle DJ. γ-Aminobutyric acid action in guinea pig ileal myenteric plexus. Eur J Pharmacol 1982; 79: 43-51.

23. Giotti A, Luzzi S, Spagnesi S, Zilletti L. GABAA and GABAB receptor-mediated effects in guinea pig ileum. Br J Pharmacol 1983; 78: 469-478.

24. Ong J, Kerr DIB. Comparison of GABA-induced responses in various segments of the guinea pig intestine. Eur J Pharmacol 1987; 134: 349-353.

25. Kerr DIB, Ong J. In: GABA outside the CNS. Erdo SL ed. Berlin: Springer, 1992: 29-44.

26. Frigo GM, Galli A, Lecchini S, Marcoli M. A facilitatory effect of bicuculline on the enteric neurones in the guinea-pig isolated colon. Br J Pharmacol 1987; 90: 31-41.

27. Minocha A, Galligan JJ. Excitatory and inhibitory responses mediated by GABAA and GABAB receptors in guinea pig distal colon. Eur J Pharmacol 1993; 230: 187-193.

28. Boeckxstaens GE, Pelckmans PA, Rampart M, Ruytjens IF, Verbeuren TJ, Herman AG, Van Maercke YM. GABAA receptor-mediated stimulation of non-adrenergic non-cholinergic neurones in the dog ileo-colonic junction. Br J Pharmacol 1990; 101: 460-464.

29. Krantis A, Kerr DIB. The effect of GABA antagonism on propulsive activity of the guinea pig intestine. Eur J Pharmacol 1981; 76: 111-114.

30. Ong J, Kerr DIB. GABAA and GABAB receptor-mediated modification of intestinal motility. Eur J Pharmacol 1983; 86: 9-17.

31. Grider JR, Makhlouf GM. Enteric GABA: mode of action and role in the regulation of the peristaltic reflex. Am J Physiol. 1992; 262: G690-G694.

GABA: Receptors, Transporters and Metabolism
ed. by C. Tanaka & N.G. Bowery

GABA$_B$ ANTAGONISTS BLOCK γ-BUTYROLACTONE-INDUCED ABSENCE SEIZURES AND COORDINATED INDUCTION OF TRANSCRIPTION FACTORS IN MOUSE BRAIN

Yoshihisa Ito, Kumiko Ishige, Masahiro Aizawa and Hideomi Fukuda

Department of Pharmacology, College of Pharmacy, Nihon University, Funabashi-shi, Chiba 274, Japan

Summary: Administration of γ-butyrolactone (GBL) induced absence-like seizures accompanied by 6-Hz spike and wave discharges (SWDs) in mice. GBL also increased nuclear CRE- and AP-1 DNA-binding activities significantly in the thalamus + midbrain and cerebral cortex but not in other regions of brain. CGP 35348, a GABA$_B$ antagonist, suppressed GBL-induced absence seizure behavior, SWDs and nuclear CRE- and AP-1 DNA-binding activities. These results suggest that GABA$_B$ receptor-mediated synaptic responses are involved in GBL-induced absence-like seizures and that the increases in nuclear CRE- and AP-1 DNA-binding activities are correlated with the seizures.

Introduction

Absence seizures are characterized by a brief period of behavioral arrest and occasional automatisms accompanied by 3-Hz spike and wave discharges (SWDs) in the electroencephalogram. These discharges may be associated with thalamocortical oscillations and GABAergic inhibition plays a critical role in synchronizing the oscillations (1). Since thalamic infusion of GABA agonists triggers or exacerbates SWDs, while GABA$_B$ antagonists block or reduce SWDs in rodent models of absence seizures (2,3), GABA$_B$ antagonists would be expected to act as possible antiepileptic drugs (4). It has been shown that administration of convulsant doses of kainic acid, metrazole or picrotoxin results in the induction of jun, c-fos and fos-related antigen immunoreactivity and also AP-1 DNA-binding activity in the brain regions such as the hippocampus and cerebral cortex (5-7). c-Fos and fos-related antigen are known to form complexes with members of the jun transcription factor family and these complexes recognize AP-1 DNA-binding elements (TGACTCA) and modulate gene transcription. It has also been reported that fos-immunoreactive nuclei are found in the thalamus of rats with GHB-induced generalized absence seizures (8). Therefore, AP-1 transcription factors could well play a role in GHB-induced generalized absence seizures. In addition to AP-1, CREB has been shown to recognize specific DNA sequence (TGACTGCA) which is closely similar to the AP-1 consensus sequence (9). However, the way of CREB induction is apparently different from that of AP-1 induction (10). In this study, we examined the involvement of the $GABA_B$ receptor and coordinated induction of transcription factors in mice with experimental generalized absence seizures induced by GBL.

Methods

Materials

Oligonucleotides (22 mer) used in this study were purchased from Boksui Brown (Tokyo, Japan). The chemicals used were [^{32}P]ATP (specific activity 6,000 Ci/mmol); ethosuximide (Sigma Chemical, USA); γ-butyrolactone (GBL, Wako Pure Chemicals, Japan) and *P*-[3-aminopropyl]-*P*-diethyoxymethyl-phosphinic acid (CGP 35348), *P*-[3-aminopropyl]-*P*-cyclohexylmethyl-phosphinic acid (CGP 46381) (CIBA-GEIGY, Switzerland).

Electrocorticogram (ECoG) recording

ECoG recording was carried out according to the method described by Zornetzer (11) with slight modifications. All recordings were carried out at least 5 days after surgery on unanesthetized and freely moving mice.

Preparation of nuclear extracts and Gel-shift DNA binding assay

Nuclear pellets and extracts were prepared as described by Gorski et al. (12) and Sakurai et al. (13) with slight modifications. Gel-shift assay was performed as described by Sonnenberg et al. (5) with slight modifications (14). DNA-protein complexes were resolved on a 4.0% non-denaturing polyacrylamide gel in buffer containing 25 mM Tris-borate (pH 8.2) and 0.5 mM EDTA. Electrophoresis was carried out in a cold room (4°C). The gel was autoradiographed with intensifying screens on diagnostic film (Kodak, USA) at -80°C.

Results

Characterization of GBL-induced absence-like seizures

Although administration of lower doses of GBL (25 and 50 mg/kg) did not show any effects on behavior or EEG pattern, higher doses of GBL (70 and 100 mg/kg, i.p.) induced generalized absence seizure behavior such as staring and arrest of activity. ECoG recording showed that these seizures were associated with 6-Hz SWDs. A typical bipolar ECoG is illustrated in Fig. 1. Pretreatment with $GABA_B$ antagonists, CGP 35348 (200 mg/kg, i.p.) or CGP 46381 (60 mg/kg, i.p.) completely suppressed the GBL-induced absence seizure behavior and SWDs. Ethosuximide (200 mg/kg, i.p.), a typical antiepileptic drug, also attenuated GBL-induced SWDs.

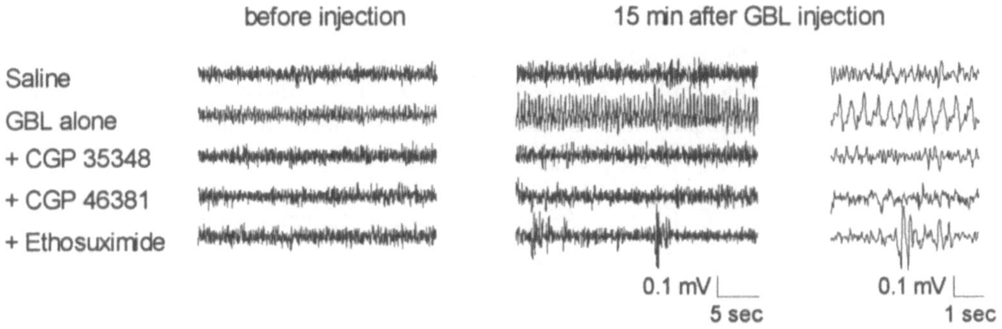

Figure 1. Representative ECoGs of mice. All ECoG patterns were recorded between right and left cortex. CGP 35348 (200 mg/kg, i.p.), CGP 46381 (60 mg/kg, i.p.) or ethosuximide (200 mg/kg, i.p.) were injected 60 min before GBL (70 mg/kg, i.p.) administration.

GBL-induced increases in nuclear CRE- and AP-1 DNA-binding activities

After administration of GBL, brains were dissected and nuclear fractions were prepared. Gel-shift assays in various brain regions revealed that administration of GBL induced nuclear CRE- and AP-1 DNA-binding activities significantly in the thalamus + midbrain and cerebral cortex but not in the cerebellum, hippocampus, hypothalamus or pons + medulla (Fig. 2). The region-specific increases in nuclear CRE- and AP-1 DNA-binding activities were also suppressed by CGP 35348 (200 mg/kg, i.p.) (Fig. 3). Although CGP 35348 did not affect DNA-binding activities of saline-treated mice, it suppressed the GBL (100 mg/kg, i.p.)-induced CRE- and AP-1 DNA-binding activities in the thalamus + midbrain and cerebral cortex.

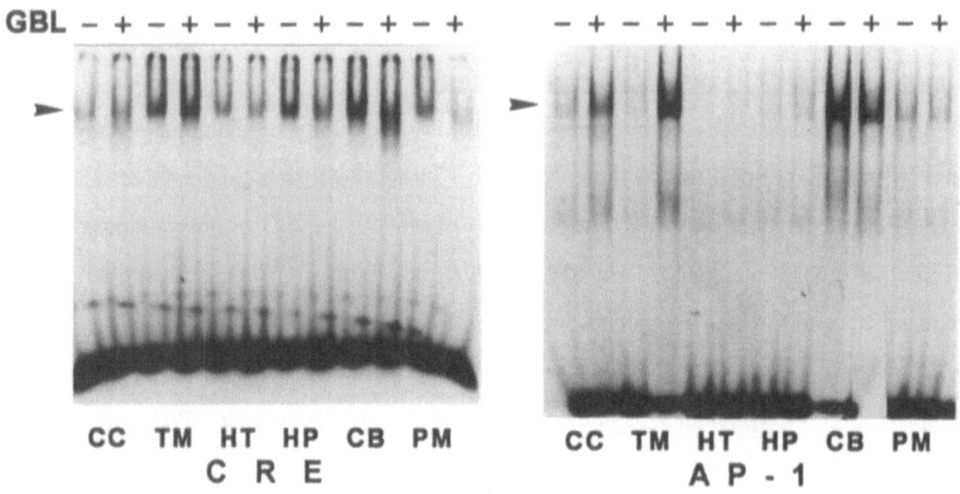

Figure 2. Effect of GBL administration on nuclear CRE- and AP-1 DNA-binding activities in various brain regions. Nuclear extracts were prepared 30 min after saline (GBL -) or GBL administration (100 mg/kg, i.p.). CC : cerebral cortex, TM : thalamus + midbrain, HT : hypothalamus, HP : hippocampus, CB : cerebellum, PM : pons + medulla oblongata.

GBL-induced nuclear CRE- and AP-1 DNA-binding activities were characterized by preincubating with various antibodies of transcription factors. In the cerebral cortex, the increased CRE-binding activity was completely supershifted by CREB antibody, but not affected by c-Jun or c-Fos antibodies. The AP-1 DNA-binding activity was blocked by c-Jun or c-Fos antibody and partially supershifted by CREB antibody (data not shown).

Discussion

Pervious results from our laboratory demonstrated that treatment of mice with GBL (70 and 100 mg/kg) induced generalized absence-like seizures with bilaterally synchronous SWDs and concomitant increases in nuclear CRE- and AP-1 DNA-binding activities in the mouse whole brain (Ishige et al., submitted). In addition, the doses that induced SWDs were the same as those that induced nuclear CRE- and AP-1 DNA-binding activities, thus suggesting that the increases in nuclear CRE- and AP-1 DNA-binding activities were correlated with the SWDs.

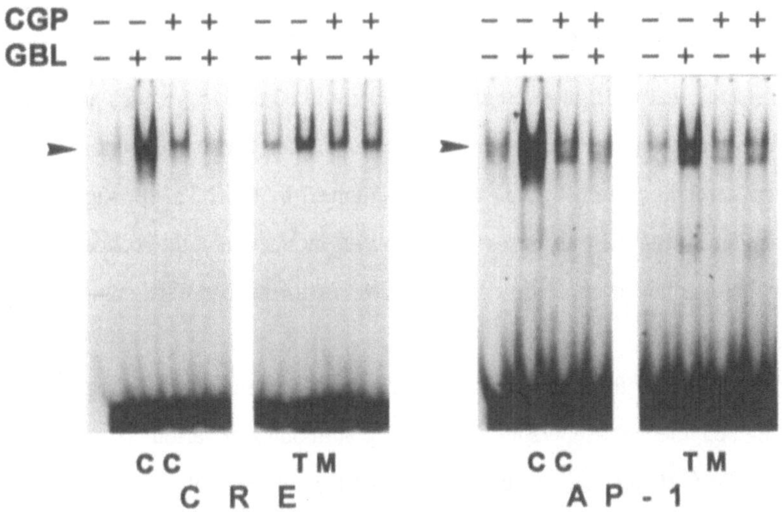

Figure 3. Antagonistic effect of CGP 35348 on GBL-induced increase in nuclear CRE- and AP-1 DNA-binding activity in the thalamus + midbrain (TM) and cerebral cortex (CC). CGP 35348 (200 mg/kg, i.p.) was injected 60 min before saline (GBL -) or GBL administration. Nuclear extracts were prepared 30 min after GBL administration (100 mg/kg, i.p.).

The major observation described here is that GBL-induced nuclear CRE- and AP-1 DNA-binding activities are region specific phenomenon. Gel-shift assays in various brain regions revealed that GBL (100 mg/kg) increased nuclear CRE- and AP-1 DNA-binding activities significantly in the thalamus + midbrain and cerebral cortex but not in the cerebellum, hypothalamus, hippocampus or pons + medulla. Certain oscillatory rhythms within the cerebral cortex are a consequence of both thalamocortical circuitry and intrinsic properties of thalamocortical neurons (1). Thalamic sensory nuclei send axonal projections to neocortex, and receive feedback projections from the cortical area. Thus, region specific increases in the DNA-binding activities in the thalamus + midbrain and cerebral cortex in absence seizure mice induced by GBL may be attributable to the thalamocortical oscillatory activity that generates the SWDs.

A GABA$_B$ antagonist, CGP 35348, has been shown to attenuate or block the absence seizure activity induced by GBL (15) or by R-baclofen (3). Particularly thalamic GABA$_B$ receptors play an important role in the control of absence seizures (3). In addition, CGP 35348 decreases the seizure frequency in lethargic (*lh/lh*) mice, a genetic model of absence seizures (16). These results suggest that $GABA_B$ receptors play a potential role in absence seizures and that an enhanced $GABA_B$ receptor-mediated synaptic responses may underlie absence seizures (15-17). Our results also demonstrated that GBL-induced absence seizure-like behavior, SWDs and nuclear CRE- and AP-1 DNA-binding activities in the mouse whole brain (Ishige et al., submitted) and the thalamus + midbrain and cerebral cortex (Fig. 3) were antagonized by CGP 35348, suggesting that $GABA_B$ receptor-mediated synaptic responses are involved in not only the behavioral symptoms and electroencephalographic events but also the increases in nuclear CRE- and AP-1 DNA-binding activities.

In order to identify trans-acting factors which are responsible to increased CRE- and AP-1 DNA-binding activities, gel-supershift assays using antibodies of various transcription factors were performed. The CRE-binding was blocked by CREB antibody, but not affected by c-Fos or c-Jun antibody. In addition, the AP-1 DNA-binding activity was blocked by c-Fos and c-Jun antibodies and partially supershifted by CREB antibody. These results suggest that CRE- binding activity is attributable to CREB. In the AP-1 DNA-binding activity, AP-1 protein (c-Fos-c-Jun complex) is a major component of the binding activity although CREB-c-Jun complex, a minor component, may also bind to this element.

The administration of metrazole or picrotoxin causes a significant increase in messenger RNAs (mRNAs) of transcription factors such as Zif/268, c-jun, jun-B and c-fos in the hippocampus and cerebral cortex (18). Convulsant doses of kainic acid induces AP-1 mRNAs, proteins and DNA binding activities in the rat hippocampus (6,7). On the other hand, no significant increase in CRE-binding activity was noted after metrazole-induced convulsions (13). Present results showed that there was a significant increase in the CRE-binding activity as well as AP-1 DNA-binding activity in GBL-induced absence-like seizure mouse. In addition, these increases did not occur in the hippocampus in these mice. These results suggest that the way of induction of transcription factors by convulsive seizure-inducing paradigms is differ from that of induction by nonconvulsive absence seizure-inducing paradigm.

In summary, our results have demonstrated that $GABA_B$ receptors play an important role in GBL-induced generalized absence seizures and that the increase in nuclear CRE- and AP-1 DNA-binding activities is correlated with these seizures.

Acknowledgments

We are grateful to Dr. K. Anzai, College of Pharmacy, Nihon University, for valuable advice. We are also grateful to CIBA-GEIGY for supplying, CGP 35348 and CGP 46381. This work was supported by Grant-in Aid for Nihon University President's Grant for Nihon University.

References

1. Striade M and Llinas R. The functional states of the thalamus and the associated neuronal interplay. Physiol Rev 1988; 68: 649-742.

2. Liu Z, Vergnes M, Depaulis A and Marescaux C. Evidence for a critical role of GABAergic transmission within the thalamus in the genesis and control of absence seizures in rat. Brain Res 1991; 545:1-7.

3 Liu Z, Vergnes M, Depaulis A and Marescaux C. Involvement of intrathalamic GABA$_B$ neurotransmission in the control of absence seizures in the rat. Neuroscience 1992; 48:87-93.

4. Upton N. Mechanisms of action of new antiepileptic drugs: rational design and serendipitous findings. Trends Pharmacol and Toxicol 1994; 15:456-463.

5. Sonnenberg JL, Mitchelmore C, Macgregor-Leon PF, Hempstead J, Morgan JI and Curran T. Glutamate receptor agonists increase the expression of Fos, Fra, and AP-1 DNA binding activity in the mammalian brain. J Neurosci Res 1989;24:72-80.

6. Pennypacker KR, McMillian MK, Douglass J and Hong JS. Ontogeny of kainate-induced gene expression in rat hippocampus. J Neurochem 1994; 62:438-444.

7. Pennypacker KR, Walczak D, Thai L, Fannin R, Mason E, Douglass J and Hong JS. Kainate-induced changes in opioid peptide genes and AP-1 protein expression in the rat hippocampus. J Neurochem 1993;60:204-211.

8. Zhang X, Ju G and Le Gal La Salle G. Fos expression in GHB-induced generalized absence epilepsy in the thalamus of the rat. NeuroReport 1991; 2:469-472.

9. Montminy MR, Gonzalez GA and Yamamoto KY. Regulation of cAMP-inducible genes by CREB. Trens Neurosci 1990; 13:184-188.

10. Lukasiuk K and Kaczmarek L. AP-1 and CRE DNA binding activities in rat brain following pentylenetetrazole induced seizures. Brain Res 1994; 643:227-233.

11. Zornetzer S. A simple and reliable chronic brain implantation technique for the mouse. Physiol Behav 1970; 5:1197-1199.

12. Gorski K, Carneiro M and Schibler U. Tissue-specific in vitro transcription from the mouse albumin promoter. Cell 1986; 47:767-776.

13. Sakurai H, Kurusu R, Sano K, Tsuchiya T and Tsuda M. Stimulation of cultured cerebellar granule cells via glutamate receptors induces TRE- and CRE-binding activities mediated by common DNA-binding complexes. J Neurochem 1992; 59:2067-2075.

14. Ito Y, Ishige K, Zaitsu E, Anzai K and Fukuda H. γ-Hydroxybutyric acid increases intracellular Ca^{2+} concentration and nuclear cyclic AMP responsive element- and activator protein 1 DNA-binding activities through $GABA_B$ receptor in cultured cerebellar granule cells. J Neurochem 1995; 65:75-83.

15. Snead OC. Evidence for $GABA_B$-mediated mechanisms in experimental generalized absence seizures. Eur J Pharmac 1992; 213:343-349.

16. Hosford DA, Clark S, Cao Z, Wilson WA, Lin F-H, Morrisett RA and Huin A. The role of $GABA_B$ receptor activation in absence seizures of lethargic (*lh/lh*) mice. Science 1992; 257:398-401.

17. Lin F-H, Cao Z and Hosford DA. Increased number of $GABA_B$ receptors in the lethargic (*lh/lh*) mouse model of absence epilepsy. Brain Res 1993; 608:101-106.

18. Saffen DW, Cole AJ, Worley PF, Christy BA, Ryder K and Baraban JM. Convulsant-induced increase in transcription factor messenger RNAs in rat brain. Proc Natl Acad Sci USA 1988; 85:7795-7799.

PHARMACOLOGICAL STUDIES ON GENERALIZED ABSENCE SEIZURE REGULATION IN GENETIC AND DRUG-INDUCED MODELS

Masahiro Aizawa, Kumiko Ishige, Yoshihisa Ito and Hideomi Fukuda

Department of Pharmacology, College of Pharmacy, Nihon University, Funabashi-shi, Chiba 274, Japan

Summary: Pharmacological profiles of generalized absence seizures in animal models were studied. Selective $GABA_B$ antagonists, CGP 35348 and CGP 46381, suppressed absence seizure behavior and spike and wave discharges (SWDs) in lethargic (*lh/lh*) mice and γ-butyrolactone (GBL)-treated mice (the γ-hydroxybutyric acid model: the GHB model) but not in stargazer (*stg/stg*) mice. Ethosuximide attenuated SWDs in all models used. In the GHB model, CGP 35348 is more potent than ethosuximide. These results suggest that $GABA_B$ receptors play a significant role in generalized absence seizures in lethargic and GHB model mice, whereas different mechanism is involved in those in stargazer mice.

Introduction

Generalized absence seizures are characterized by a brief period of behavioral arrest and occasional automatisms accompanied by a characteristic electroencephalographic pattern defined by 3-Hz spike and wave discharges (SWDs) (1). γ-Hydroxybutyric acid (GHB) and γ-butyrolactone (GBL), a prodrug of GHB, have been shown to induce absence-like seizures in rats (1). Genetic mutant mice such as lethargic (*lh/lh*) (2) and stargazer (*stg/stg*) (3) have also been shown to elicit absence-like seizures accompanied by 6-Hz SWDs. Intrathalamic infusion of baclofen triggers or exacerbates SWDs, while $GABA_B$ antagonists suppress or attenuate the SWDs (4). Although genetic and drug-induced models of absence seizures were characterized independently, drug sensitivities in various models have not been compared in detail. In this study, we examined further the role of $GABA_B$ receptors in generalized absence seizure regulation in these animal models.

Materials and Methods

Animals

Male ddY, male and female lethargic (*lh/lh*) and stargazer (*stg/stg*) mice were used at 8-10 weeks old.

Electrocorticogram (ECoG) recording

Two silver ball electrodes (right and left surface cortical electrodes) and one screw-type electrode (reference electrode, anterior nasal sinus) were implanted into mice under pentobarbital anesthesia. All recordings were carried out at least 5 days after surgery on unanesthetized and freely moving mice.

Results

ECoG recordings of lethargic, stargazer and GBL (70 mg/kg)-treated mice (the GHB model) revealed 6-Hz spike and wave discharges (SWDs, Fig. 1) accompanied by simultaneous arrest of movement with the duration of the discharge. Selective $GABA_B$ receptor antagonists, CGP 35348 (200 mg/kg) and CGP 46381 (60 mg/kg), suppressed absence seizure behavior and SWDs in lethargic and GHB model mice but CGP 35348 did not affect SWDs in stargazer mice (Figs. 1 and 2).

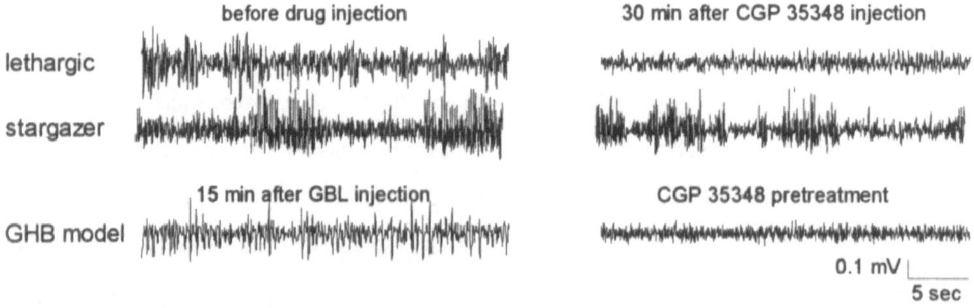

Figure 1. Effect of CGP 35348 on SWDs in lethargic, stargazer and GHB model mice. ECoG patterns before and after CGP 35348 (200 mg/kg) administration are shown. In the GHB model, a dose of GBL (70 mg/kg) was given. Representative bipolar ECoGs from left and right hemispheres are shown. All drugs were injected intraperitoneally.

Ethosuximide (200 mg/kg) also attenuated SWDs in lethargic, stargazer and GHB model mice (Fig. 2). In the GHB model, the antiepileptic effect of this compound was slightly weaker than that of the same dose of CGP 35348. MK-801 (0.5 mg/kg), a noncompetitive NMDA receptor antagonist, had no effect on SWDs in lethargic and GHB model mice; however, it completely suppressed those in stargazer mice (Fig. 2).

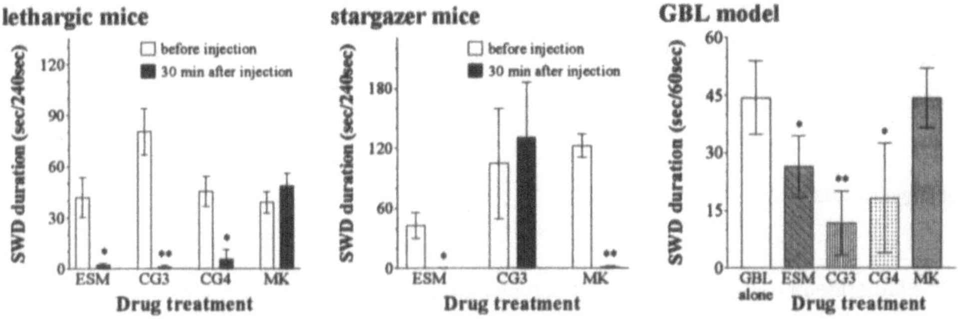

Figure 2. Effects of ethosuximide (200 mg/kg), CGP 35348 (200 mg/kg), CGP 46381 (60 mg/kg) and MK-801 (0.5 mg/kg) on SWD duration of lethargic, stargazer and the GHB model mice. Each value represents the mean ± S.E.M. for three to six different experiments. ESM, ethosuximide; CG3, CGP 35348; CG4, CGP 46381; MK, MK-801. *p<0.05, **p<0.01 as compared with the value before drug injection in lethargic and stargazer mice and the value in GBL alone in the GHB model.

Discussion

Oscillatory burst firing of thalamocortical neurons may underlie the generation of SWDs in absence seizures (5). Activation of $GABA_A$ receptors on the neurons produces short-lasting chloride-dependent IPSPs, whereas activation of $GABA_B$ receptors produces long-lasting potassium-dependent IPSPs. These late $GABA_B$-dependent IPSPs have been shown to prime thalamocortical cell (TC cell) for burst firing by activating low-threshold calcium currents which can be blocked by ethosuximide (Fig. 3) (6). Present results

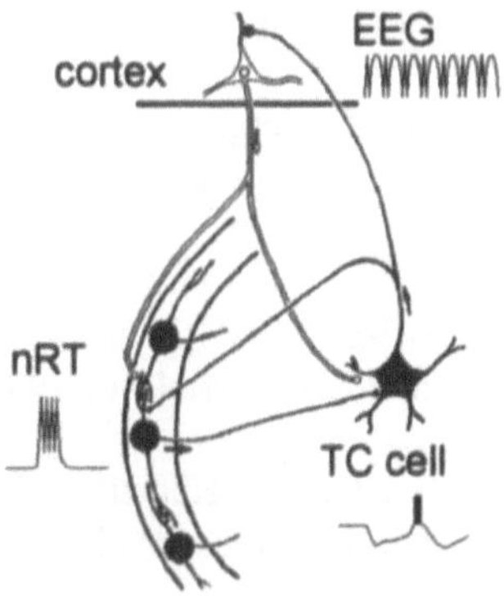

Figure 3. Schematic diagram of thalamocortical circuitry.

demonstrate that selective $GABA_B$ antagonists such as CGP 35348 and CGP 46381 suppress absence seizure behavior and SWDs as potent as and more potent than ethosuximide does in lethargic and GHB model mice, respectively. CGP 35348 has also been shown to suppress absence seizure behavior and SWDs in Genetic Absence Epilepsy Rats from Strasbourg (GAERS) (7). Our results combined with other findings suggest that $GABA_B$ receptors well play a role in genesis of SWDs in lethargic and GHB model mice as well as GAERS. On the other hand, SWDs in stargazer but not those in lethargic or the GHB model mice are suppressed by MK-801, suggesting that different mechanism, possibly neocortical and hippocampal networks (4), is involved in SWDs in stargazer mice.

Acknowledgment

We are grateful to CIBA-GEIGY for supplying CGP 35348 and CGP 46381.

References

1. Snead OC III. Pharmacological models of generalized absence seizures in rodents. J Neural Transm 1992; 35: 7-19.

2 Hosford DA, Clark S, Cao Z, Wilson WA , Lin F-H, Morrisett RA and Huin A. The role of $GABA_B$ receptor activation in absence seizures of lethargic (*lh/lh*) mice. Science 1992; 257: 398-401.

3. Qiao X and Noebels JL. Developmental analysis of hippocampal mossy fiber outgrowth in a mutant mouse with inherited spike-wave seizures. J Neurosci 1993; 13: 4622-4635.

4. Liu Z, Vergnes M, Depaulis A and Marescaux C. Involvement of intrathalamic $GABA_B$ neurotransmission in the control of absence seizures in the rats. Neuroscience 1992; 48: 87-93.

5. Striade M and Llinas R. The functional states of the thalamus and the associated neuronal interplay. Physiol Rev 1988; 68: 649-742.

6. Coulter DA, Hugeuenard JR and Prince DA. Characterization of ethosuximide reduction of low-threshold calcium current in thalamic neurons. Ann Neurol 1989; 25: 582-593.

7. Marescaux C, Liu Z, Bernasconi R and Vergnes M. $GABA_B$ receptors are involved in the occurrence of absence seizures in rats. Pharmacol Commun 1992; 2: 57-62.

Acknowledgements

We cordially thank all the participants of the symposium for coming to Kyoto. In addition, we would like to express our sincere thanks to the following organizations for their support:

International Society for Neurochemistry
Japan Foundation for Neuroscience and Mental Health
The Pharmaceutical Manufacturers Association of Tokyo
Osaka Pharmaceutical Manufacturers Association

Asahi Chemical Co. Ltd.
Astra Japan Ltd.
Banyu Pharmaceutical Co. Ltd
Bayer Yakuhin Ltd.
Chugai Pharmaceutical Co. Ltd
Ciba-Geigy (Japan) Ltd.
Ciba-Geigy Ltd.
Daiichiseiyaku Co. Ltd.
Dainippon Pharmaceutical Co. Ltd.
Eisai Co. Ltd.
Fujisawa Pharmaceutical Co. Ltd.
Fuso Pharmaceutical Industries Ltd.
Grelan Pharmaceutical Co. Ltd.
Hishiyama Pharmaceutical Co. Ltd.
Iwaki Seiyaku Co. Ltd.
Janssen-Kyowa Co. Ltd.
Kaken Pharmaceutical Co. Ltd.
Kanebo Ltd.
Kowa Shinyaku Co. Ltd.
Kyorin Pharmaceutical Co. Ltd.
Kyowa Kakko Kogyo Co. Ltd.
Marion Merrell Dow, K.K.
Maruho Co. Ltd.
Maruishi Pharmaceutical Co. Ltd.
Meiji Seika Kaisha Ltd.
Merck and Co. Inc.
Mochida Pharmaceutical Co. Ltd.
Mohan Medicine Research Institute
Nihon Pharmaceutical Co. Ltd.
Nihon Schering K.K.
Nippon Boehringer Ingelheim Co. Ltd.
Nippon Kayaku Co. Ltd.
Nippon Shinyaku Co. Ltd.
Nipponzoki Pharmaceutical Co. Ltd.
Ono Pharmaceutical Co. Ltd.
Otsuka Pharmaceutical Co. Ltd.
Otsuka Pharmaceutical Factory Ltd.
Pfizer Pharmaceuticals Inc.
Rohto Pharmaceutical Co. Ltd.
Roussel Morishita Co. Ltd.
Sandoz Pharmaceuticals Ltd.
Sankyo Co. Ltd.
Santen Pharmaceutical Corporation Ltd.
Sato Pharmaceutical Co. Ltd.
Sawai Pharmaceuticals
Shin Nippon Pharm. Inc.
Shionogi & Co. Ltd.
Showa Yakuhin Kako Co. Ltd.
SS Pharmaceutical Co. Ltd.
Sumitomo Pharmaceuticals Co. Ltd.
Taiho Pharmaceutical Co. Ltd.
Taisho Chemical Co. Ltd.
Takeda Chemical Industries Ltd.
Tanabe Seiyaku Co. Ltd.
Teikoku Hormone M.F.G. Co. Ltd.
The Green Cross Corporation
Tokyo Tanabe Co. Ltd
Torii & Co. Ltd.
Towa Pharmaceutical Co. Ltd.
Toyama Chemical Co. Ltd.
Tsumura & Co.
Upjohn Pharmaceuticals Ltd.
Wakamoto Pharmaceuticals Co. Ltd.
Yamanouchi Pharmaceutical Co. Ltd.
Yoshitomi Pharmaceutical Industries Ltd.
Zeneka K.K.

Keyword Index

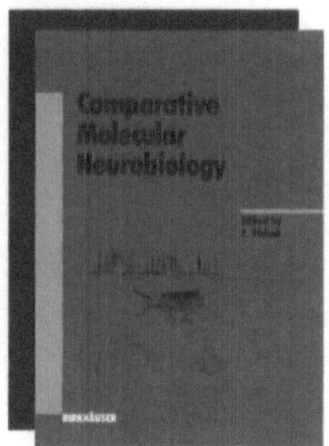